Using Environments to Enable Occupational Performance

Using Environments to Enable

Occupational Performance

edited by

Lori Letts, PhD, OT Reg. (Ont.)
Assistant Professor
School of Rehabilitation Science
McMaster University
Hamilton, Ontario

Patty Rigby, MHSc, OT Reg. (Ont.)
Assistant Professor
Department of Occupational Therapy, Faculty of Medicine
University of Toronto

Research Coordinator for Occupational Therapy
Bloorview MacMillan Children's Centre
Toronto, Ontario

Debra Stewart, MSc, OT Reg. (Ont.)
Assistant Clinical Professor
School of Rehabilitation Science

Co-Investigator
CanChild Centre for Childhood Disability Research
McMaster University
Hamilton, Ontario

An innovative information, education, and management company
6900 Grove Road • Thorofare, NJ 08086

As the main authors are based in Canada, this book uses Canadian spellings of some words. There are a number of theoretical models and assessments described throughout the book. For consistency purposes, the editors have chosen to present all names of models and assessents in lowercase letters.

The work SLACK publishes is peer reviewed. Prior to publication, recognized leaders in the field, educators, and clinicians provide important feedback on the concepts and content that we publish. We welcome feedback on this work.

Printed in the United States of America

Library of Congress Cataloging-in-Publication Data

Using environments to enable occupational performance / [edited by] Lori Letts, Patty Rigby, Debra Stewart.
    p. ; cm.
  Includes bibliographical references and index.
  ISBN 1-55642-578-3
  1. Occupational therapy. 2. Occupational therapy--Environmental aspects.
  [DNLM: 1. Occupational Therapy--methods. 2. Holistic Health. 3. Sociology, Medical. WB 555 U845 2003] I. Letts, Lori. II. Rigby, Patty. III. Stewart, Debra
  RM735.U85 2003
  615.8'515--dc21

                              2003005515

Published by: SLACK Incorporated
               6900 Grove Road
               Thorofare, NJ 08086 USA
               Telephone 856-848-1000
               Fax 856-853-5991
               www.slackbooks.com

Contact SLACK Incorporated for more information about other books in this field or about the availability of our books from distributors outside the United States.

For permission to reprint material in another publication, contact SLACK Incorporated. Authorization to photocopy items for internal, personal, or academic use is granted by SLACK Incorporated provided that the appropriate fee is paid directly to Copyright Clearance Center. Prior to photocopying items, please contact the Copyright Clearance Center at 222 Rosewood Drive, Danvers, MA 01923 USA; phone: 978-750-8400; website: www.copyright.com; email: info@copyright.com.

For further information on CCC, check CCC Online at the following address: http://www.copyright.com.

Last digit is print number: 10   9   8   7   6   5   4   3   2   1

TABLE OF CONTENTS

Instructors: *Using Environments to Enable Occupational Performance Instructor's Manual*
is also available from SLACK Incorporated. Don't miss this important companion to
Using Environments to Enable Occupational Performance. To obtain the *Instructor's
Manual,* please visit http://www.efacultylounge.com.

ACKNOWLEDGMENTS

A book does not come together without the help of a number of people, and this one is no exception. We want to thank each of the contributing authors for their efforts to submit, revise, and resubmit their chapters for this book. They helped to shape the book immeasurably, and the examples they share demonstrate the significant role that occupational therapists can have in enabling occupational performance using environmental interventions. The reviewers of the book also provided us with a number of insights and suggestions that helped strengthen the content. Donna Marfisi deserves our special thanks for her help in formatting all of the submissions consistently, checking references, and reformatting figures and tables. The members of the Occupational Therapy Environment Research Group originally conceived the idea for this book. The other members of the group, Mary Law, Barbara Acheson Cooper, and Susan Strong, have been very supportive of our efforts to turn the group's idea into reality. Finally, we thank the clients, groups, and communities with whom we have worked. They have inspired us to look at the environment as a source of change and opportunity, and have helped us to understand how the environment can be used to optimize occupational performance.

ABOUT THE EDITORS

Lori Letts, PhD, OT Reg. (Ont.) is an Assistant Professor within the School of Rehabilitation Science at McMaster University. She received her Bachelor of Science degree in Occupational Therapy from the University of Western Ontario in 1987. In 1991, she received a Master of Arts with a joint degree in Gerontology and Regional Planning and Resource Development. She received her PhD in Environmental Studies at York University, Canada. Her research and practice interests include aging, environment (theory, assessment, intervention), health promotion, community rehabilitation, evidence-based practice, program evaluation, and participatory research.

Patty Rigby, MHSc, OT Reg. (Ont.) is an Assistant Professor in the Department of Occupational Therapy, Faculty of Medicine, University of Toronto. She is also the Research Coordinator for Occupational Therapy at Bloorview MacMillan Children's Centre in Toronto. She completed her undergraduate studies in occupational therapy at the University of Alberta in 1976, and in 1991 received a Master of Health Sciences degree from McMaster University. Her research and practice interests include enabling childhood occupations such as play and school productivity, assistive technology for children, environment (theory, assessment, and intervention), and cost-utility evaluation of assistive technology.

Debra Stewart, MSc, OT Reg. (Ont.) is an Assistant Clinical Professor in the School of Rehabilitation Science, and a Co-Investigator at *CanChild* Centre for Childhood Disability Research at McMaster University. She received her Bachelor of Science degree in Occupational Therapy from the University of Toronto in 1976, and has worked clinically in the field of pediatrics for many years. In 1998, she received her Master of Science degree in Design, Measurement, and Evaluation at McMaster University. Debra's research interests include person-environment relations in occupational therapy, evidence-based practice, and the experiences of young people with disabilities in transition from adolescence to adulthood.

Contributing Authors

Carol Anderson, MA, OTR, CHT
Clinical Manager of Hand Therapy Services
Baylor College of Medicine
Houston, TX

Sue Baptiste, MHSc, OT Reg. (Ont.)
Associate Professor
School of Rehabilitation Science
McMaster University
Hamilton, Ontario, Canada

Catana Brown, PhD, OTR, FAOTA
Associate Professor
Department of Occupational Therapy Education
University of Kansas Medical Center
Kansas City, KS

Gunilla Carlsson, PhD, Reg. OT
Department of Clinical Neuroscience,
Division of Occupational Therapy
Lund University
Lund, Sweden

Barbara Acheson Cooper, PhD, DipP&OT
Professor Emerita
School of Rehabilitation Science
McMaster University
Hamilton, Ontario, Canada

Kristen Day, PhD
Associate Professor
Department of Urban and Regional Planning
University of California, Irvine
Irvine, CA

Karen Whalley Hammell, PhD, MSc, OT(C), DipCOT
Researcher and Writer
Oxbow, Saskatchewan, Canada

Wendy Hildenbrand, MPH, OTR
Clinical Instructor
Department of Occupational Therapy Education
University of Kansas Medical Center
Kansas City, KS

Lorrie Huggins, MA, EdS, BScOT
General Manager, Training, Development, and Research
YMCA of the Greater Toronto Area
Toronto, Ontario, Canada

Susanne Iwarsson, PhD, Reg. OT
Associate Professor
Department of Clinical Neuroscience
Division of Occupational Therapy
Lund University
Lund, Sweden

Mary Law, PhD, OT Reg. (Ont.)
Professor and Associate Dean
School of Rehabilitation Science
McMaster University
Hamilton, Ontario, Canada

Mary Muhlenhaupt, BS OT, OTR/L, FAOTA
Graduate Certificate, Related Services in Inclusive Environments
Independent Practice
Valley Forge, PA

Occupational Therapy Dept., Child and Family Studies Research Programs
Thomas Jefferson University
Philadelphia, PA

Kathy Parker, MS, OTR
Clinical Instructor
Department of Occupational Therapy Education
University of Kansas Medical Center
Kansas City, KS

Linda Petty, BSc (OT), OT Reg. (Ont.)
Certificate, High Technology Vision Aids
Clinical Specialist
Adaptive Technology Resource Centre
University of Toronto
Toronto, Ontario, Canada

Karen Rebeiro, MScOT, BScOT, OT Reg. (Ont.)
Clinical Researcher
Northeast Mental Health Centre
Sudbury, Ontario, Canada

Melisa Rempfer, PhD
Assistant Professor
Department of Occupational Therapy Education
University of Kansas Medical Center
Kansas City, KS

Laurie Ringaert, MSc, BMR-OT, BSc
Director
The Center for Universal Design
Raleigh, NC

Jean Spencer, PhD, OTR, FAOTA
Professor, School of Occupational Therapy
Texas Woman's University
Houston, TX

Agneta Ståhl, PhD
Professor in Public Transport Planning
Department of Technology and Society
Division of Traffic Planning
Lund University
Lund, Sweden

Susan Stark, PhD, OTR, FAOTA
Instructor
Washington University School of Medicine,
Program in Occupational Therapy
St. Louis, MO

Susan Strong, MSc, BSc, OT Reg. (Ont.)
Assistant Clinical Professor
Research Associate at the Work Function Unit
School of Rehabilitation Science
McMaster University

Staff Occupational Therapist
Community Schizophrenia Services
St. Joseph's Healthcare
Hamilton, Ontario, Canada

Gail Whiteford, PhD, MHSc (OccTherapy), BAppSc (OccTherapy)
Professor and Head, School of Community Health
Charles Sturt University
Albury, NSW, Australia

Ann Wilcock, PhD, GradDipPH, DipCOT, BAppScOT
Professor
School of Health Sciences
Occupational Science and Therapy
Deakin University
Geelong, Australia

PREFACE

The Environment Research Group at McMaster University began in 1991 as a forum for six occupational therapists to discuss research and practice issues related to the environment. As a group, we came from diverse perspectives, including practice areas of pediatrics, gerontology, and mental health. At the time, several of us were pursuing graduate studies in areas of architecture, urban and regional planning, environmental studies, and clinical health sciences. After conducting reviews of the literature in occupational therapy more broadly, we concentrated on reviewing environmental assessment measures and how others conceptualized person-environment relationships. We found few models explaining what we as therapists and researchers had come to understand as the importance of occupations and relationships among individuals, occupations, and environments. Therefore, we developed an occupational therapy model of practice, the person-environment-occupation (PEO) model (Law et al., 1996). We have continued to work together in various combinations over the past 10 years to conduct research centred on persons and the occupations they fulfill within their everyday environments. As a group, we identified a need for a resource that focused on how occupational therapists can use environments in their interventions. We decided to pursue publication of a book that would present interventions with environment as a focus from all areas of occupational therapy, with three members of our group acting as its editors.

The focus of this book is on how environments can be used by occupational therapists to enable occupational performance. Environments are broadly defined to include physical, social, cultural, and institutional components. As occupational therapy practice has evolved, increasing recognition has been given to the importance of making adjustments to all aspects of environments to improve clients' performance. Occupational therapists have a long history of being involved in home modifications, but even that has changed over time. There are now opportunities to branch into new areas of intervention, such as industry, health promotion, public planning, and technology.

The consideration of environments and their influence on health and function is demonstrated globally through the adoption of the International Classification of Functioning, Disability and Health (ICF) (WHO, 2001), as well as by the disability rights and consumer movements. Developments within our own profession have also influenced our understanding about the relationship of environments to health and occupational performance. Client-centred practice has become an accepted (and in many cases expected) approach to working with clients, and this requires occupational therapists to consider the whole picture of a client's daily experiences. Theory in occupational therapy has now evolved to incorporate the influence of the environment on occupational performance. Increasingly, occupational therapists are emphasizing occupation as a holistic, dynamic, and interactive concept.

Reflecting these trends, assessments of environments have received increased attention in occupational therapy literature (Cooper, Letts, Stewart, Strong, & Rigby, 2001; Letts et al., 1994). Therapists can now access environmental assessments that help them to consider the influence that physical, social, and institutional environments have on the occupational performance of clients.

Until now, there has been little focus on how occupational therapists can take full advantage of environments in their interventions. This book provides occupational therapy students and occupational therapists with ways to think about how environments fit into their practice. Additionally, it provides a number of examples that demonstrate approaches to occupational therapy intervention involving environmental changes. Members of the research group have three main goals for the book:

1. To promote a broader focus within occupational therapy practice from changing the person to changing occupational performance by focusing on environments, occupations, and the person.

2. To provide concrete examples of how occupational therapists use environments in their interventions to enable clients' occupational performance.

3. To integrate new information and ideas about environments with current knowledge about health, occupation, and occupational performance.

The underlying purpose of this book is to demonstrate how occupational therapy practice has changed over time in response to increased knowledge, new experiences, and research. The incorporation of environment as a key component of health and occupational performance is a prime example of the evolution of occupational therapy practice. This evolution started with a shift in views, or paradigms, about the relationship between the environment, health, and occupation. These views are now represented by theories and models of practice that incorporate the concept of the environment and its relationship to health and occupation. Occupational therapists around the world are incorporating these theories and models into daily practice within different systems, with different people, and at different levels of the environment. The first section of the book provides an overview of the conceptual underpinnings that guide the book.

In identifying contributors for this book, one of our goals was to bring together a breadth of diverse perspectives related to thinking about the environment in occupational therapy practice. That diversity is represented in terms of the countries in which the authors live and work, the types of clients, settings, and environments represented, and the models that the authors draw upon to plan and implement their interventions. A range of examples is included across Sections Two and Three of the book. Section Two provides examples of occupational therapists working to address environments at societal and community levels, while Section Three is focused on the level of communities, groups, and individuals.

It is not our intent to provide readers with recipes on "how to" incorporate environments into intervention. Rather, the book provides tools to help readers think about environments, and apply the ideas in their own areas of practice. The instructor's manual provides responses to some study questions and includes a number of suggested learning activities to further support student occupational therapists in thinking about using environments in their future practice. The book demonstrates the importance of thinking about occupational performance and using environments to optimize clients' occupational performance. The book highlights the ways in which environments are used in occupational therapy intervention and we hope this will inspire us all to use environments to enable our clients' occupational performance.

Mary Law, PhD, OT Reg. (Ont.); Barbara Acheson Cooper, PhD, DipP&OT;
Lori Letts, PhD, OT Reg. (Ont.); Patty Rigby, MHSc, OT Reg. (Ont.);
Debra Stewart, MSc, OT Reg. (Ont.); and Susan Strong, MSc, BSc, OT Reg. (Ont.)
The Environment Research Group, School of Rehabilitation Science,
McMaster University, Hamilton, Ontario, Canada

REFERENCES

Cooper, B. A., Letts, L., Stewart, D., Strong, S., & Rigby, P. (2001). Measuring environmental factors. In M. Law, C. Baum, & W. Dunn (Eds.), *Measuring occupational performance: Supporting best practice in occupational therapy* (pp. 229-256). Thorofare, NJ: Slack Incorporated.

Law, M., Cooper, B., Strong, S., Stewart, D., Rigby, P. & Letts, L. (1996). The person-environment-occupation model: A transactive approach to occupational performance. *Can J Occup Ther*, 63, 186-192.

Letts, L., Law, M., Rigby, P., Cooper, B. A., Stewart, D., & Strong, S. (1994). Person-environment assessments in occupational therapy. *Am J Occup Ther*, 48, 608-618.

World Health Organization. (2001). *International classification of functioning, disability and health*. Geneva: Author.

FOREWORD

The concept of environment has been in the occupational therapy literature since Meyer (1922) first described its importance in the *Philosophy of Occupation Therapy*. Not until the 1970's was the environment conceptualized as supporting clients' competencies; at that time several authors proposed that practitioners manipulate the environment to challenge and support the individual in achieving their treatment objectives (Dunning, 1972; Kiernat, 1972). By the 1980's the first occupational therapy environmental model was published— *The Ecological Systems Model* (Howe & Briggs, 1982)—and occupational therapists were beginning to write about the environment (Barris, 1982; Barris, Keilhofner, Levine, & Neville, 1985). However, the practice was not yet ready to move to an occupational performance focus. Beginning in 1990, a number of occupational performance models emerged that have introduced students and practitioners to the environment as a central focus of occupational therapy practice. We also have seen environmental assessments emerge to help guide the clinical reasoning process of practitioners.

At the same time the environment has become central to the discussion of function and health. Prior to the early 1990's, models of disability viewed pathology and disability interchangeably and excluded the environment as a factor contributing to disability, even though architectural barriers were beginning to be recognized as factors that limited persons with mobility problems from moving about their communities. In the past decade, many countries have developed models of health that clearly identify the role of the environment in fostering participation and quality of life. One example is the enabling-disabling process model in the report *Enabling America* (Brandt & Pope, 1997). This model describes the rehabilitation process as both restoring the individual's function and employing environmental strategies that remove barriers that limit performance. More recently, the WHO (2001) released the ICF. The model in the ICF shifts the view of the indicators of health from one based on mortality rates of populations to one focused on how people live with health conditions and how the individual can achieve a productive, fulfilling life. The ICF addresses the social aspects of disability that include the social and physical environment as central to the person's function. These policy initiatives provide opportunities for occupational therapists to become central to the delivery of client-centred care that enables occupational performance using environmental strategies.

The profession is ready for this book, *Using Environments to Enable Occupational Performance*. Progress has been made in applying environmental interventions; however, there has not been a text dedicated to helping students and practitioners understand the underlying theories, concepts, and interventions that can be employed by occupational therapy practitioners to enhance the performance of our clients. This is such a text.

I have had the opportunity to review this book and am excited about what it offers educators and practitioners who are seeking to use evidence and state-of-the-art interventions. The editors have brought together an international group of scholars and practitioners who are doing seminal work in environment and occupation. The content adds to the profession's base of knowledge—but more important, the content presents practitioners and students with models that can be used to improve the lives of the people occupational therapists serve.

Carolyn Baum, PhD, OTR/L
St. Louis, Missouri
Washington University School of Medicine

References

Barris, R. (1982). Environmental interactions: An extension of the model of occupation. *Am J Occup Ther, 36,* 637-644.

Barris, R., Keilhofner, G., Levine, R., & Neville, A. (1985). Occupation as an interaction with the environment. In G. Kielhofner (Ed.), *A model of human occupation: Theory and application* (pp. 42-62). Baltimore: Williams and Wilkins.

Brandt, E. N., & Pope, A. M. (1997). Executive summary. In E. N. Brandt & A. M. Pope (Eds.), *Enabling America: Assessing the role of rehabilitation science and engineering* (pp. 1-23). Washington, DC: National Academy Press.

Dunning, H. (1972). Environmental occupational therapy. *Am J Occup Ther, 26,* 292-298.

Howe, M. C., & Briggs, A. K. (1982). Ecological systems model for occupational therapy. *Am J Occup Ther, 36,* 322-327.

Kiernat, J. M. (1972). Promoting community awareness of architectural barriers. *Am J Occup Ther, 26,* 10-12.

Meyer, A. (1922). The philosophy of occupation therapy. *Archives of Occupational Therapy, 1*(1), 1-10.

World Health Organization. (2001). *International classification of functioning, disability and health.* Geneva: Author.

SECTION ONE

Occupational Therapy and Environment: Conceptual Underpinnings

The introductory section of this book deals with paradigms, theories, and models that are the underpinnings to the practice of occupational therapy and the use of environments to enable occupational performance. Chapter 1 focuses on the broad paradigms and theories about health and the environment that influence occupational therapists' views and practice. Chapter 2 describes specific theories and models used by occupational therapists that explain the environment and person-environment-occupation relationships. These overviews provide the context for the remainder of the book, which describes a wide variety of applications of current theories and models in using environments in occupational therapy practice.

The Environment: Paradigms and Practice in Health, Occupational Therapy, and Inquiry

Debra Stewart, MSc, OT Reg. (Ont.); and Mary Law, PhD, OT Reg. (Ont.)

CHAPTER OBJECTIVES

* To describe concepts of paradigm, theory, and models of practice in relation to our understanding of health, occupation, and the environment.
* To explore paradigms of health, occupation, and inquiry in relation to our changing views of the environment and person-environment relations.
* To encourage occupational therapists to incorporate emerging views about health, occupation, and the environment into their practice.

INTRODUCTION

The ways in which occupational therapists view the world have direct influence on daily practice, including assessments and interventions. A generally accepted view of the world is called a *paradigm*. A paradigm has been described by Guba (1990) as "a basic set of beliefs that guides action, whether of the everyday garden variety or action taken in connection with a disciplined inquiry" (p. 17).

Table 1-1 defines the terms *paradigm, theories, and models*, which will be used throughout this book. An example of adult learning is offered in this table to enhance understanding of the concepts. Table 1-2 provides a summary of the key characteristics of three paradigms, which represent a range of views of how the world is constructed and how study and inquiry are influenced by these views.

Paradigms change as people gain knowledge and become more sophisticated in their thinking (Guba, 1990). Individuals base their ideas on existing evidence, which is also con-

Table 1-1

Definitions and Examples of Concepts:
Paradigms, Theories, and Models

Concept	Definitions	Examples
Paradigm	• A basic set of beliefs or views of the world • Paradigms guide our actions (Guba, 1990) • Global visions (Keilhofner, 1992)	• Adult learning. Our views of adult learning incorporate beliefs about active participation and life-long learning. The adult learner is viewed as an individual who has characteristics of self-direction and autonomy (Merriam & Caffarella, 1991).
Theories	• Sets of interrelated concepts, facts and assumptions that explain people and things (Cooper & Saaranin-Rahikka, 1986)	• Different theories about adult learning assume that key components of learning include life experiences, self-reflection, self-direction and action (Merriam & Caffarella, 1991).
Models	• Ways of organizing different concepts from theories and paradigms to explain phenomena • Ways to implement our global vision in practice (Keilhofner, 1992) • Conceptual models explain "why" we work as we do; practice models explain "how" we work (Reed & Sanderson, 1999)	• Problem-based or Case-based educational programs represent models of education built on the paradigm of adult learning, which has been adopted by some universities to educate student occupational therapists (Baptiste, 2003; Salvatori, 1999).

tinually growing and changing. In the past few decades many disciplines have been part of several paradigm shifts, as traditional views have been challenged and new paradigms have emerged. For example, in the field of science and inquiry, traditional, objective views of the world considered the laws of nature to be set, and there was one "truth" out there to be found. This led to the adoption of the scientific method as the best way to study science and nature. These views have been challenged recently, as people have learned more about the way the world works. Scientists have found that nature and science are not static, and therefore, inquiry must be more dynamic and flexible. New views of "truth" and "reality" acknowledge that humans are significantly influenced by their own social and mental constructions, and that the relationship between the inquirer and the inquired is interactive. These views are leading to new types of scientific inquiry, including greater use of qualitative methods of research.

Through most of the 20th century, health care has been dominated by a medical, reductionistic paradigm that views the person and the environment as separate elements. These views echo society's belief that a person is distinct from the environment. This approach also tends to compartmentalize different aspects of the environment (e.g., physical versus social environments). As knowledge about the nature of the world and person-environment relationships has changed, and evidence has mounted about these relationships, new models of

Table 1-2
Paradigms of Inquiry: Three Approaches

Question	Positivism	Post-Positivism	Constructivism
What is the nature of "reality?"	Reality really exists "out there." Cause and effect laws exist.	Reality exists but cannot be fully appreciated or understood.	Reality is constructed, is dependent on context, and is socially based. There are multiple realities; no ultimate truth.
What is the nature of the relationship between the inquirer (the knower) and the inquired (the known)?	The inquirer must be distant and non-interactive, neutral, mainly an observer. Bias and confounding factors are excluded.	Inquirer and inquired (subject and object) are still viewed as separate.	Inquirer and inquired determine findings together—can't be separated. Findings are the creation of the process of interaction between the two.
What methods should we use to learn or find knowledge?	Empirical experimentalism. Manipulative tests under controlled conditions.	Objective observational techniques, do inquiry in more natural settings.	Individual constructions are elicited and compared, and consensus is reached.
What are some examples of this form of inquiry?	Randomized controlled trials.	Quantitative methods— e.g., surveys. Qualitative methods— e.g., focus groups.	Largely qualitative— e.g., semi-structured interviews; participant observation.

Adapted from Guba, 1990; Law et al. (1992).

practice are emerging in different health care disciplines that incorporate the view that person and environment are interactive. In occupational therapy, ideas or conceptualizations about occupation and occupational performance represent a re-emerging paradigm, and several models have been developed to put these views and beliefs into practice. Although the models differ in terms of definitions and explanations of various elements, they all support the view of occupational performance as the outcome of an interactive or transactional relationship between person, occupation, and environment.

This chapter explores the paradigm shifts that have occurred, or are occurring, in the fields of health and occupational therapy, specifically in relation to concepts of the environment and person-environment relations. Changes in the way occupational therapists study these concepts are presented as worldviews that also influence a discipline's approach to research and inquiry.

Table 1-3

Comparison of Medical and Social Views of Health and Environment

Concepts	Biomedical Paradigm	Social Paradigm
Basic assumptions	• Specific disease entities are associated with specific biological processes	• Many factors "outside of the body" interact with the person to influence health and well-being
Key concepts of health	• Cause and effect mechanisms • Normal and abnormal functioning • The problem is within the person (body) • Body and mind are separate	• Disease varies temporally, culturally, and geographically • Determinants of health include social, environmental and behavioral elements • Body and mind are inseparable
Views of the environment	• Person and environment are separate entities • Risk factors in the environment may contribute to the person's problem	• Environment and person are interactive • Many environmental factors influence health: physical, social, institutional, cultural, societal, etc.
Views of disability	• An observable deviation from biomedical norms • The direct result of a disease, trauma, or other health condition	• Disability is a social construct, influenced by the world, which is built and designed by people
Model(s) of practice related to the paradigm	• Medical model of care • Acute care models of practice	• Socio-political models • Health promotion • Independent Living Movement

PARADIGMS OF HEALTH: VIEWS OF THE ENVIRONMENT

Paradigms of health have evolved over time, and these views have influenced the way we deliver health care services. There are literally thousands of historical reviews of the many perspectives from which health and health care have been approached over this past century. Two primary paradigms are described below, the biomedical and social paradigms, with a focus on how the environment is described and represented in each approach. Table 1-3 summarizes these two approaches. Views about disability are also explored, as they are closely connected to health paradigms and practices. Examples of current models of health and disability will be used to demonstrate how these paradigms influence practice.

The Biomedical Paradigm

The biomedical paradigm assumes that there are specific disease entities that are associated with specific biological processes. Cause and effect is a paramount concept in this paradigm. Etiology is viewed as biologically specific, indicating that health and disability emanate from the person (Eyles & Woods, 1983; Jones & Moon, 1987).

There are several key assumptions in this paradigm:
1. *Disease* is defined as a deviation from normal biological functioning.
2. It is an individualistic view of health—the problem rests with the person.
3. There is a causal explanation of disease and illness. Cause and effect is a linear process, with a specific pathology within the body resulting in a disease.
4. Health and disease are viewed in a reductionistic way, reducing the body to parts or systems. Mind and body are separate entities.
5. Medicine is a special section of society, objective and neutral in its study and treatment of health and disease.

Using a biomedical perspective, the environment is considered to be a separate entity from the person. Risk factors in the environment such as parasites, pollution, and daily stresses may contribute to the process of disease, but the disease itself rests within the person. Within this paradigm, *health* is defined as the absence of disease (Reed & Sanderson, 1999). Disability is viewed as an observable deviation from biomedical norms of structure or function that directly results from a disease, trauma, or other health condition.

The biomedical paradigm is represented in several models of health that have existed throughout the 20th century. The best known is the medical model, also known as the biomedical model. Within the medical model of health care practice, doctors and other health care workers (including occupational therapists) are viewed as experts in the study and treatment of health, disease, and disability. The role of "scientific medicine" is to identify sources of disease (risk factors) and to provide therapeutic and preventive measures to fix the problem. The approach is reductionistic as health and function are reduced to inner mechanisms of the body and mind (Kielhofner, 1992). Preventive measures may include modifications to the physical environment, such as the use of assistive devices, but they seldom involve changes in the sociocultural or institutional environments, as they are not considered to be the source of the problem.

The biomedical model has promoted knowledge and understanding of the inner mechanisms of the human body and disease processes. The scientific approach has facilitated the medical system's ability to discover and test medications and other forms of the treatment for specific diseases. However, the person as a whole being in context of the daily environment has been neglected in this approach to health and disease. A major criticism of the biomedical model is the lack of attention paid to environmental influences on health, as the model places most of the focus of disease and disability on the person (the body and mind) alone.

The Social (Ecological) Paradigm

The social (ecological) paradigm developed in response to criticisms of the biomedical paradigm. There is a wide range of understandings of this paradigm. A social or ecological perspective to health considers the many factors "outside of the body" that can influence or be influenced by an individual (Jones & Moon, 1987). Determinants of health include a wide range of environmental and behavioral factors that interact with personal variables (Taylor, 1990). Some theorists believe that environmental factors, including housing, employment, neighborhood resources, and safety, may be more important in predicting health behavior and outcomes than individual characteristics (Cohen, Scribner, & Farley, 2000). Ecological theories of health underpin the public health movement, which has traditionally targeted the physical conditions in which people live, and offers a preventive approach to the health of communities (Ashton & Seymour, 1988).

According to the social paradigm, health interventions can occur at many levels, including family, neighborhood, workplace, institutions, communities, and society at large (Cohen, et al., 2000). A shift toward this thinking has occurred during the latter part of the 20th century, with the acknowledgment of the importance of social and ecological determinants of health. The concept of health promotion, as it is currently understood, emerged in the 1970's out of the work of the World Health Organization (WHO). Principles for health promotion were developed in the Ottawa Charter for Health Promotion (WHO, 1986). Literature on health promotion emphasizes the importance of social relations in health and well-being (Berkman, 1995).

The social paradigm of health is perhaps most clearly articulated in relation to views about disability. Disease and disability are considered human constructs, based on social and moral judgements of what is normal and abnormal (Jones & Moon, 1987). Although different terms are used in the literature to identify this emerging view ("socio-political", "interactive," and "environmental"), the concepts are similar; health conditions, including disability, are influenced by the world (the environment), which is built and designed by people. Barriers in the social and physical environment create the disadvantages that people with disabilities experience (Bickenbach, Chatterji, Badley, & Ustun, 1999; Hahn, 1984). Disability is not viewed as a personal characteristic, but rather as a gap between personal capability and environmental demand (Verbrugge & Jette, 1993). The rise of the consumer disability movement and the work of individual scholars who have disabilities have contributed to our understanding of this paradigm. We are now recognizing the importance of strategies that focus on changing attitudes, policies, and laws that disadvantage people with disabilities (Bickenbach et al., 1999).

An example of the influence of a social view of health in the field of disability is the Independent Living Movement (ILM). This movement developed in opposition to the institutionalization of people with disabilities (DeJong, 1979). The basic principles of the ILM include consumer control over one's life in the community, and the right to self-determination (Crewe & Zola, 1984). Within the ILM, problems are viewed in terms of environmental barriers rather than personal impairments (DeJong, 1979).

The social view of health is a holistic approach that encourages health care professionals to address person and environment issues together. It moves professionals away from a linear, cause-effect explanation of disease and disability, as they recognize the dynamic interactions that take place between person and environment. The challenge for health care providers now is to put these new concepts and beliefs into practice. This may appear to some to be a daunting task, given the multiple elements and interactions involved in health and disability.

Paradigm to Practice: A Global Perspective of Health and Environment

Different models of health and disability have evolved over the years. A prime example of this evolution is the development and revision of a global model of health and human functioning by the WHO. The first International Classification of Impairments, Disabilities, and Handicaps (ICIDH) was published by WHO in 1980, and was subsequently revised several times to become the new International Classification of Functioning, Disability and Health (ICF), which was approved by the World Health Assembly in May 2001. Influenced by changing paradigms, the evolving global model of health depicts how views of the environment in relation to health, and person-environment relationships, have changed over time.

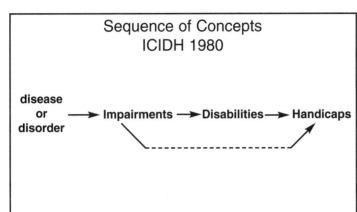

Figure 1-1. International Classification of Impairments, Disabilities, and Handicaps (WHO, 1980). (Reprinted with permission from WHO [1980]. International classification of impairments, disabilities and handicaps. Geneva: Author.)

The graphic model of the first ICIDH (WHO, 1980) is presented in Figure 1-1. This model introduced the use of the term "handicap," which was described as social disadvantages that a person with an impairment or disability experiences (WHO, 1980). Some people considered this to be a global move toward a social view of health; however, there is no reference in the first ICIDH to the environment, and therefore the model does not fit clearly with a social paradigm. The first ICIDH was a linear, cause-and-effect model of health, wherein disability and handicap were viewed as the direct result of an impairment within the body of a person.

Critics of the first ICIDH quickly identified ambiguous language in the definitions of disability and handicap, and the lack of environmental factors in the model. In 1993, the WHO began a revision process for the ICIDH, and spent the next 9 years seeking consultation from people with disabilities, service providers, and policy makers around the world. The result of this process is the new ICF (Figure 1-2). The WHO describes this model as an integration of the traditional medical model and emerging social models of health (WHO, 2001). The new ICF model acknowledges an interactive relationship between person and environment. Although there are still concerns about the model and some of the terms, particularly within the disability community, the ICF represents a step forward in world views of health and disability with the inclusion of environmental factors. The work of the WHO since 1980 is an example of how paradigms or views of health and the environment change over time and influence our practice models. A more detailed critique of the ICF can be found in Chapter 3.

Changing Paradigms in Occupational Therapy: Views of the Environment

The profession of occupational therapy has also witnessed several paradigm shifts in the 20th century. At the beginning of this century, occupational therapy developed from a moral treatment paradigm, which had a strong humanitarian view of health and illness. The first paradigm of the profession focused on occupation (Kielhofner, 1992; Reed & Sanderson, 1999; Wilcock, 1998a), viewing health as closely linked to being occupied, in harmony with nature (Meyer, 1922). Health was reflected in the balance between work, rest, and leisure, and in the habits that organized a person's use of time (Meyer, 1922). Other important concepts within this first paradigm of occupation were the unity of mind and body and the importance of the environment in the therapeutic process. The occupational therapy envi-

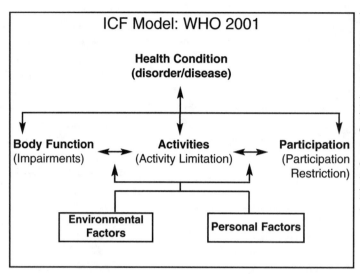

Figure 1-2. International Classification of Functioning, Disability and Health (ICF) (WHO, 2001). (Reprinted with permission from World Health Organization [WHO]. [2001]. International classification of functioning, disability and health. Geneva: Author.)

ronment was viewed as a carefully-managed environment that provided patients with the opportunity to have their needs met, develop healthy habits, and strive to lead meaningful lives (Meyer, 1922; Slagle, 1922).

In response to the development of the biomedical paradigm of 20th-century medicine, the profession of occupational therapy reoriented its views about health and illness to fit with the scientific and reductionistic beliefs of the time. Views of the individual changed over time to focus on the internal functioning of the person, based on scientific knowledge of the inner mechanisms of the body. Occupational therapists explained health, illness, and disability in terms of normal and abnormal functioning, and treatments focused on the "disorder" within the person (Baum & Baptiste, 2002). Activity was used for a particular therapeutic reason, to change and/or compensate for the disorder. There was little mention of the environment and no reference to person-environment relationships in this biomedical paradigm.

Beginning in the 1960's, many occupational therapy scholars began to question the narrow views of the biomedical paradigm for the profession. They noted that the problems people with disabilities were facing in society were not simply the result of disorders of the person. A new paradigm of occupation has been growing since Mary Reilly first challenged occupational therapists to consider the occupational nature of human beings (Yerxa, 1998). This emerging paradigm recognizes the importance of the interaction between person and environment for development, health, and well-being. The concept of occupational performance has been adopted by many occupational therapists to represent current views of the interaction of person, environment, and occupation (Chapparo & Ranka, 1997; Christiansen & Baum, 1997; Law et al., 1996; Reed & Sanderson, 1999). It is an evolving concept that demonstrates how occupational therapists are building on earlier views of occupation with new knowledge and experiences about the relationship between health and the environment.

Occupational therapists also recognize that the interaction between person, environment, and occupation is mutual or reciprocal in nature, and dynamic in process (Law et al., 1996; Wilcock, 1998a; 1998b).

The emerging paradigm of occupation and occupational performance has influenced the practice of occupational therapy in the past few decades in numerous ways. The profession's beliefs in the uniqueness of the individual has led occupational therapists to adopt a client-centred approach to practice (Canadian Association of Occupational Therapists [CAOT], 1997; Law, 1998; Sumsion, 1999), in which clients and their families have the ultimate responsibility for decisions. One of the key concepts of client-centred occupational therapy practice is a focus on the person-environment-occupation relationship (Law, 1998). Occupational therapists see clients in their daily environments of home, school, community, and workplace as they recognized the importance of the interaction between person and environment during daily occupation. Expanded views of health have facilitated a realignment of who occupational therapists associate with in their day-to-day work. They now collaborate with individuals and groups in educational, social, and recreational systems as well as health care, and have developed partnerships with consumer groups. They are promoting concepts of health promotion, in particular the idea of creating supportive environments for the people with whom they are working (Letts, Fraser, Finlayson, &Walls, 1993). Occupational therapists have also developed new practice models that incorporate theories of person-environment relations (CAOT, 1997; Christiansen & Baum, 1997; Dunn, Brown, & McGuigan, 1994; Kielhofner, 1992; Law et al., 1996). Chapter 2 describes some of these models of practice in detail.

Current worldviews of health, disability, and the environment are represented in occupational therapy practice frameworks and standards developed by some national associations and the World Federation of Occupational Therapists. For example, the *Occupational Therapy Practice Framework* (AOTA, 2002) in the United States includes cultural, physical, social, personal, spiritual, temporal, and virtual contexts for therapists to consider in assessment, occupational analysis, and selection of intervention strategies. In the *Profile of Occupational Therapy Practice in Canada* (CAOT, 2002), the environment includes "social, political, cultural, economic, physical, and geographic conditions which impact on occupational performance" (p. 12). Also, the World Federation of Occupational Therapists has revised the minimum standards for the education of occupational therapists (Hocking & Ness, 2002) to include important elements of the environment and the person-occupation-environment relationship with health.

Paradigm Shift in Research and Inquiry

A profession's views of the world also influence how concepts of health and person-environment relations are studied. Within a biomedical paradigm, the major focus of health research has been to uncover the causes of ill health and disease, and test treatment techniques and drugs that can "fix" the underlying disorder (Wilcock, 1998a). With the emphasis on a scientific basis of medicine, many researchers have considered the experimental method to be the only acceptable form of study. The randomized controlled trial (RCT) became the gold standard for health research in the 20th century (Sackett, Haynes, Tugwell, & Guyatt, 1991).

In the past few decades, researchers found that quantitative methods of study did not always answer the complex questions that were emerging in medicine and health. They began to explore the use of qualitative approaches to research that had been traditionally used in the social sciences. Experts now feel that qualitative methods can complement or be used in conjunction with quantitative methods of research to inform health care professionals about the interactive and dynamic nature of person-environment relationships, and concepts of quality of life and well-being (Pope & Mays, 2000). This is not to say that one method of study is better than another, but rather the range of options available to researchers has broadened to facilitate the study of complex issues and new knowledge.

The shift in thinking to a more social view of health is promoting a strong interest in studying the social determinants of health and the relationship between people and environments. The social paradigm promotes the study of the disablement process at political and social levels (Bickenbach et al., 1999), and this has assisted in the development of participatory action research and other forms of study that include persons with disabilities in planning and carrying out research programs.

A participatory approach to inquiry is being used by occupational therapists to increase awareness of the resources and strengths of persons with disabilities. The ultimate goal is to improve the quality of life for persons with disabilities. Participatory action research (PAR) involves people with disabilities articulating the problems and participating in the process of defining, analyzing, and solving the problem (Cockburn & Trentham, 2002). There is a strong action component to research that fits well with occupational therapy's enabling orientation (CAOT, 1997; Cockburn & Trentham, 2002; Letts, 2003). Occupational therapists are involving consumers as user-experts in the design and implementation of programs and resources, including environmental modifications, advocacy, and community development (Law, 1997; Wilcock, 1998b). Several examples of a participatory approach to research and program planning are described in this book.

These new approaches to viewing and studying occupation and health are the underpinnings of occupational science, which began in 1989. Within this discipline of inquiry, occupation is described as "chunks of activity... which are named in the lexicon of the culture" (Yerxa et al., 1990, p. 5), and requires an interaction between an individual and the environment. Yerxa (1998) believes that the study of occupational therapy can promote a new view of health as possession of a repertoire of skills that enables people "to achieve their valued goals in their own environments" (p. 417). This view recognizes that health is for all people, including those with chronic impairments and disabilities, and encourages occupational therapists to integrate environment and person-environment relations into daily practice.

In summary, paradigm shifts are occurring in occupational therapy practice and research in parallel with changing views of health, occupation, and the environment. Changes are also occurring in education, and this can create a dissonance between what is studied and taught, and what is practiced. There is a need for occupational therapy educators, researchers, students, and practitioners to have an open dialogue about approaches to assessment and intervention. All occupational therapists have responsibility for their professional development to ensure that new knowledge is effectively disseminated and used. Up-to-date practice in all areas includes using evidence about the relationships and interactions between people and environments. The chapters in this book provide occupational therapists with current evidence to support using environments to enable occupational performance.

CONCLUSION

Paradigms represent our worldviews. This chapter demonstrates how views of the environment in relation to health, occupation, and inquiry/study have changed through the 20th century. Worldviews influence our practice in occupational therapy and other health care disciplines. New understandings and beliefs about the environment and person-environment relations are promoting the development of practice models that incorporate a social view of health and disability. As a result, occupational therapists now acknowledge the importance of environmental factors in the lives of the people with whom they work, and they are changing assessment and treatment practices to reflect this. Research methods are expanding with the use of qualitative and participatory methods of inquiry to study complex issues of health and occupation in partnership with persons with disabilities and consumer groups.

This book represents a stage of applying knowledge, views, and beliefs about environment and person-environment-occupation relations to the practice of occupational therapy. We believe that occupational therapists are ready for this book. The profession's views about health, disability, and occupation have changed to promote a greater understanding of the strong role of the environment and the transactional influence of person-environment-occupation relationships in the lives of all people. The chapters in this book demonstrate innovative ways that occupational therapists are using the environment to enable occupational performance. These new approaches to practice and inquiry encourage all occupational therapists to continue to challenge their views of the world and to leave their minds open to new understanding and application of person-environment-occupation relations within the profession's unique perspective of health and disability.

Study Questions

1. What paradigms have influenced the practice of occupational therapy in the past few decades?

2. What are the main differences between the medical model and the social model of health in relation to the environment?

3. How are occupational therapists incorporating current views about health, occupation, and the environment into practice and research?

4. How can occupational therapy educators, researchers, students, and clinicians work together to influence positive, evidence-based changes in practice?

REFERENCES

American Occupational Therapy Association (AOTA). (2002). Occupational therapy practice framework: Domain and process. *Am J Occup Ther, 56*, 609-639.

Ashton, J., & Seymour, H. (1988). The setting for a new public health. In J. Ashton & H. Seymour (Eds.), *New public health: The Liverpool experience* (pp. 15-39). New York, NY: Open University.

Baptiste, S. (2003). *Understanding problem-based learning: A self-directed journey*. Thorofare, NJ: SLACK Incorporated.

Baum, C., & Baptiste, S. (2002). Reframing occupational therapy practice. In M. Law, C. Baum, & S. Baptiste (Eds.), *Occupation-based practice: Fostering performance and participation* (pp. 3-15). Thorofare, NJ: SLACK Incorporated.

Berkman, L. F. (1995). The role of social relations in health promotion. *Psychosom Med, 57*, 245-254.

Bickenbach, J. E., Chatterji, S., Badley, E. M., & Ustun, T. B. (1999). Models of disablement, universalism and the international classification of impairments, disabilities, and handicaps. *Soc Sci Med, 48*, 1173-1187.

Canadian Association of Occupational Therapists (CAOT). (1997). *Enabling occupation: An occupational therapy perspective*. Ottawa, ON: CAOT Publications ACE.

Canadian Association of Occupational Therapists (CAOT). (2002). *Profile of occupational therapy practice in Canada* (2nd ed.). Ottawa, ON: CAOT Publications ACE.

Chapparo, C., & Ranka, J. (Eds.) (1997). *Occupational performance model (Australia)*. Monograph 1. Lidcome, Australia: Occupational Performance Network.

Christiansen, C., & Baum, C. (Eds.) (1997). *Occupational therapy: Enabling function and well-being* (2nd ed.). Thorofare, NJ: SLACK Incorporated.

Cockburn, L., & Trentham, B. (2002). Participatory action research: Integrating community occupational therapy practice and research. *Can J Occup Ther, 69*, 20-30.

Cohen, D. A., Scribner, R. A., & Farley, T. A. (2000). A structural model of health behavior: A pragmatic approach to explain and influence health behaviors at the population level. *Prev Med, 30*, 146-154.

Cooper, B. A., & Saaranin-Rahikka, H. (1986). Interrelationship of theory, clinical models, and research. *Physiotherapy Canada, 38*, 97-100.

Crewe, N., & Zola, I. (1984). *Independent living for physically disabled people*. San Francisco: Jossey-Bass.

DeJong, G. (1979). Independent living: From social movement to analytic paradigm. *Arch Phys Med Rehabil, 60*, 435-446.

Dunn, W., Brown, C., & McGuigan, A. (1994). Ecology of human performance: A framework for considering the effect of context. *Am J Occup Ther, 48*, 595-607.

Eyles, J., & Woods, K. J. (1983). Perspectives on health and health care. In J. Eyles & K. J. Woods (Eds.), *The social geography of medicine and health* (pp. 15-30). London: Croom Helm.

Guba, E. G. (1990). *The paradigm dialog*. Newbury Park, CA: Sage.

Hahn, H. (1984). Reconceptualizing disability: A political science perspective. *Rehabilitation Literature, 45*, 362-365, 374.

Hocking, C., & Ness, N. E. (2002). World federation of occupational therapists revised minimum standards for the education of occupational therapists—2002. Retrieved November 1, 2002 from www.wfot.org.au/.

Jones, K., & Moon, G. (1987). The social context of disease, health and medicine. In K. Jones & G. Moon (Eds.), *Health, disease, and society* (pp. 1-33). London: Routledge.

Kielhofner, G. (1992). *Conceptual foundations of occupational therapy*. Philadelphia, PA: F. A. Davis.

Law, M. (1997). Changing disabling environments through participatory action research. In S. Smith, D. Willms, & N. Johnson (Eds.), *Nurtured by knowledge: Participatory action research* (pp. 34-58). New York: Aspen.

Law, M. (Ed.). (1998). *Client-centered occupational therapy*. Thorofare, NJ: SLACK Incorporated.

Law, M., Cooper, B., Letts, L., Rigby, P., Stewart, D., & Strong, S. (1992). *The environment: A critical review of person-environment relations and environmental assessments*. Unpublished document. Hamilton, ON: McMaster University, School of Rehabilitation Science.

Law, M., Cooper, B., Strong, S., Stewart, D., Rigby, P., & Letts, L. (1996). The person-environment-occupation model: A transactive approach to occupational performance. *Can J Occup Ther, 63*, 9-23.

Letts, L. (2003). Occupational therapy and participatory research: A partnership worth pursuing. *Am J Occup Ther, 57*, 77-87.

Letts, L., Fraser, B., Finlayson, M., & Walls, J. (1993). *For the health of it! Occupational therapy within a health promotion framework*. Toronto, ON: CAOT Publications ACE.

Merriam, S. B., & Caffarella, R. S. (1991). *Learning in adulthood. A comprehensive guide*. San Francisco, CA: Jossey-Bass.

Meyer, A. (1922). The philosophy of occupational therapy. *Archives of Occupational Therapy, 1*, 1-10.

Pope, C., & Mays, N. (Eds.) (2000). *Qualitative research in health care* (2nd ed.). London: BMJ Books.

Reed, K. L., & Sanderson, S. N. (Eds.). (1999). *Concepts of occupational therapy* (4th ed.). Baltimore, MD: Lippincott, Williams & Wilkins.

Sackett, D. L., Haynes, R. B., Tugwell, P., & Guyatt, G. H. (1991). *Clinical epidemiology: A basic science for clinical medicine* (2nd ed.). Philadelphia, PA: Lippincott, Williams & Wilkins.

Salvatori, P. (1999). Meaningful occupation for occupational therapy students: A student-centred curriculum. *Occup Ther Int, 6*, 207-223.

Slagle, E. C. (1922). Training aides for mental patients. *Archives of Occupational Therapy, 1*, 11.

Sumsion, T. (Ed.) (1999). *Client-centered practice in occupational therapy: A guide to implementation*. London: Churchill Livingstone.

Taylor, M. (1990). Geographic perspectives on national health challenges. *The Canadian Geographer, 34*, 334-338.

Verbrugge, L. M., & Jette, A. M. (1993). The disablement process. *Soc Sci Med, 38*, 1-14.

Wilcock, A. (1998a). *An occupational perspective of health.* Thorofare, NJ: SLACK Incorporated.

Wilcock, A. (1998b). Reflections on doing, being and becoming. *Can J Occup Ther, 65*, 248-256.

World Health Organization (WHO). (1980). *International classification of impairments, disabilities and handicaps.* Geneva: Author.

World Health Organization (WHO). (2001). *International classification of functioning, disability and health.* Geneva: Author.

World Health Organization (WHO). (1986). *Ottawa charter for health promotion.* Geneva: Author.

Yerxa, E. J. (1998). Health and the human spirit for occupation. *Am J Occup Ther, 52*, 412-418.

Yerxa, E. J., Clark, F., Frank, G., Jackson, J., Parham, D., Pierce, D., et al. (1990). An introduction to occupational science, a foundation for occupational therapy in the 21st century. *Occupational Therapy in Health Care, 6*, 1-17.

Chapter 2

Using Environments to Enable Occupational Performance

Environment and Occupational Performance: Theoretical Considerations

Patty Rigby, MHSc, OT Reg. (Ont.); and Lori Letts, PhD, OT Reg. (Ont.)

CHAPTER OBJECTIVES

* To review person-environment theories that have influenced occupational therapy theory development.
* To describe occupational therapy theories that incorporate environment, with particular emphasis on the person-environment-occupation model of occupational performance.
* To describe how the remainder of the text is organized based on occupational therapy views of environment.

INTRODUCTION

In recent years, the focus of occupational therapy has broadened. Occupational therapists are working in increasing numbers in a variety of community roles, within and outside of the health care system. In part, this is due to the fact that occupational therapists are articulating a common focus—on occupations that matter to clients, on occupational science, and on occupation as a determinant of health. Occupational therapists have also begun to place more attention on client-centred practice. We see individuals within the context of where they live, work, and play. These changes have resulted in a greater awareness of influences of environments on the performance of occupations. Environment has become an integral part of occupational therapy practice models, many of which are grounded in theories of person-environment relationships.

The purpose of this chapter is to provide a brief overview of current models of practice in occupational therapy, and to demonstrate how the environment is conceptualized. Many of these theories were developed from theories of person-environment relations. These theories, which do not often incorporate occupation, are useful to occupational therapy and will be briefly described as well.

THEORIES OF PERSON-ENVIRONMENT RELATIONSHIPS

The influence of environment on human behavior has long captured the interest of many disciplines, including environmental psychology, anthropology, social science, architecture, and now occupational therapy. In previous occupational therapy publications, a number of key person-environment theories have been reviewed (Law et al., 1994; Law et al., 1997). Table 2-1 provides an overview of some of the major theories of person-environment relations that have influenced theory development in occupational therapy.

Three common threads can be identified across the different environmental theories relevant to occupational therapy: how the person is conceptualized, how environments are considered, and the nature of person-environment relationships. Although the emphasis varies, characteristics of the person that are considered include physical status, mental/cognitive status, attitudes and personal beliefs, needs and preferences, knowledge, and social skills. The person might be considered at the level of the individual (Bronfenbrenner, 1977) or at the level of community or group (Moos, 1980). People are generally understood to filter environmental input through their perceptions, which means that people may interpret the same environmental characteristics differently. The emphasis varies in how the environment is understood, but it includes physical, social, cultural, economic, and organizational aspects. Perception (i.e., how environments are perceived) is an important component of some theories (Bandura, 1986). *Press*, or demand, is frequently described in explaining the environment. Generally, press represents characteristics of the environment that place some demand on the person to respond. For example, the physical environment may demand specific levels of ability (e.g., walking certain distances down an institution's hallway to manage functional activities like eating or bathing).

In almost all of the person-environment theories considered here, person-environment interactions or transactions are described. In addition, the relationships between people and their environments are considered to be dynamic (ever-changing in both directions), complex, and interdependent. Multiple outcomes are considered, yet there is always a specific outcome of interest such as fit, perceived self-efficacy, health, and congruence.

The types of theories described in Table 2-1 provide a framework for occupational therapists to incorporate environment into occupational therapy. Although different occupational therapy theories have relied on different person-environment theorists, as a whole person-environment theories provide useful frameworks to consider how people interact with and in their environments. Occupational therapists, in developing theories, have found ways to begin with those theories, and have incorporated concepts of occupation and occupational performance into them.

OCCUPATIONAL THERAPY THEORIES

A number of universal (or overarching) occupational therapy models have been developed over the past few decades to frame occupational therapy practice. Although each

Table 2-1

Overview of Person-Environment Theories

Theorists	Disciplines	Characteristics of Person-Environment Relationship
Bandura (1986)	Psychology	• People are described by six cognitive capacities. • Perceptions of the environment are emphasized. • Outcome is perceived self-efficacy (i.e., person's perception of his or her ability to be successful in light of abilities and environmental characteristics).
Barker (1968); Schoggen, 1989)	Ecological Psychology	• Person-environment relationships are the unit of study. • The goal is to understand community environments. • Environments include physical and geographical characteristics, and also the typical patterns of behaviors that occur within the location. • Behavior settings are emphasized with most emphasis on the physical environment.
Bronfenbrenner (1977)	Developmental Psychology	• Focused on interdependencies between people and their social settings. • People interact with their environments at various levels: groups, social structures, societal institutions.
Gibson (1977); Gibson (1988)	Ecological Psychology	• Person and environment cannot be separated. • People learn more about their environments in trying to achieve goals. • Adaptation occurs when the people are able to modify their performance or their environments to reach a goal.
Lawton (1982); Lawton & Nahemow, (1973)	Environment Psychology & Gerontology	• Adaptive behavior and positive affect occur when there is a good "fit" between the competence of the person and the press of the environment. • As competence declines, the person is more sensitive to slight changes in the environment.
Moos (1980)	Psychology	• People interact with environments to provide stability. • Environments are made up of physical, institutional, human aggregate, and social climate factors. • Personal characteristics (generally at the congregate rather than individual level) include health and functional status, and sociodemographics. • People's perceptions of their abilities and the environmental systems influence whether they attempt to adapt to environments and in turn this influences stability.

model has unique features and theoretical roots, the common emphasis is the understanding of occupational behaviors and performance through the identification and analysis of elements and/or conditions that influence occupational performance. In these models the environment is the context for occupational performance, and is believed to influence occupational performance. There is some variance in how the environment is described, the assumptions the models make about the environment, the descriptions of person-environment relationships, and influences on occupational performance as shown in Table 2-2.

Most of these models view the environment broadly and describe cultural, social, physical and temporal characteristics. A few models also describe the setting or place as a context for occupational performance, including the home, community, school, or place of employment (Kielhofner & Forsyth, 1997; Law et al., 1996). Law (1991) also proposed that environmental factors are influential at various levels including the person, household, community, and broader society.

There is general agreement that the relationship between persons and environments is complex and influences or mediates a person's behavior and his or her performance of any occupation. Occupational performance or occupational behavior is considered to be the outcome of this relationship. This view is consistent with the views of environmental behavior theorists who propose that behavior is the outcome of the person-environment relationship. The unique contribution of the occupational therapy theories and models is the inclusion of the person's roles and occupations (including activities and tasks) as integral to understanding the person-environment relationship.

Several models maintain that occupational behavior is a response to environmental demands, press, or expectations (Christiansen & Baum, 1997; Dunn, Brown, & McGuigan, 1994; Hagedorn, 2000; Kielhofner, 1997; Schkade & McClung, 2000). Conditions and/or cues in the environment can shape people's responses. For example, in rain and snow conditions people generally drive more cautiously and dress differently than they would when it is sunny and warm. The concept of person-environment-occupation fit is also described in several models (Christiansen & Baum, 1997; Hagedorn, 2000; Law et al., 1996). Fit is viewed as a congruence or balance between what the person brings to the transaction and the environmental demands or resources available.

The environment in these occupational therapy models is viewed as a resource, enabler or facilitator of occupational performance, as well as a feature that can present barriers or excessive demands that hinder or prevent occupational performance. The environment is also considered to be amenable to change through modification or adaptation. The implications for occupational therapists when applying any of these models is to assess the influences of environment on occupational performance and to consider how the environment can be used (accessed, modified, adapted) to enable occupational performance.

The Person-Environment-Occupation Model

One of the overarching models in occupational therapy that incorporates environment is the person-environment-occupation (PEO) model (Law et al., 1996). It is described here in more detail to illustrate the ways in which occupational therapists can use theory to guide environmental interventions. Law and colleagues (1996) developed the PEO model to provide a clear, broad framework for using environment in occupational therapy practice. One of the major goals in developing this model was to emphasize the many possibilities and opportunities for accessing environmental resources and reducing environmental barriers to optimize occupational performance. The developers also felt that it was important to have

Table 2-2

**Comparison of How Environment is Viewed
by Various Models That Guide Occupational Therapy Practice**

Model	Characteristics of Environment	Major Assumptions About Environment	Person-Environment Relationship and Influences on Occupational Performance
Canadian model of occupational performance (CAOT, 1997)	• Defined as having cultural, institutional (including political, economic and legal aspects), physical, and social elements. • Includes community, provincial, national, and international factors.	• Occurs outside of the person and elicits responses from him or her. • Individuals ascribe meaning to environments around them, which can change over time and vary between persons. • Environments are influenced by the behaviors of person(s).	• Occupational performance is the result of the dynamic relationship of person, environment, and occupation (p. 45). • Environmental conditions influence a person's occupational performance. • Environment provides the context for occupations.
Competent occupational performance in the environment (Hagedorn, 2000)	• Context includes physical and social content. • Environment is also viewed from each person's perspective of proximity, accessibility, and significance.	• Characteristics of the environment demand action from the person. • The demand characteristics of the environment can be analyzed and altered to create a better fit with the person and occupation.	• Competent occupational performance is enabled when there is a fit, or dynamic balance between person, environment, and occupation. • Competence in performance happens when the person adaptively responds to the demands of the environment. • Competence is situation specific and may not be generalizable from one setting to the next.

Table 2-2, continued

Comparison of How Environment is Viewed by Various Models That Guide Occupational Therapy Practice

Model	Characteristics of Environment	Major Assumptions About Environment	Person-Environment Relationship and Influences on Occupational Performance
The ecology of human performance model (Dunn, Brown, & McGuigan, 1994)	• Context includes physical, social, cultural, and temporal environmental features, and also includes the phenomenological experience of person with the environment.	• Person is viewed through contextual lens; persons interact with context to perform tasks. • Environmental cues and supports are used by a person to support performance of tasks. • It varies from person to person how well the environmental cues and/or supports are used, depending upon the person's skills and abilities to access these.	• Ecology (the interaction between person and environment) influences human behavior and performance. • Context is viewed as dynamic (non-linear) and interactional. • Task/activity performance cannot be understood outside of context. • Assessment tools and intervention(s) are contextually focused.
Model of human occupation (Kielhofner, 1997; Kielhofner & Forsyth, 1997)	• Physical and social elements combine to form meaningful settings for occupational behavior. • Typical settings for occupational behavior include the home, neighborhood, school, workplace, or community.	• Environment influences occupational behavior—it can afford opportunities for performance and/or press for specific behaviors. • The synergy of affordances and presses channels behavior. • Opportunities and presses from the environment are usually viewed very differently by the person with a disability compared with the typically functioning person.	• The human system interacts with the environment to produce occupational behavior. • What is viewed as an opportunity for one person may be a source of frustration for another, particularly for the person with a disability. • Assessment involves looking at the environmental influences—how this affords or presses occupational behavior. • Intervention—therapists can change aspects of environments to support or precipitate a change in occupational behavior.

Table 2-2, continued

Comparison of How Environment is Viewed by Various Models That Guide Occupational Therapy Practice

Model	Characteristics of Environment	Major Assumptions About Environment	Person-Environment Relationship and Influences on Occupational Performance
Occupational adaptation model (Schkade and McClung, 2001)	• The physical, social, and cultural subsystems of the occupational environment; these provide context-specific expectations for occupational roles and adaptive behaviors. • Occupational environments are those that call for an occupational response—they are contexts in which occupations occur (work, play, leisure, and self-care).	• The environment demands mastery from the person, through facilitators or barriers. • The demands for mastery from the occupational environment are equally as important as the person's occupational role expectations and desire for mastery. The interaction is a press for mastery toward competence in occupational functioning.	• The person and occupational environment interact together to create a press for mastery. • This interaction produces a dynamic tension; the person desires to confront challenges adaptively and masterfully; the environment demands that the person responds adaptively and masterfully (p. 14). • The outcome of this relationship is an occupational response.
Person-environment-occupation model (Law et al., 1996; Strong et al., 1999)	• Broadly defined to include cultural, socioeconomic, institutional, physical, and social domains. • Each domain is considered from the unique perspective of the person, household, neighborhood, and community.	• Environment provides the context for occupational performance; it influences performance but is also influenced by performance. • A person's environments are continually shifting and changing over time and space, and as these change, the behavior necessary for occupational performance also changes; the environment can either enable or constrain performance. • Environment is considered to be more amenable to change than the person.	• Behavior is influenced by and cannot be separated from contextual influences. • Occupational performance is the outcome of the transaction of the person, environment, and occupation. • Assessment includes looking at environmental conditions and influences (positive or negative). • Intervention can target the environment as a way to optimize PEO fit and occupational performance.

Table 2-2, continued

Comparison of How Environment is Viewed
by Various Models That Guide Occupational Therapy Practice

Model	Characteristics of Environment	Major Assumptions About Environment	Person-Environment Relationship and Influences on Occupational Performance
Person-environment-occupational performance model (Baum & Christiansen, 1997)	• Physical, cultural, social, or societal conditions; environmental conditions are either objective or are perceived by person.	• Environment creates demands or expectations for occupational behavior.	• Occupational performance is outcome of complex interactions between person and the environment in which he or she carries out tasks and roles. • Performance is facilitated by environmental enablers. • Attention should be paid to the individual's environment and the potential to modify environment and/or access environmental enablers during assessment and intervention.

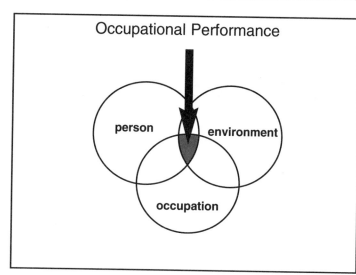

Figure 2-1. The person-environment-occupation model. (From Law, M., Cooper, B., Strong, S., Stewart, D., Rigby, P., & Letts, L. [1996]. The person-environment-occupation model: A transactive approach to occupational performance. *Can J Occup Ther*, 63, 9-23. Reprinted with permission from CAOT Publications ACE.)

an occupational therapy model that could be easily understood by those both inside and outside of the profession, while the scope of occupational therapy practice expands into areas such as ergonomics, health promotion, and occupational justice.

With this model, occupational therapists can easily explain what is contributing to an occupational performance issue or problem, and their unique contributions to assessment, intervention, and evaluation. Occupational therapists can use the graphic illustration of the model (Figure 2-1) to show the relationship of the person engaged in occupation(s) within the context of the environment. The therapist can go on to explain that performance problems occur when there is a lack of fit within this relationship.

During the development of this model, the developers explored how the various ideas and models that emerged from the study of person-environment relationships could be applied to occupational therapy (Law et al., 1996). Many of these models have been discussed earlier in this chapter. The concept of person-environment fit described in the ecological model of aging (Lawton, 1986) and the ideas about matching challenges of activities with the skills of a person as described in the model of flow (Csikszentmihalyi, 1990) helped to shape the ideas toward the conceptualization of PEO fit.

In Lawton's model (1986) the outcomes of a good fit are adaptive behavior and positive affect. When the competence of the person is too low in relation to the demands of the environment, the person can experience negative affect and maladaptive behavior, which could be in the form of feeling stressed and/or experiencing failure of performance, as shown in Figure 2-2. This can be illustrated in the example of a frail elderly woman whose competence with descending stairs has deteriorated. She lives alone in her own home, which has a few stairs at each entrance. The environmental demands or press of the stairs no longer match her competence in mobility, and she is afraid to descend the stairs without help. Consequently, she becomes isolated, withdrawn, and weakened by not leaving her home for weeks at a time.

The model of flow (Csikszentmihalyi, 1990) is similar conceptually. The experience of flow is a positive experience that provides the person with satisfaction. Flow happens when there is a good fit between the person's skills and the challenges or opportunities for action provided by the activity in which the person engages. Like Lawton's model, when the challenges are too great and do not match the person's level of skill, the quality of the experi-

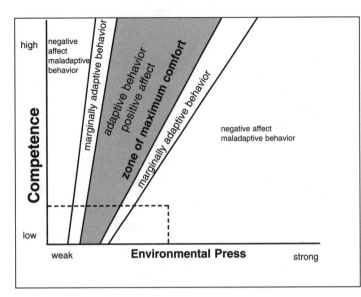

Figure 2-2. Lawton's ecological model of aging. The intersection of the two dashed lines demonstrates poor person-environment fit. (Reprinted with permission from Lawton, M. P. & Nahemow, L. [1973]. Ecology and the aging process. In C. Eisdorfer & M. P. Lawton (Eds.). Psychology of adult development and aging. Copyright 1973 by the American Psychological Association.)

ence of doing the activity is less positive and the person could feel anxious or discouraged. Similarly, if the challenges are low, but the person's skills are high, the activity may not sustain the person's interest and the doing of the activity is not enjoyed. For example, a school-aged child becomes easily bored and disinterested during play with an activity geared for his preschool-aged sibling, because it doesn't provide sufficient challenge.

The theory of person-environment fit as described in Lawton's model (Lawton, 1986) and the notion of person-activity fit as described in Csikszentmihalyi's model of flow (Csikszentmihalyi, 1990) provide conceptual foundations for the PEO model. While the outcome of fit in Lawton's model is affect and behavior, and the outcome of person-activity fit in Csikszentmihalyi's model is flow, or the experience of engagement in the activity and enjoyment, the outcome of a good PEO fit is optimal occupational performance. The nature of the relationship between the person (whether an individual, family or group), the occupation of interest, and the environmental setting in which the occupation will be performed is also different from Lawton's and Csikszentmihalyi's models and is seen as transactional in the PEO model. Transactional relationships imply that the relationship among person, environment, and occupation, and the outcome of occupational performance are viewed as interwoven and inseparable. The examination of occupational performance involves analysis of the goodness of fit for the PEO relationship.

The better the fit, the more optimal the occupational performance. This can be observed in Figure 2-3, which demonstrates how fit increases the overlap of PEO, as depicted in the three overlapping spheres. The area of overlap is labeled occupational performance. When there is less congruence or a poor PEO fit, there is less optimal or poor occupational performance.

The degree of PEO fit is influenced by the passage of time. Priorities and interests can vary over the course of a day, a week, or a lifetime and this will influence the degree of PEO fit at any given time. The variations in PEO fit across a lifespan are illustrated in Figure 2-4. This can be understood through the example of a woman engaged in the occupation of grooming. She may perform this occupation at a more optimal level during her adolescence when this occupation takes on great importance as a part of peer socialization and belonging. The performance of grooming may have less importance for the same woman when she

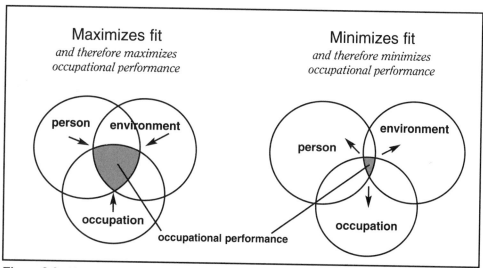

Figure 2-3. Changes in occupational performance as a consequence of variations in person-environment-occupation fit. (from Law, M., Cooper, B., Strong, S., Stewart, D., Rigby, P., & Letts, L. [1996]. The person-environment-occupation model: A transactive approach to occupational performance. *Can J Occup Ther, 63*, 9-23. Reprinted with permission from CAOT Publications ACE.)

starts a family and has less time for herself, and her occupational performance in grooming may be less optimal then. After a passage of time, the same woman may simplify her grooming routines, change her expectations for her own grooming, and experience less environmental press about grooming in order to achieve better PEO fit in grooming.

In the PEO model, the person is understood to be dynamic, motivated and ever-developing, and is a composite of mind, body, and spiritual qualities (Law et al., 1996). People assume a variety of roles, each of which vary in their importance depending upon such things as the person's developmental level, or the environmental presses at a given time. For example, the role of student would typically be prominent for an adolescent. Yet, the role of football teammate may assume greater importance during play-offs, and consequently the occupations of the student role may get less attention.

Occupations are viewed as self-directed tasks and activities in which the person engages to meet intrinsic needs for self-maintenance, expression and fulfillment, within a variety of roles. Environment is broadly defined in the PEO model and the cultural, socioeconomic, institutional, social, and physical elements are given equal importance (Law et al., 1996). A matrix was also proposed to view each of these elements from the perspective of the individual, household, or community (Law, 1991). For example, the cultural environment within one household may apply to the traditions arising from the ethnicity and religion of a family, while in another household the culture is more about the common values and beliefs shared by a group of single, ambitious, professional roommates. These two household cultures may vary greatly, and their influence on occupational performance is quite likely to be different.

Through analysis of the PEO relationship, the factors influencing occupational performance are clearly identified (Strong et al., 1999). This knowledge can be used by the occupational therapist, together with the client, to plan interventions to optimize the PEO fit and the outcome of occupational performance.

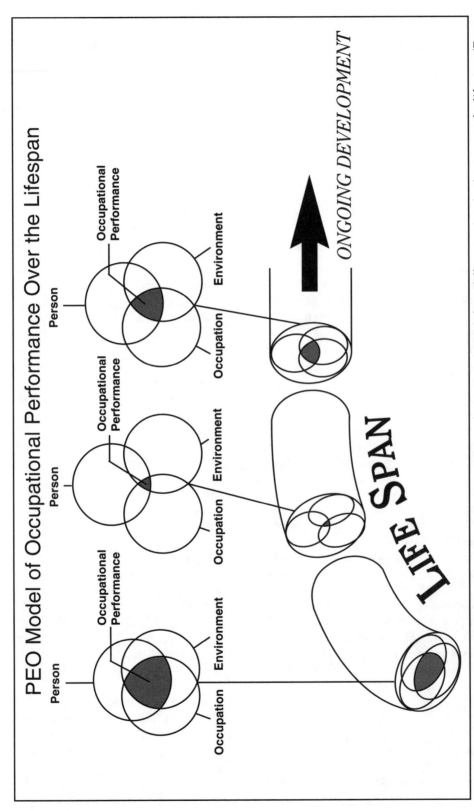

Figure 2-4. Changes in occupational performance as a consequence of variations in PEO fit at different points in time across the lifespan. (From Law, M., Cooper, B., Strong, S., Stewart, D., Rigby, P., & Letts, L. [1996]. The person-environment-occupation model: A transactive approach to occupational performance. *Can J Occup Therapy, 63,* 9-23. Reprinted with permission from CAOT Publications ACE.)

Lyle: An Example of PEO Application in Occupational Therapy

The following example demonstrates how readily the PEO model can be applied in occupational therapy practice. It also shows how easily environments can either be changed or utilized to improve occupational performance.

Lyle is a 22-year-old bank teller who was involved in a motorcycle accident after which his right (dominant) arm was amputated just below the elbow. Following an intensive rehabilitation program Lyle was ready to return home and to his job at the bank. Lyle had become quite proficient using a prosthesis during his rehabilitation, but doesn't like using it, as he is embarrassed by its appearance. He is anxious about returning to work as he does not think that he can manage the specific tasks of a bank teller with one hand, and he is nervous about dealing with the public and how they may react to his disability.

One week following Lyle's discharge from the rehabilitation centre, the occupational therapist, Sue, completed a job-site analysis and found that there was a poor PEO fit for Lyle in performing the tasks required of a bank teller. Most of the tasks required bimanual skills or the use of the right hand, many of which Lyle could not perform. Lyle was also very self-conscious using his prosthesis. Sue introduced a few simple modifications to the physical environment, such as creating a work-space that minimized the bimanual demands and accommodated left-handed function, which enabled Lyle to succeed in performing some of the basic tasks. Lyle found this encouraging. However, Lyle's performance was slow and he had difficulty manipulating the money.

Sue met with Lyle and his bank manager and used the PEO model to facilitate collaborative problem-solving to address Lyle's return to work issues. Sue first showed Lyle and his manager how the loss of Lyle's right hand changed how he was now performing the tasks of a bank teller, as shown in Figure 2-5a. She then showed them Figure 2-5b, to demonstrate how modifying the environment and accessing resources in the environment, in addition to changing some of the demands of the job (the occupation) could improve Lyle's job performance. These explanations using the figures of the PEO model provided a catalyst for Lyle and his manager to consider how the environment could be used or changed to improve the fit with Lyle's current physical abilities. Sue and Lyle's manager also wanted to consider the psychosocial demands of the bank teller job and how these contributed to PEO fit for Lyle. As a group they discussed various options, such as shifting Lyle's role to that of helping customers with their Internet banking by guiding them over the phone. He would then work in a virtual environment in which he would not need to physically manipulate the money, nor interact directly with the public until he feels more able to manage the demands of his previous role of bank teller. Lyle was enthusiastic about trying the new role and found that he could meet the job demands by accessing a few technology aids recommended by Sue.

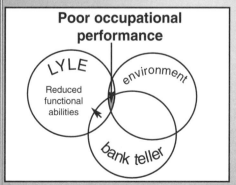

Poor occupational performance

Figure 2-5a. The occupational therapist performs a jobsite analysis and observes that Lyle now has poor PEO fit in performing the occupation of a bank teller. (From Law, M., Cooper, B., Strong, S., Stewart, D., Rigby, P., & Letts, L. [1996]. The person-environment-occupation model: A transactive approach to occupational performance. *Can J Occup Ther, 63*, 9-23. Adapted with permission from CAOT Publications ACE.)

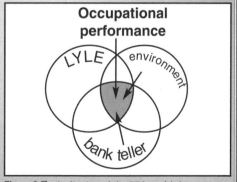

Occupational performance

Figure 2-5b. Application of the PEO model demonstrates how interventions that change Lyle's work environment with modified computer technology, thus allowing him to perform the bank teller role in a new way enables him to successfully return to work.(From Law, M., Cooper, B., Strong, S., Stewart, D., Rigby, P., & Letts, L. [1996]. The person-environment-occupation model: A transactive approach to occupational performance. *Can J Occup Ther, 63*, 9-23. Adapted with permission from CAOT Publications ACE.)

Theory and Environmental
Interventions in Occupational Therapy

As has been demonstrated in this chapter so far, there are many ways to think about environments and how occupational therapists can use environments to enable occupational performance. This text offers many ideas for therapists to use environments in their interventions. It is useful to incorporate a framework into thinking about environments, not as a means to limit the ways that environments can be used, but to help occupational therapists understand and communicate with one another. The PEO model (Law et al., 1996) has received particular attention in this chapter, as one example of how environment can be incorporated into the concepts and practice of occupational therapy. Throughout the remainder of the text, the PEO model or another occupational therapy model is used to frame the discussion.

One of the main purposes of this book is to provide occupational therapists and student occupational therapists with suggestions that they can use when working with clients to optimize their client's occupational performance, and to do so using the environments of the client. This book provides many examples of how environments can be used by occupational therapists to enable occupational performance of clients. This is a new resource for occupational therapists, as no other books that we know of focus exclusively on providing students and practitioners with specific examples and suggestions for interventions that use environments. We have not included an exhaustive or comprehensive collection of environmental interventions, but have selected innovative, cutting-edge practices that have had success in changing environments to enable occupational performance. Section Two presents examples of environments used at their broadest level. The applications are focused on communities and society, rather than specific groups or individuals as clients. We felt that this was an important starting place, because we want to encourage readers to consider environments at their broadest levels. The remaining chapters of the book focus on groups and individuals. Features of the environment that receive attention in the book vary between physical, cultural, social, political, and technological. Settings include schools, clinics, workplaces, homes, and institutions. The perspectives offered by contributors to this book are diverse, and demonstrate the breadth of resources and opportunities that can be accessed from environments to support the occupational performance of the varied clientele receiving occupational therapy services.

Conclusion

An occupational therapist's goal is to promote optimal occupational performance with his or her clients. Recent theoretical developments in occupational therapy have built on theories of person-environment relationships to incorporate environment more explicitly into theories of occupational performance. The challenge now is to ensure that occupational therapists are using the environment optimally in practice. This book provides a number of illustrative examples to challenge occupational therapists to use the environment in intervention to enable occupational performance.

Study Questions

1. Identify three common threads that cross the environmental theories that are described in this chapter.

2. Pick two of the occupational therapy models described in this chapter and examine how the views about the influence of environment on occupational performance are similar and different.

3. Use the concepts from the PEO model to describe how your own occupational performance has varied across a passage of time for a specific occupation that is of importance to you. What influenced the PEO fit for you at two different points in time?

REFERENCES

Bandura, A. (1986). *Social foundations of thought and action: A social cognitive theory.* Englewood Cliffs, NJ: Prentice-Hall.

Barker, R. G. (1968). *Ecological psychology: Concepts and methods for studying the environment of human behavior.* Stanford, CA: Stanford University.

Bronfenbrenner, U. (1977). Toward an experimental ecology of human development. *Am Psychol, 32,* 513-531.

Canadian Association of Occupational Therapists (CAOT). (1997). *Enabling occupation: An occupational therapy perspective.* Ottawa, ON: CAOT Publications ACE.

Christiansen, C., & Baum, C. (1997). Person-environment-occupational performance: A conceptual model for practice. In C. Christiansen & C. Baum (Eds.), *Occupational therapy: Enabling function and well-being* (2nd ed., pp. 46-70). Thorofare, NJ: SLACK Incorporated.

Csikszentmihalyi, M. (1990). *Flow: The psychology of optimal experience.* New York: Harper & Row.

Dunn, W., Brown, C., & McGuigan, A. (1994). The ecology of human performance: A framework for considering the effect of context. *Am J Occup Ther, 48,* 595-607.

Gibson, E. (1988). Exploratory behavior in the development of perceiving, acting and acquiring of knowledge. *Annu Rev Psychol, 39,* 1-41.

Gibson, J. (1977). The theory of affordances. In R. Shaw, & J. Bransford (Eds.), *Perceiving, acting and knowing* (pp. 67-82). Hillsdale, NJ: Erlbaum.

Hagedorn, R. (2000). *Tools for practice in occupational therapy: A structured approach to core skills and processes.* London: Churchill Livingstone.

Kielhofner, G. (1997). *Conceptual foundations of occupational therapy.* Philadelphia, PA: F. A. Davis.

Kielhofner, G., & Forsyth, K. (1997). The model of human occupation: An overview of current concepts. *British Journal of Occupational Therapy, 60,* 103-110.

Law, M. (1991). The environment: A focus for occupational therapy. *Can J Occup Ther, 58,* 171-179.

Law, M., Cooper, B. A., Stewart, D., Letts, L., Rigby, P., & Strong, S. (1994). Person-environment relations. *Work: Journal of Assessment, Prevention and Rehabilitation, 4,* 228-238.

Law, M., Cooper, B., Strong, S., Stewart, D., Rigby, P., & Letts, L. (1996). The person-environment-occupation model: A transactive approach to occupational performance. *Can J Occup Ther, 63,* 186-192.

Law, M., Cooper, B.A., Strong, S., Stewart, S., Rigby, P., & Letts, L. (1997). Theoretical contexts for the practice of occupational therapy. In C. Christiansen & C. Baum (Eds.), *Occupational therapy: Enabling function and well-being* (2nd ed., pp. 72-102). Thorofare, NJ: SLACK Incorporated.

Lawton, M. P. (1982). Competence, environmental press, and the adaptation of older people. In M. P. Lawton, P. G. Windley, & T. O. Byerts (Eds.), *Aging and the environment: Theoretical approaches* (pp. 33-59). New York: Springer.

Lawton, M. P. (1986). *Environment and aging* (2nd ed.). Albany, NY: Plenum.

Lawton, M. P., & Nahemow, L. (1973). Ecology and the aging process. In C. Eisdorfer & M. P. Lawton (Eds.), *The psychology of adult development and aging* (pp. 619-674). Washington, DC: American Psychological Association.

Moos, R. (1980). Specialized living environments for older people: A conceptual framework for evaluation. *Journal of Social Issues, 36,* 75-94.

Schkade, J., & McClung, M. (2000). *Occupational adaptation in practice: Concepts and cases.* Thorofare, NJ: SLACK Incorporated.

Schoggen, P. (1989). *Behavior settings: A revision and extension of Roger G. Barker's ecological psychology.* Stanford, CA: Stanford University.

Strong, S., Rigby, P., Law, M., Cooper, B., Letts, L., & Stewart, D. (1999). Application of the person-environment-occupation model: A practical tool. *Can J Occup Ther, 66,* 122-133.

SECTION TWO

Using Environments to Enable Occupational Performance at the Community and Societal Level

Section Two focuses on the broadest applications of environment and person-environment relations to the practice of occupational therapy. This includes the broad concept of community as a geographical entity, closely associated with society. The section covers aspects of the environment including institutional, social, cultural, and physical environments. These chapters challenge occupational therapists to consider how societies and the structures within them influence occupational performance, and how their skills can be applied in interventions at this broad level.

Changing Institutional Environments to Enable Occupation Among People With Severe Physical Impairments

Karen Whalley Hammell, PhD, MSc, OT(C), DipCOT

CHAPTER OBJECTIVES

* To challenge occupational therapists to work in participation with disabled people toward institutional changes (political, legal, economic, social policy) that will enable occupational choice and performance and enhance quality of living for those with severe physical impairments.
* To encourage occupational therapists to identify those elements of institutional environments that oppress and marginalize disabled people.
* To describe a qualitative research study demonstrating the relationships between people with high spinal cord injuries, their occupations, their environments, and their experiences of quality in living.

INTRODUCTION

"The reasonable man adapts himself to the world; the unreasonable one persists in trying to adapt the world to himself. Therefore, all progress depends on the unreasonable man [sic]."

(George Bernard Shaw, 1903, p. 260)

Occupational therapy has traditionally been framed within a medical model, viewing disability as an individual problem to be addressed on an individual basis, perceiving any problems confronting individuals to be a consequence of their impairments, and attempting to teach each person to adapt themselves to the world (their environment; Law, 1991). This approach has long been rejected by disability activists and academics who have instead proposed a *social model* of disability (DeJong, 1979; Hahn, 1984; Oliver, 1983). The social model claims that the problems confronting people with impairments are the consequence of social oppression and disabling environments.

Although controversial, use of the term "disabled people" within this chapter (rather than "people with disabilities") is a deliberate strategy to acknowledge the perspectives of disabled people around the world who claim this term as a political statement: they are people who are disabled by society's response to their differences (Driedger, 1989; Morris, 1993; Oliver, 1990).

Recent occupational therapy theory has identified the need to consider occupational performance as a complex equation of person, environment, and occupation (Canadian Association of Occupational Therapists [CAOT], 1991; Law et al., 1996, 1997). Occupational therapists recognize the futility of interventions focused solely upon individuals, arguing that we must seek to improve the circumstances of all disabled people by working for environmental changes (CAOT, 1991; Hammell, 1995; Law, 1991).

The Canadian model of occupational performance describes the *environment* as comprising cultural, physical, social, and institutional elements (CAOT, 1997). The *institutional* environment collectively describes the economic, legal, and political elements (policies, procedures, practices, and services) of an environment that exert a strong influence on everyday life (CAOT, 1997). Regrettably, while embracing the challenge to address physical environments, occupational therapists have an inauspicious record in the struggle to change social policies that might benefit disabled people, tending to reserve advocacy in the political and institutional arenas for issues pertaining to our own professional interests (Jongbloed & Crichton, 1990).

This chapter explores those institutional changes that enhanced the person-occupation-environment (PEO) fit (Law et al., 1996) for a group of people with severe physical impairments. The residential Institution will be capitalized in this chapter to differentiate it from the institutional dimensions—political, legal, economic, and social policy—of the environment. The primary aim of the chapter is to encourage occupational therapists to form closer alliances with disabled people and work to change oppressive environments at the societal level.

CHALLENGING INSTITUTIONAL ENVIRONMENTS: POLITICAL AND THEORETICAL CONTEXTS

If we believe that occupational therapy can contribute to the process of creating a non-disabling society we must enter into a dialogue with disabled people to discover effective ways of doing this (Hunt, 1996). Believing that theories are part of the political environment (that theorizing about others can never be politically "neutral"), the present chapter draws both upon theoretical perspectives and models designed by disabled people and their occupational therapists. It is appropriate that a profession aspiring to client-centredness acknowledges the perspectives of those about whom we have seen fit to theorize, determining the degree to which "our" theories respond to "theirs."

Classified by Experts: Medical Models and Deviations from the "Norm"

The WHO developed the International Classification of Impairments, Disabilities and Handicaps (ICIDH) and later the International Classification of Functioning, Disability and Health (ICF) to try to provide a framework for classifying and measuring the consequences of injuries and diseases. Within these models, *impairments* (deviations from socially defined norms) are the root cause for any subsequent problems encountered in everyday lives. Badley (1998) outlined the potential benefits of these systems of classification, for clinical record keeping, computer coding, assessing deviations from normality, and determining eligibility for services and programs. Such benefits are for professionals and bureaucrats, rather than with those people being classified.

By definition, the WHO taxonomies are not needs-assessments but systems for classifying differences of assessing and labelling deviations from an assumed norm (Law, 1992). As Badley (1998, p. 21) explained, "the assessment of disability requires a judgment of what is normal." Critics contend that societal and institutional practices that segregate certain categories of people from others (e.g., by ability, ethnicity, or color) contribute to marginalization by establishing value and inferiority (Charlton, 1998). Marginalization, it is argued, results from classificatory systems whereby certain sorts of people are compared, categorized, hierarchized, excluded, separated, and Institutionalized (Foucault, 1977). While the intent of classifying, measuring, and statistically analyzing divergence from assumed norms may, in some societies, be benign, even enlightened, the consequences for those classified as less than normal can be devastating (including denial of medical interventions, Institutional confinement and, for example, in Nazi Germany, death) (Foucault, 1977; Gallagher, 1990; Pfeiffer, 1998; Shapiro, 1993).

Throughout medical and occupational therapy literature the WHO classifications tend to be used without reference to alternate models or theories, reflecting an unchallenged assumption that this model, or way of thinking is correct or "right" (Northway, 1997). Indeed, Barnes and Mercer (1997) observe that the opposition expressed by disabled people to the ICIDH and ICF classifications has been in complete contrast to their broad and uncritical acceptance by researchers. The very premise that underpins the WHO classifications has been denounced by disability theorists who note that assessments of ability or of "categories of functioning" (the units of classification) require a value judgement of what constitutes normality. They contest the right of "experts" to reserve for themselves the privilege of classifying other people (the "subjects of classification"; Brattemark, 1996, p. 4) and challenge the suggestion that any problems experienced by those deemed "different" should be blamed on their differences (Douard, 1995; Pfeiffer, 1998). Nonetheless, due to their broad acceptance by health care professionals, the WHO taxonomies comprise an important part of the ideological and institutional environment—or context of societal practices— within which disability is currently experienced.

Self-Definition: The Social Model of Disability

Disability theorists reject the idea that disability is an inevitable consequence of individual impairments, viewing disability as a social phenomenon. Their social model of disability claims that it is society that disables physically impaired people; disability is something imposed upon people with impairments through societal practices (Oliver, 1996; although this does not deny the inherently negative dimensions of certain impairments). Hence the Winnipeg-based, global, cross-disability coalition of disabled people—Disabled Peoples' International (DPI)—adopted a two-fold classification of "impairment" and "disability" that had been developed by the British Union of the Physically Impaired Against Segregation (UPIAS) in 1975, 4 years before the ICIDH was published (Barnes, 1998; Oliver, 1996).

Importantly, the social model rejects a causal linkage between impairment and disability, claiming that disability is a consequence of social oppression (Oliver, 1996). Charlton (1998) argues that this experience of oppression includes *marginalization* (classification as deviant from assumed norms, economic and social deprivation, lack of access to employment or to the full rights accorded other citizens, segregation, and confinement), *exploitation* (the social process by which work is organized to enact relations of power and perpetuate economic inequality, including structural barriers—such as lack of accessible transportation—that serve to exclude disabled people from paid labor) and *powerlessness*, that derives from a lack of decision-making power and the inability to enact choices (Young, 1990). Northway (1997, p. 738) observes that, "oppression may arise not just because society actively seeks to disadvantage some groups of people but rather because of the effects of societal norms, laws, and unchallenged assumptions." Oppression may result from both commission and omission.

Reflecting the philosophy that underpins the social model, the ILM developed as a social movement opposed to the compulsory Institutionalization of disabled people (DeJong, 1979). The ILM has an agenda consisting of some basic principles: consumer control over service delivery, the right to take risks, self-determination and the right of disabled people to direct their own lives, and individual choice of service delivery to enable and empower them to pursue their own lifestyles (Oxelgren, Harker, Hammell, & Boyes, 1992). Clearly, the ILM's principles support concepts of client-centred practice to which many occupational therapists aspire (CAOT, 1991; Rebeiro, 2000; Sumsion, 1999) and demonstrate the translation of theory into practice (model to movement).

Institutional Environments: Corresponding Theories

Fundamentally then, the social model of disability that has been articulated by disabled theorists locates the problems experienced by disabled people within civil, political, economic, social, and cultural environments (Johnstone, 1998). Responding to this paradigm that shifts focus from individuals with impairments to the society that disables them, Jongbloed and Crichton (1990) challenged occupational therapists to change from their individualistic approach to disability to a socio-political model and "be less willing to accept individual explanations for problems that are essentially economic, social or political" (p. 32). Richardson (1997) claims, "once professionals working with disabled people accept the barriers [social] model it behooves them to form alliances with disabled people, assisting them with citizenship rights, with managing personal support, liaison with other services, and advocacy in support of the removal of social and physical barriers in the local environment" (p. 1270). If practice is informed by theory, such collaboration will more easily be achieved if occupational therapy theories correspond with those articulated by disabled people.

Recent changes to the Canadian model of occupational performance (CAOT, 1997) acknowledge the role of economic, social, and political contexts in erecting barriers for disabled people (Jongbloed & Crichton, 1990), identifying the various elements of the institutional environment as follows:

> "Societal institutions and practices, *including policies, decision-making processes, procedures, accessibility and other organizational practices. Includes* economic components *such as economic services, financial priorities, funding arrangements, and employment support;* legal components *such as legal processes and legal services; and* political components *such as government-funded services, legislation, and political practices*" (CAOT 1997, p. 46, emphasis added).

It is encouraging to note that since 1997 Canadian occupational therapists have been promoting an understanding of disabling environments that is in tune with that outlined by disabled theorists for the past quarter-century. However, little occupational therapy literature has addressed dimensions of economic, legal, and political environments as these relate to client issues (Sumsion, 1997).

THE IMPACT OF INSTITUTIONAL ENVIRONMENTS: EVIDENCE FROM RESEARCH

This section uses evidence from a qualitative study to explore the impact of institutional environments on perceptions of quality in living among people with severe physical impairments.

The Study

Much research has demonstrated, counter-intuitively, that perceptions of quality in living following spinal cord injury do not correlate with level of function, degree of impairment, or level of physical independence (e.g., Fuhrer, 1996; Fuhrer, Rintala, Hart, Clearman, & Young, 1992). Clearly, these findings challenge rehabilitation's traditional attempts to enhance quality of life through maximizing each individual's physical function and would seem to challenge the logic of the medical model.

In response to a paucity of research into the experience of living with high spinal cord injuries, a qualitative study was undertaken to explore perceptions of life satisfaction among people with high, complete tetraplegia and the factors they identified as contributing to enabling or constraining their lives (Hammell, 1998).

Men and women were invited to participate in the study if they had complete, traumatic spinal cord injuries between C1 and C4, were at least two years post-injury, aged between 20 and 50, and living in the community in the Lower Mainland of British Columbia (BC) or on Vancouver Island. Everyone who was invited to participate agreed to do so. Participants were both men (11) and women (4), of mixed ethnicity. Age at injury ranged from 12 to 44 years, with an average of 24 years of age. The number of years since injury varied from 4 to 28 years.

Semi-structured interviews were undertaken to explore the experience of living with complete paralysis below the neck (Hammell, 2000a) and were supplemented by a reflexive diary and field notes. In an attempt to reflect occupational therapy's client-centred philosophy and to respond to the demands of disability activists for accountable, collaborative, "non-parasitic" research (Stone & Priestley, 1996), there was some involvement of participants in planning the research, analyzing the data, and identifying locations for findings to be published and made available. Thematic analysis of the data was undertaken and the findings were related to contemporary theories of biographical disruption and occupation (Bury, 1982, 1991; CAOT, 1997).

The Findings

Unanimously glad to be alive at the time of the study, several participants reported lives that were more focused, less superficial, and more meaningful and rewarding than prior to injury. The themes that emerged from the data were overlapping and interdependent and described a process of re-establishing a view of the self as able and valuable following injury. Primarily, the participants indicated that the exercise of choice and the opportunity to take control of one's own life are central contributors to perceptions of quality in living. They assessed ability and autonomy not by how many activities they could accomplish without assistance but by how much control they had over their lives.

After injury, the participants felt they had lost so many abilities that they perceived themselves to have no value. Feeling helpless, useless, and unable to contribute to society, life itself held little value. Assisted suicide, for some, appeared a seductive option. Once they were able to gain control of their lives through enacting choices and directing their every-day lives, they redefined themselves as able and competent. Indeed, many no longer per-ceived themselves to be disabled. Autonomy was strongly linked with the experience of a life worth living.

Supporting the proposal advanced by Yerxa et al. (1989, p. 5) that "to engage in occupa-tions is to take control," the participants described the importance of using time in ways deemed personally meaningful to attaining a view of a competent, "able self" (McCuaig & Frank, 1991) and to experiencing quality in living. They emphasized the importance of "doing something" and included within this theme five elements: the need to keep busy, to have something to wake up for, to explore new opportunities, to envision future time engaged in valued activities, and to contribute reciprocally to others (Hammell, in press).

Contextual factors were identified that enabled or constrained everyday lives. These included those "micro" resources—family members, friends, intimates, and assistants—whose relationships affirmed the worth and value of the injured person and whose practical and emotional support enabled the exploration and exploitation of opportunities.

The "macro" context of factors that shaped the experience of high tetraplegia was dis-cussed primarily in terms of deinstitutionalization and as an example of the impact of insti-tutional environments on occupational satisfaction among people with severe physical impairments. This will be the focus of the following sections.

Institutionalization: Inhibiting Occupational Performance

In the absence of policies enabling self-managed community living, many of the study participants had experienced prolonged confinement in long-term care facilities ("Institutions") following injury, some for many years. Prior to the 1990s, people with high tetraplegia were sent from acute care facilities to a ward that had been established to cope with ventilator-dependent survivors of the polio epidemic of the early 1950s and most of the residents had been there for several decades. As Colin explained (all names have been changed to respect privacy), "there's polio patients that have been there for 40 years and still living in the same bed!"

Acknowledging that the intent of the policy of Institutionalizing people with high spinal cord injuries was "to take care of us", they said the reality had been less benevolent. David tried to explain what it was like to live in an Institution, "Well, I mean—hell, really." Beth concurred, "We were basically considered second- or third-class citizens and we were subject to a lot of denigration and abuse... I wanted to leave the [Institution] that I was in so badly that it became less important to me that I had a spinal cord injury than where I was." Owen explained, "They didn't need actual bars on the windows 'cause nobody was going to escape, but it was like a prison in every other sense."

Not solely identified as an experience of powerlessness, the experience of Institutionalization had been one of impoverished occupational choices. The experience of occupational deprivation associated with Institutional living was described as, "sheer boredom... the days were very long and very boring" (David); "I watched TV 10, 12 hours a day, and I really wasn't thinking and was completely wasting my life" (Luke). Alan asked himself, "'How am I going to fill 1 hour, let alone 24?' And I did have the answer then: I could sleep for 18 hours, eat for 2, and that makes 20, so then I could just put up with the other 4." He said, "I felt so useless, I seemed so dependent. I *felt* so dependent."

Institutionalization and Occupation: Summary

It is claimed that Institutional confinement is characterized by powerlessness, low self-esteem, depersonalization, loss of choices and options, lack of fruitful use of time, inflexibility of routines and menus, desexualization, regulation and restriction, recipient status, lack of legal rights, limited social contact, lack of opportunity to engage in productive occupations, subordination, and negative, disrespectful, and belittling staff attitudes (Dijkers, 1998; Lorimer, 1984). These claims were substantiated by the participants in this study.

After existing in the Institution for over a decade Alan observed, "I would go home for 3 or 4 months a year, and so we thought, 'well, why would we need an Institution?'" The movement toward deinstitutionalization had begun.

Deinstitutionalization: Enabling Occupational Performance

The previous section briefly outlined the relationships between people with high spinal cord injuries, their occupations, and their perceptions of quality of living in a specific institutional context. Unwilling victims of specific social policies and practices, discriminatory funding arrangements, restrictive government-funded services, and a lack of legal rights, these people discovered that with the advent of injury came also compulsory confinement within an Institution in which they were expected to live and die.

The importance of occupation in reconstructing a life worth living cannot be over-stated. Studies have suggested that perceptions of quality in living following spinal cord injury are closely correlated with length of life and that, beyond simply enhancing life's quality, engagement in fulfilling occupations may influence survival itself (Krause, 1991; Krause & Kjorsvig, 1992).

The next section explores the relationship between the same people with high spinal cord injuries, their occupations, and their perceptions of quality of living in a different institutional context. Changes in social policies, economic, political, and legal environments—which some of the study participants themselves had initiated—enabled them to move into the community using direct funding to pay for their own personal assistants. Once staff members were attached to people rather than buildings, these people with high tetraplegia were suddenly liberated to explore and exploit all of the opportunities afforded by community living.

Since leaving Institutional custody, the participants' lives have come to resemble those of other members of society. Several have married, and some have fathered children. Some have attained university degrees or are currently studying for academic or professional qualifications. Some are employed full-time in the competitive labor market. Others volunteer or pursue artistic, literary, or business endeavours. They eat at restaurants; attend concerts, sporting events, and parents' meetings; travel locally and globally; and, importantly, they follow their own agendas.

Living in the community, the participants described experiences of life satisfaction that I interpret as ranging from tolerable: "I guess it's pretty rewarding—I keep pretty busy" (Erik) and "I'm up and down but overall I'm glad I'm living" (Jal) to very high. Alan claimed, "life is rich... life is really valuable" and Beth agreed, "I have an exceptional quality of life." While these four comments are representative of the 15 study participants, David was especially effusive, "I wouldn't trade my life for anybody's now, you know, although for a few years there, I sure would have!"

Crewe (1996) suggests that people who make a good adjustment to the sudden onset of spinal cord injury are those who can redefine their values, broaden the range of things that are cherished and decrease the emphasis on physique as a measure of the self. This hypothesis was exemplified by the people in this study, who said, "You have to throw away the way you looked at life" (David). Alan explained, "The person you were pre-injury is the person you are going to be post-injury, but with different values."

Values inform priorities; but time use reflects not only values and priorities. Time use also reflects opportunities and resources. The study participants needed to marshal a wide range of services, resources, and supportive social networks. On a daily basis they utilized mouthsticks (for turning pages and operating some electrical devices); computer input devices, both at home and in the workplace (such as Morse code sip/puff, voice control, and mouse replacements); environmental control systems; telephone systems; accessible work stations; powered wheelchairs (operated by sip/puff and enabling power tilt and/or recline); pressure relief cushions, and lift-equipped vans. However, the opportunities they had maximized—attending college or university, working on artistic and business projects, being a community member, having children—had all been dependent upon leaving residential Institutions; an opportunity that was predicated upon change at the institutional level of the environment.

CHANGING INSTITUTIONAL PRACTICES AND LEAVING INSTITUTIONS

David: "They just *warehoused* us in an Institution and we fought *very* hard to get out of that and they said at first you'll never do it. You'll live the rest of your life in an Institution. And for me and a few other fellows, we thought that wasn't good enough. We wanted more of a quality of life."

Karen: "So, what was it like for you to leave the Institution?"

Nicholas: "A burst of freedom. I was beside myself. I couldn't believe it was true."

Li: "It's a *great* change. You get a lot more freedom. You're running your own life and there's no set schedule where you have to do certain things at certain times... it's back more to the way things were, when you can come and go and do whatever you like, eat whatever you like, eat when you like... you've got a lot more freedom."

Alan: "The quality of life living in the community is immeasurably greater."

It has been observed that marginalized minority groups are analyzed by those wielding more power, "not as citizens, or even people, but as problems to be solved or confined" (Said, 1978, p. 207). Members of the more powerful, dominant group in society take for granted the social policies of which they are the beneficiaries (for example, maternity benefits, schools, pensions); "special programs" are those devised for marginalized others. In speaking of policy initiatives that benefit a defined group, there is a temptation to further marginalize them as a problem to be solved; indeed, as has been shown, this study population was until recently a problem to be confined. Although one could agree wholeheartedly with Young (1990) that: "To promote social justice... social policy should sometimes accord special treatment to groups" (p. 158); it is difficult to discuss special treatment without implying that this is not a right but a favor to be bestowed or withheld according to the whim or "generosity" of the majority. A man who has a spinal cord injury commented, "People sometimes ask me why society should bother with us and I always tell them it's because it's our society as well as theirs" (Slatter, as cited in Hurley, 1983, p. 137).

Steps to Freedom: The First Step

When six young men with high tetraplegia—five of whom were ventilator dependent—moved into an apartment in Vancouver using direct funding from the government to hire their own staff, they became the first group of people with such profound injuries in North America to achieve self-managed care (BC Rehab Society, 1990). This achievement demanded a blurring of boundaries separating provincial ministries and the negotiation and development of innovative programs with government agencies. The project was not without opponents, as Alan explained, "It was seen by the Ministry of Health as impossible; seen by [the Institution] as impossible! And the CEO at [the Institution] at the time was *totally* against it... so we did everything without their assistance."

The group of disabled men examined cost effectiveness and decided that they could probably rent a large apartment and hire their own assistants for roughly half the cost of being in an Institution. An independent study subsequently determined that the actual cost was less than 50% that of Institutional care (BC Rehab Society, 1990). Initiated by people with tetraplegia and supported by the BC division of the Canadian Paraplegic Association (CPA), planning and implementation demanded the involvement of those in positions of power. Much initial effort was expended in challenging those who resisted change, including government officials, hospital staff, medical personnel, and community housing groups; "professional attitudes about the care and *control of care* of those who were severely disabled and ventilator-dependent, had to be changed" (BC Rehab Society, 1990, p. 17, emphasis added).

Planning and implementation were coordinated by the six future tenants and three CPA counselors, with eventual support from an occupational therapist, nursing unit manager, physiotherapist, health care worker, and director of social services from the Institution; and representatives from the long-term care division of the Vancouver Health Department and the community home care agency.

This shared-care arrangement was a huge success, but because the staff members were required to stay on site, any resident wishing to leave the apartment had to make arrangements to be accompanied by family or friends; or run the substantial risk of going out alone. This placed restrictions on opportunities for occupations. "And people were going, 'You're going to spend the rest of your life here' and I'm going, 'No, NO!' I mean, to me that was just a stepping stone—'I'm going to the next stage of my life'" (David).

Steps to Freedom: A Giant Leap

Alan was the first to move out of the apartment and on to the next stage of his life. In so doing, he ventured into what he admits was "virgin territory." He knew exactly what he needed: care assistants to be tailored to individuals rather than tied to buildings, and funding for services going directly to the user to administer. This would give the disabled person autonomy and control over the provision of personal assistance that would not be restricted solely to personal care needs but would extend to every daily task with which help was required.

Again, changes were implemented that were initiated by the men themselves and that enabled complete autonomy. Within this system of direct funding, each person can design a program of support to meet their own needs. For example, someone who lives with a spouse or other family member willing to provide care might only need 8 hours of support a day. Those people who live alone are able to design the support to meet their needs; up to 24 hours of care a day may be required by people who are ventilator dependent. Funding is tailored accordingly. Alan observed, "I manage my life and I direct it. I'm the boss, I just do it differently. Rather than, as I did before, from brain to hand, it's now more, um, brain to mouth, to communicate, so it's perhaps a bit more involved, but I'm in control."

The ability to lead a self-managed lifestyle by directing their own personal assistants contributed to a tremendous sense of autonomy, as illustrated by the following exchange, that was characteristic of many others. I asked, "Can you describe for me a typical day; sort of where you go, and what you do, and how long it takes you?" To which Erik responded, "Well, normally I guess we get started at 5 in the morning, 'cause I'm out of here at quarter to 7. Then I just go across the road, catch the bus over to [my office] which is about a half-hour bus ride; so I'm at work at 7:30. Then I pretty much work 'til 5 and then just take the bus home and then I'm just finishing up my last couple of courses, so I come home and do a couple of hours work, uh, have something to eat, and go to bed—not too exciting!" Achieving autonomy by directing his own care, it is difficult to discern the imprint of profound impairment upon his life, or to recall that Erik is completely paralyzed below the neck.

Despite the evident success of the Vancouver experience of self-managed care for severely physically impaired people, this is not a model that has spawned a domino effect of policy changes across the continent despite studies demonstrating cost-effectiveness, client satisfaction, and reductions in preventable medical complications (BC Rehab Society, 1990; Beatty, Richmond, Tepper, & DeJong, 1998; Mattson-Prince, 1997); and in view of the apparent fit between the deinstitutionalization of young people and current political rhetoric of "care in the community." Unions protest any move toward contracting out of work, and their professional associations have skilled lobbyists working to preserve power (Hammell, 2000b). The status quo benefits many in the majority population.

Many community supports are necessary to enable autonomous living for people with high lesion tetraplegia including (but not limited to) accessible housing, access to transportation, financial support, medical supplies, and equipment. However, it was evident that the ability to manage their own daily affairs by employing and directing their own personal care assistants was the critical determinant in enabling full participation in community living. The policy change that had enabled them to do so was the single most important innovation impacting their lives. Discharge from the rehabilitation centre to the community is now the norm, such that newly injured people with high tetraplegia in BC are no longer sent to Institutions (unless they so choose). This suggests that these disabled pioneers created a new perspective on ability within disability that has permeated not only the ideology of the medical professions but through every level of government in BC. (This is also the reason why some of the participants' voices have been absent in this chapter. Benefitting from the path forged by others, they had been discharged from rehabilitation directly to self-managed, community living).

ENABLING OCCUPATION THROUGH
SOCIETAL INSTITUTIONS AND PRACTICES

Societal institutions and practices comprise "policies, decision-making processes, procedures, accessibility, and other organizational practices" (CAOT, 1997, p. 46). The most significant policy change to enable occupational performance and quality of living was that of allowing the redirection of funds from Institutions to individuals, permitting people with severe impairments to employ and pay staff of their own choosing. However, several of the study participants described continual difficulties with gaining access to basic services—such as ventilator repairs—that they identified as being the result of "power games" rather than policy. This experience was not universal among the group, nor isolated (or universal) among those receiving welfare benefits; this tended to support their premise that the problem was located among individual gate keepers rather than with the policies they were authorized to enact. Beth explained, "I have to work so *damned hard* at keeping my life as it is... I have a feeling that people have good intentions—but they *cannot* see how a person with a [high] spinal cord injury can live a life that's in anyway satisfactory—or that they should be given any opportunity to do that." Through the exercise of power, organizational procedures and practices may subvert the policies they are empowered to enact.

Economic Components

Economic components include "economic services, financial priorities, funding arrangements, and employment support" (CAOT, 1997, p. 46). The majority of disabled people live in poverty and face considerable barriers to participation in the paid workplace (Barnes, 1991; Charlton, 1998; Crichton & Jongbloed, 1998). Clearly, confinement within an Institution effectively prohibits participation in the competitive labor market. Existing income "systems" are a patchwork of separate financial and tax programs, such that the amount an individual with an impairment receives will depend primarily on how he or she became disabled (Prince, 1991), people with injuries being deemed more worthy of financial support than people with diseases.

Individualized funding arrangements that might enable self-managed care have not been a burning priority for Canadian or American governments. Each jurisdiction makes different arrangements for personal care support, adding to the discrepant life experiences of people with the same degree of impairment living in different regions of the same country. Very different direct payment schemes operate in various countries of the European Union, for example, Austria, Britain, Denmark, France, and Germany (Ungerson, 1997).

Legal Components

Legal components include: "legal processes and legal services" (CAOT, 1997, p. 46). Legislation grounded in the principles of human rights and social justice has only recently required that access to facilities and services be available for all, not just certain citizens. Although the Americans With Disabilities Act of 1990 was heralded as a way to end discrimination in both employment and public accommodations, Tanzman (1991) notes that in the absence of personal assistance services, people with severe physical impairments are unable to access their civil rights. Legislated requirements had provided the study participants with various housing options (cooperative apartments, subsidized self-contained apartments or houses, in addition to family homes); the right to workplace access (e.g., physical access, modified work-spaces, adapted computer technology); to education (e.g., physical access, note-taker services, computer-generated examinations); and to public transportation. Accessible hotels, planes, trains, ferries, and buses were used both for personal excursions and for work-related business and were all the result of legislated initiatives.

Political Components

Political components include "government-funded services, legislation, and political practices" (CAOT, 1997, p. 46). Government-funded services had been persuaded to reflect an espoused commitment to honouring the rights of all citizens to self-determination and the right (in the absence of criminal conviction) to live outside an Institution. Evidence suggests that it is considerably cheaper to maintain people with high tetraplegia in their own homes (with direct funding to employ personal assistants) than to warehouse them in Institutions (Zejdlik & Forwell, 1993). As Katherine explained, "over and above the fact that it's a much higher quality of life, it's cheaper!" However, arguing that self-directed care and community living is a right, necessitated by principles of social justice, then as a society we *ought* to adhere to this basic right irrespective of whether it costs less money to do so.

CONCLUSION

"Disability itself does not determine the quality of our lives. Rather, it is the resources available to us which make all the difference. If we have, or can get, the housing and personal care we need, if we have friends and family who value us, occupations (within and outside the home) that we enjoy, then there will be joy in our lives" (Morris, 1989, p. 190).

This chapter drew upon a study into perceptions of life with complete paralysis to demonstrate the relationships between a group of people, their occupations, and their experiences of quality in living, thereby demonstrating a clear link with occupational therapy theory.

Not solely interconnected, occupational performance and quality of living were contingent upon changes to social policy that enabled these people to access the sorts of lives previously denied them on the basis of their impairments. The participants identified the need to escape the restricted and impoverished environment of the Institutions in which some had resided for many years, describing feelings of incarceration and imprisonment, helplessness as opposed to autonomy, and occupational deprivation. They had been (as they explained) unable to contribute to others, unable to participate in education or employment, unable to form relationships, and unable to enact adult roles; they experienced feelings of uselessness and lack of value. Since forcing institutional changes, some of these pioneers have married, had children, completed university degrees, travelled the globe, volunteered, undertaken paid employment, and have, in very many different ways, used their time, energies, and resources to participate in personally meaningful occupations.

Innovative changes in political, economic, and social policy ("institutional") structures enabled occupational performance and the attainment of quality of living. By contesting prevailing perceptions of people with high tetraplegia as dependent, helpless, and in need of care directed by others, these people demonstrated their abilities to manage their own care and in so doing, have changed the options and opportunities for every other person sustaining a high spinal cord injury in BC.

How might these findings be linked to occupational therapy theory and practice? The research provides evidence-based support for the Canadian model of occupational performance (CAOT, 1997) and its depiction of the inter-relationship between individuals and their institutional environments. Further, the strong emphasis placed by the participants on self-determination and the need to be in control of their own lives supports a client-centred approach to occupational therapy practice, characterized by a sharing of information to inform choices, respect for clients and the choices they make, individualized intervention, and a focus upon the relationships between the client, environment, and occupations (Law & Mills, 1998). Clearly, however, occupational therapy practice must more closely match theoretical rhetoric, expanding the focus on environmental intervention to include social, legal, economic, and political arenas. Building ramps is a necessary but inadequate response to the oppression faced by disabled people!

Should this approach be generalized to other populations and settings? Pentland, Krupa, Lynch, and Clark (1992) explored the community needs of three populations: people with physical, developmental, and psychiatric impairments. They identified shared needs and visions among these diverse populations: deinstitutionalization to non-segregated living, consumer-directed care, full participation in society, increased community-based supports, and the opportunity to engage in more meaningful roles and activities. This sounds familiar. Although this chapter has focused upon a defined group of people with specific and severe physical impairments, the issues these people raised—and the impact of particular configurations of institutional environments in either enabling or constraining their occupational choices and performance—are shared by other client groups. The life stories glimpsed in this chapter prompt consideration of the ways in which environments may be experienced as disempowering and oppressive, or alternatively, enabling and inclusive for all people.

APPLICATIONS: STRATEGIES FOR CHANGE

"Citizens who have disabilities experience oppression in many aspects of their lives. The causes of oppression include poverty, other people's attitudes, and the systems of publicly and privately funded support services comprising—laws, policies and regulations; state and private sector funding bodies; agencies which provide services. These systems operate in ways that deny control to those they are intended to serve."
(The Seattle 2000 Declaration on Self-determination & Individualized Funding, as cited in BC Coalition of People with Disabilities, 2000, p. 4).

More than a decade has passed since Jongbloed and Crichton (1990) challenged occupational therapists to become more involved in advocacy with disabled people. Despite their call for an expanded vision of occupational therapy services, it is evident that the profession is still more closely aligned with those of physical therapists, speech and language pathologists, and nurses than with those of architects, lawyers, politicians, or with disability activists. Thus, while paying lip service to the notion of intervention focused on institutional environments (CAOT, 1997), we remain mired in the medical model, squarely focused upon "health care services" and one-to-one clinical interactions and thus are at odds both with our own theoretical model and the social model of disability.

Research has demonstrated that occupational therapists' attempts to enact a client-centred philosophy are subverted by organizational processes and work environments that impose conflicting philosophies and demands (Townsend, 1998). Nonetheless, these are organizational processes in which some occupational therapists themselves collude. This prompted Townsend to call for occupational therapists to challenge and change institutional organizations of power, choosing not to become victims of our own work environments. Having recently broadened our focus from attempting to change individuals to changing environments, "we cannot forget to include within that focus the institutional environment in which our own practice occurs" (Hammell, 2001a, p. 418). Perhaps we will need to challenge and change our own institutional environments, give up power and acknowledge the political nature of our role (Law, 1991) before we can work effectively (and with legitimacy) with clients and others to resolve problems in those institutional environments that serve to oppress disabled people and that conspire to reduce the autonomy and quality in their lives.

Working in Participation

How might occupational therapists work in a participatory manner with disabled people to effect change in macro, institutional environments? To become advocates—to speak for others—and to work in partnership with others requires a clear understanding of their needs and perspectives. We must first gain a clear understanding of the barriers presented by the institutional environment before we can work together, or advocate, for a more just and fair society.

As academics and clinicians we need to forge closer alliances with disabled people, speaking not only to each other in our own theoretical language but with those of whom we speak. We cannot claim allegiance to client-centred philosophy when we are not paying attention to what disabled people are saying. We must not confuse the debate and dialogue that we have among ourselves with debate and dialogue with and among disabled people (Germon, 1998). A large body of theoretical and analytical literature has been generated by disabled and disability theorists yet reference to this wealth of consumer-generated knowledge (e.g. the journal *Disability and Society*, and books such as: Albrecht, Seelman, & Bury, 2001; Barnes, Mercer & Shakespeare, 1999; Barton, 1996, 2001; Barton & Oliver, 1997; Corker & French, 1999; Charlton, 1998; Davis, 1997; Oliver, 1990, 1996; Priestley, 2001; Shakespeare, 1998; Swain, Finkelstein, French, & Oliver, 1993) are virtually absent in occupational therapy texts and journals (where reference to the work of disabled people is usually confined to anecdotal biographies).

Occupational therapists claim a client-centred orientation to practice, yet rarely appear to involve clients in deciding what research should be undertaken. This generates an evidence base that is grounded in a therapist-centred view of the world and provides little insight into the issues, needs, and problems of clients (Hammell, 2001b). Disabled people have criticized researchers for failing to fully acknowledge the role of environmental factors in constraining the lives of people with impairments, producing research perceived to be irrelevant to their needs (Oliver, 1997). If we are to work together with disabled people to change institutional environments, we may first need to give up our own power and work to change the relationships within which research is undertaken and the issues facing disabled people are identified and understood.

Occupational therapists have traditionally worked only on weekdays, and then only during social hours. While working social hours may be viewed as a privilege (and congruent with professional power) it does not fit well with a philosophy of client-centredness (Hammell, 1995; Wigham & Supyk, 2001). Furthermore, it precludes contact with those disabled people who are also employed and prohibits collaboration with disabled people who meet to exchange views or plan strategies in the evenings or at weekends. Advocacy could usefully begin with efforts to move toward flexible working hours, enabling occupational therapists to work in partnership for change at hours best suited to disabled people.

Collaboration with disabled people working for institutional change reflects other forms of client-centred work. There is recognition that disabled people are the experts concerning their lives and the issues they face. There are continual and concerted efforts to comprehend disabled people's perspectives. There is a collaborative process to identify needs, share information, and plan strategies. Disabled people and their families are present at all decision-making tables as equal partners in policy development and planning.

Mary Law's visionary Muriel Driver Lecture might supply a useful mantra for our future practice: "[Occupational therapists] will give up power, acknowledge the political nature of our role, and work together with clients and others to resolve environmental problems" (Law, 1991, p. 178).

DEDICATION

This chapter is dedicated to the lives of my friends Jil Harker, of Regina, SK and Mary Williams, of Vancouver, BC who both died while I was writing it. Following their own high spinal cord injuries, Jil and Mary dedicated their energy towards improving the lives of other disabled people, advocating—most particularly—for deinstitutionalization and the right to self-managed community living.

ACKNOWLEDGMENTS

I acknowledge, with gratitude, the financial support I received for the duration of this study from: The University of British Columbia, Graduate Fellowship, the Rick Hansen Man in Motion Foundation, and the Social Sciences and Humanities Research Council of Canada. Primarily, however, I wish to thank the 15 remarkable people with high lesion tetraplegia who welcomed both me and my study and generously shared their stories with me.

Study Questions

1. This chapter has attempted to provoke and challenge occupational therapists to participate actively with disabled people in changing oppressive institutional environments. From your experience of working with occupational therapists and reading the profession's books and journals, can you provide examples demonstrating how occupational therapists have already been actively engaged in the process of creating social change?

2. What institutional barriers would be faced by someone with a physical impairment (who resided in an Institution) in engaging in adult roles and occupations?

3. As individuals and as a profession, how might occupational therapists work in collaborative partnerships with disabled people in achieving societal change?

4. It has been argued that occupational therapists espouse a client-centred orientation to practice yet refer only to literature produced by their own theorists in speaking about disability. How might students, academics, and practitioners seek to incorporate the theories and perspectives of disabled people into their education and practice?

5. What employment environments would best enable occupational therapists to work collaboratively with disabled people to create institutional change? Are traditional, facility-based employment settings compatible with this approach to practice? What workplace changes could occupational therapists initiate to enhance the fit between their theory and practice?

REFERENCES

Albrecht, G. L., Seelman, K. D., & Bury, M. (2001). *The handbook of disability studies*. London: Sage.

Badley, E. M. (1998). Classification of disability. In M. A. McColl & J. E. Bickenbach (Eds.), *Introduction to disability* (pp. 19-28). London: WB Saunders.

Barnes, C. (1991). *Disabled people in Britain and discrimination*. London: Hurst.

Barnes, C. (1998). The social model of disability: A sociological phenomenon ignored by sociologists? In T. Shakespeare (Ed.), *The disability reader. Social science perspectives* (pp. 65-78). London: Cassell.

Barnes, C., & Mercer, G. (1997). *Doing disability research*. Leeds: Disability Press.

Barnes, C., Mercer, G., & Shakespeare, T. (1999). *Exploring disability: A sociological introduction*. Cambridge: Polity.

Barton, L. (1996). *Disability and society: Emerging issues and insights*. London: Longman.

Barton, L. (2001). *Disability, politics and the struggle for change*. London: David Fulton.

Barton, L., & Oliver, M. (1997). *Disability studies: Past, present and future*. Leeds: Disability Press.

BC Coalition of People With Disabilities. (2000). If update Fall, 4-6.

BC Rehab Society. (1990). *Keys to freedom. Resource manual for the development of self-managed community living alternatives for quadriplegic persons*. Vancouver, BC: Author.

Beatty, P. W., Richmond, G. W., Tepper, S., & DeJong, G. (1998). Personal assistance for people with physical disabilities: Consumer-direction and satisfaction with services. *Arch Phys Med Rehabil, 79*, 674-677.

Brattemark, M. (1996). International classification of impairments, disabilities and handicaps (ICIDH). *Helios, 16*, 4.

Bury, M. (1982) Chronic illness as biographical disruption. *Sociology of Health and Illness, 4*,167-182.

Bury, M. (1991). The sociology of chronic illness: a review of research and prospects. *Sociology of Health and Illness, 13*, 451-468.

Canadian Association of Occupational Therapists (CAOT). (1991). *Occupational therapy guidelines for client-centred practice.* Toronto, ON: CAOT Publications ACE.

Canadian Association of Occupational Therapists. (1997). *Enabling occupation: An occupational therapy perspective.* Ottawa, ON: CAOT Publications ACE.

Charlton, J. I. (1998). *Nothing about us without us. Disability, oppression and empowerment.* Berkeley, CA: University of California.

Corker, M., & French, S. (1999). *Disability discourse.* Buckingham, UK: Open University.

Crewe, N. M. (1996). Gains and losses due to spinal cord injury: Views across 20 years. *Topics in Spinal Cord Injury Rehabilitation, 2,* 46-57.

Crichton, A., & Jongbloed, L. (1998). *Disability and social policy in Canada.* North York, ON: Captus.

Davis, L. J. (1997). *The disability studies reader.* London: Routledge.

DeJong, G. (1979). Independent living: From social movement to analytic paradigm. *Arch Phys Med Rehabil, 60,* 435-446.

Dijkers, M. (1998). Community integration: Conceptual issues and measurement approaches in rehabilitation research. *Topics in Spinal Cord Injury Rehabilitation, 4,* 1-15.

Douard, J. W. (1995). Disability and the persistence of the "normal." In S. K. Toombs, D. Barnard, & R. A. Carson (Eds.), *Chronic illness. From experience to policy* (pp. 154-175). Bloomington, IND: Indiana University.

Driedger, D. (1989). *The last civil rights movement. Disabled peoples' international.* London: Hurst and Co.

Foucault, M. (1977). *Discipline and punish* (A. Sheridan Smith, Trans.). New York: Random House.

Fuhrer, M. J. (1996). The subjective well-being of people with spinal cord injury: Relationships to impairment, disability and handicap. *Topics in Spinal Cord Injury Rehabilitation, 1,* 56-71.

Fuhrer, M. J., Rintala, D. H., Hart, K. A., Clearman, R., & Young, M. E. (1992). Relationship of life satisfaction to impairment, disability and handicap among persons with spinal cord injury living in the community. *Arch Phys Med Rehabil, 73,* 552-557.

Gallagher, H. (1990). *By trust betrayed. Patients, physicians and the license to kill in the Third Reich.* New York: Henry Holt.

Germon, P. (1998). Activists and academics: Part of the same or a world apart? In T. Shakespeare (Ed.), *The disability reader: Social science perspectives* (pp. 247-255). London: Cassell.

Hahn, H. (1984). Reconceptualizing disability: A political science perspective. *Rehabilitation Literature, 45,* 362-365, 374.

Hammell, K. W. (1995). *Spinal cord injury rehabilitation.* London: Chapman and Hall.

Hammell, K. W. (1998). *From the neck up: Quality in life following high spinal cord injury.* Unpublished Doctoral Thesis, The University of British Columbia, Vancouver, Canada.

Hammell, K. W. (2000a). High level injury: Self-managed care and quality of life. *Total Access, 1,* 31-32.

Hammell, K. W. (2000b). Representation and accountability in qualitative research. In K.W. Hammell, C. Carpenter, & I. Dyck (Eds.), *Using qualitative research: A practical introduction for occupational and physical therapists* (pp. 59-71). Edinburgh: Churchill Livingstone.

Hammell, K. W. (2001a). Applying the client-centred philosophy [Letter to the Editor]. *British Journal of Occupational Therapy, 64,* 418-9.

Hammell, K. W. (2001b). Using qualitative research to inform the client-centred evidence-based practice of occupational therapy. *British Journal of Occupational Therapy, 64,* 228-234.

Hammell, K. W. (in press). Using qualitative evidence to inform theories of occupation. In K. W. Hammell & C. Carpenter (Eds.), *Evidence-based rehabilitation: Informing practice through qualitative research.* Oxford: Harcourt.

Hunt, J. (1996). Joining in the dialogue set by disabled people [Letter to the Editor]. *British Journal of Occupational Therapy, 59,* 243.

Hurley, G. (1983). *Lucky break?* Hampshire, UK: Milestone.

Johnstone, D. (1998). *An introduction to disability studies.* London: David Fulton.

Jongbloed, L., & Crichton, A. (1990). A new definition of disability: Implications for rehabilitation practice and social policy. *Can J Occup Ther, 57,* 32-38.

Krause, J. S. (1991). Survival following spinal cord injury: A fifteen-year prospective study. *Rehabilitation Psychology, 36,* 89-98.

Krause, J. S., & Kjorsvig, J. M. (1992). Mortality after spinal cord injury: A four-year prospective study. *Arch Phys Med Rehabil, 73,* 558-563.

Law, M. (1991). The environment: A focus for occupational therapy. *Can J Occup Ther, 58*, 171-180.

Law, M. (1992). Michel Foucault's historical perspective on normality and restrictive environments. *Can J Occup Ther, 5*, 193-203.

Law, M., Cooper, B., Strong, S., Stewart, D., Rigby, P., & Letts, L. (1996). The person-environment-occupation model: A transactive approach to occupational performance. *Can J Occup Ther, 63*, 9-23.

Law, M., Cooper, B., Strong, S., Stewart, D., Rigby, P., & Letts, L. (1997). Theoretical contexts for the practice of occupational therapy. In C. Christiansen & C. Baum (Eds.), *Occupational therapy: Overcoming human performance deficits.* (2nd ed., pp. 72-102). Thorofare, NJ: SLACK Incorporated.

Law, M., & Mills, J. (1998). Client-centred occupational therapy. In M. Law (Ed.), *Client-centered occupational therapy* (pp. 1-18). Thorofare, NJ: SLACK Incorporated.

Lorimer, E. A. (1984). Learned helplessness as a framework for practice in long-term care environments. *Australian Occupational Therapy Journal, 31*, 62-67.

Mattson-Prince, J. (1997). A rational approach to long-term care: Comparing the independent living model with agency-based care for persons with high spinal cord injuries. *Spinal Cord, 35*, 326-331.

McCuaig, M., & Frank, G. (1991). The able self: Adaptive patterns and choices in independent living for a person with cerebral palsy. *Am J Occup Ther, 45*, 224-234.

Morris, J. (1989). *Able lives. Women's experience of paralysis.* London: Spinal Injuries Association and the Women's Press.

Morris, J. (1993). *Independent lives? Community care and disabled people.* Basingstoke, UK: Macmillan.

Northway, R. (1997). Disability and oppression: Some implications for nurses and nursing. *J Adv Nurs, 26*, 736-743.

Oliver, M. (1983). *Social work with disabled people.* London: Macmillan.

Oliver, M. (1990). *The politics of disablement.* Basingstoke, UK: Macmillan.

Oliver, M. (1996). *Understanding disability. From theory to practice.* London: Macmillan.

Oliver, M. (1997). Emancipatory research: Realistic goal or impossible dream? In C. Barnes & G. Mercer (Eds.), *Doing disability research* (pp. 15-31). Leeds, UK: The Disability Press.

Oxelgren, C., Harker, J., Hammell, I., & Boyes, B. (1992). *A new beginning. Attendant services through individualized funding.* Regina, SK: South Saskatchewan Independent Living Centre.

Pentland, W., Krupa, T., Lynch, S., & Clark, C. (1992). Community integration for persons with disabilities: Working together to make it happen. *Can J Occup Ther, 59*, 127-130.

Pfeiffer, D. (1998). The ICIDH and the need for its revision. *Disability and Society, 13*, 503-523.

Priestley, M. (2001). *Disability and the life course. Global perspectives.* Cambridge: Cambridge, UK: University.

Prince, M. J. (1991). *The disability income system in Canada: A literature review and policy analysis.* Report prepared for the Community Services Team of the Premier's Advisory Council for Persons with Disabilities, Province of British Columbia.

Rebeiro, K. (2000, March/April). Reconciling philosophy with daily practice: Future challenges to occupational therapy's client centred practice. *OT Now*, 12-14.

Richardson, M. (1997). Addressing barriers: Disabled rights and the implications for nursing of the social construct of disability. *J Adv Nurs, 25*, 1269-75.

Said, E. (1978). *Orientalism.* London: Routledge.

Shakespeare, T. (1998). *The disability reader: Social science perspectives.* London: Cassell.

Shapiro, J. P. (1993). *No pity: People with disabilities forging a new civil rights movement.* New York: Times Books.

Shaw, G. B. (1903). *Maxims for revolutionists. Man and superman.* London: Penguin.

Stone, E., & Priestley, M. (1996). Parasites, pawns and partners: disability research and the role of non-disabled researchers. *British Journal of Sociology, 47*, 699-716.

Sumsion, T. (1997). Environmental challenges and opportunities of client-centred practice. *British Journal of Occupational Therapy, 60*,53-56

Sumsion, T. (1999). *Client-centred practice in occupational therapy: A guide to implementation.* Edinburgh: Churchill Livingstone.

Swain, J., Finkelstein, V., French, S., & Oliver, M. (1993). *Disabling barriers—Enabling environments.* London: Sage Publications & The Open University.

Tanzman, M. R. (1991). The campaign for a national personal assistance for independent living program. *SCI Psychosocial Process, 4*, 98-99.

Townsend, E. (1998). *Good intentions overruled: A critique of empowerment in the routine organization of mental health services.* Toronto, ON: University of Toronto.

Ungerson, C. (1997). Social politics and the commodification of care. *Social Politics, 362-381.*

Wigham, S., & Supyk, J. (2001). Should occupational therapists work shifts? *British Journal of Occupational Therapy, 64,* 151-152.

Yerxa, E. J., Clark, F., Frank, G., Jackson, J., Parham, D., Pierce, D., et al. (1989). An introduction to occupational science: A foundation for occupational therapy in the 21st century. *Occupational Therapy in Health Care, 6,* 1-17.

Young, I. M. (1990). *Justice and the politics of difference.* Princeton NJ: Princeton University.

Zejdlik, C., & Forwell, S. (1993). *A PhD in life. Living with SCI.* Vancouver: BC Rehab & British Columbia Paraplegic Association.

Occupation, Health Promotion, and the Environment

Ann Wilcock, PhD, GradDipPH, DipCOT, BAppScOT; and
Gail Whiteford, PhD, MHSc (OccTherapy), BAppSc (OccTherapy)

CHAPTER OBJECTIVES

* To enable occupational therapists to recognize the multidimensional and environmental nature of health promotion.
* To assist occupational therapists to become advocates for World Health Organization (WHO) targets by focusing practice toward occupation that promotes health and wellbeing.
* To describe how occupational therapists might address or mediate the underlying social and political determinants of health to advance occupationally focused and enabling practice.

INTRODUCTION

In this chapter the relationship between occupation and health promotion is considered in the context of social and political environments. Specifically, the chapter addresses how, through a health promotion framework, occupational therapists can enable people to engage in health-giving occupations of personal, social, and cultural meaning within their communities. To do that, it is essential to understand the impact of environmental factors on what is possible, or what needs to be the focus of change. This highlights the necessity for occupational therapists to become more socially responsive and politically aware and active. The Ottawa Charter for Health Promotion (WHO, 1986), as it relates to these issues, is revisited and the chapter concludes with a series of recommendations for current and future occupational therapy practice.

The rise of the health promotion movement has occurred in recent times (Jones, 1997). Jones' chronology includes the following key events: the 1974 Lalonde report, 1977 Health For All by the year 2000 launched at the 30th World Health Assembly; the 1978 Declaration of Alma Ata; the 1986 Ottawa Charter for Health Promotion, the 1992 Sundsvall (Sweden) conference on supportive environments, and the 1991 revised WHO Health For All targets. As a relatively new movement, health promotion has, in a short period of time, reached significant populations of people and contributed to noteworthy gains in health status within pockets across the globe. But where that has happened has also been a consequence of ecological and socio-political environments.

Throughout history, health challenges and strategies to improve health have been related to prevailing environments. For example, industrialization in 19th-century Britain created health challenges for armies of workers, while ensuring a viable workforce was of central interest to the new captains of industry (Peersman, 2001). Industrialization, in fact, produced horrendous health conditions even though it was a time when industriousness itself was valued highly. Not least, large groups of people living in newly created urban and manufacturing environments created a need to focus attention on sanitation. Public health statistics of that time demonstrate that the implementation of social and environmental strategies to improve sanitation contributed more significantly to the health status of the country than have the medical and technological advances of recent decades (Wilcock, 1998).

It was largely the case that the era, although epitomized by scientific and technological advances and occasional heroic attempts to promote better health for the underprivileged masses, had many negative impacts on health, on how it was viewed, and on subsequent responses to it. For example, some perceptions about ill, disabled, poor, or culturally different people were being characterized as different or "other" (Whiteford & Barns, 1999). The time also encouraged adoption of conceptualizing the body as a machine.

Biomedicine, which had its foundations in the latter conceptualization, rose to prominence in the 20th century. It has remained the dominant health paradigm, despite retrospective analyses that point to its mechanistic approach to the body and mind, and others of similar ilk, as failings. Although not addressing the complex health issues of communities and populations, for at least the last half-century biomedicine has reaped the rewards of greater resources than interventions to prevent illness and promote health.

Although it is generally accepted that health promotion arose as a defined movement in the 1970s, some authors suggest that the movement in itself was not new, rather a re-visitation of the public health concerns of the 1800s (Green & Kreuter, 1991). Those acknowledged the relationship between illness, poverty, and housing through such developments as the Public Health Act in Britain in 1842 (Peersman, 2001). In addition to the phenomenon of a "renaissance" of interest in the relationships between those health-related environmental factors, more recent antecedents have been linked to the rise of health promotion. These include the burgeoning problem of smoking-related illnesses (and the costs associated therein) that were identified in the USA in the 1960s, and the increasing costs of work-related injuries (Stokols, 2000). The lifestyle focus of the 1970s has also been identified as a cultural context that influenced the development of health promotion (Peersman, 2001). In some respects its largely individualistic orientation has been controversial in population health circles. On a more global scale, the growing identification of health inequalities as being socioeconomically determined (Minkler, 2000) alongside the growing recognition of the threat of civil unrest and war to health of populations (perhaps in aftermath of the war in Vietnam) shaped the philosophical orientation of health promotion (Jones, 1997).

HEALTH AND OCCUPATION: EXPLORING THE RELATIONSHIP

In exploring the issue of promoting better health by occupational means it is useful to recall that all people are products of their environment as well as their inherited constitution. Both impact each other. While individuals can sometimes benefit by a change of environment, the effect can differ according to individual needs of an inborn nature. According to the rules of health known as the *Regimen Sanitatis*, change of environment has been a medical consideration and prescription for health for at least 2,000 years (Wilcock, 2001).

The nature of health has also been the subject of earlier chapters. Here, a few ideas that have been found useful from an occupational health promotion perspective are discussed briefly to set the scene. The last four decades have witnessed an intense focus on health and what constitutes it. Over this period of time it has been considered from multiple perspectives and with a range of foci, from the individual and biological to the collective and spiritual. Definitions of health moved away from simplistic notions relating to the absence of disease to encompass the inherent dynamism and holism of human well-being. Such broadened definitions range from the often cited WHO statement of the mid-20th century, that health is a state of complete physical, mental, and social well-being (WHO, 1946), to the suggestion that it is a "neutral idea relating to non-pathological physical functioning and the fulfilment of ordinary social roles" (Kelly & Charlton, 1995, p. 83).

Jones (1997) makes an interesting suggestion that all definitions of health contain concepts that refer to what identifies a state of health, whose responsibility it is to maintain that state and how health is viewed socially. Similarly, Fox (1999) maintains that all definitions of health "have a politics associated with them; all try to persuade us to adopt a particular perspective on the person who is healthy or ill" (p. 213). Those ideas imply the extent to which environmental contexts and socio-political influences impact on the experience of health.

Kass (1981), in an earlier examination of the complexity of the issues involved, suggested a more individually focussed notion, although he thought it difficult to define health precisely. He offered the simple yet encompassing explanation of health as being when the organism as a whole is working well. To this simple definition, which he described as a natural standard or norm, he added a more complex rider with environmental overtones, that health is a state of being revealed through activity. Kass's linkage of a state of being with a state of doing anticipates the notion of people as "occupational beings." Although earlier than Kelly & Charlton's (1995) similar notion that health is related to fulfilment of social roles, Kass's definition also implies that health is related to the activities associated with such roles.

Those environmental notions of health having a political nature and dependency on social and occupational factors provide a background to consider how people, in general, describe their own health. Blaxter (1990) found that many of a sample of 9,000 adults in the United Kingdom described a relationship between what they or others did and the state of their health. Such respondents provided descriptions about positive health, many of which had environmental as well as social and occupational implications. These included perceptions of physical and psychosocial fitness, energy levels, and functional ability. Younger adults were more likely than middle-aged or older respondents in the study to define health as healthy behavior, stressing the role of "bad habits" in the causation of disease. They associated health with not drinking or smoking, "virtuous" eating patterns, and exercise. For many young men even not staying in bed seemed related to positive healthiness. When they discussed physical fitness, young men tended to point to sporting ability, strength, and ath-

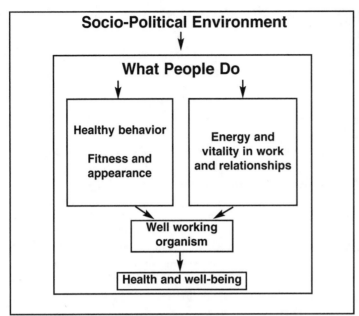

Figure 4-1. Environmental notions of health.

letic prowess, while young women often described aspects of appearance such as body size, complexion, eyes, and hair (Blaxter, 1990, pp. 20-23). The choice of such concepts points to the influence of popular media and commercial interests, probably more than medical advice or natural awareness of the issues. It also raises the idea that if those are, indeed, factors in the health experience, that socio-political environments are very important in creating social climates for better health.

Among older respondents in Blaxter's study the pattern of descriptions about health differed from those of younger adults. Interestingly, energy was "the word most frequently used by all women and older men to describe health" (1990, p. 25). Respondents referred to either physical energy or psychosocial vitality, or combined them. They used words such as being full of life or "get up and go," being lively and alert, and about not being tired or listless (p. 25). For women of any age and older men ideas about energy and vitality embraced enthusiasm about work. Women also defined health as "doing everything easily," "feeling like conquering the world," "being keen and interested," "lots of get up and go," and "having the energy to be with other people" (pp. 25-27). Indeed, women often defined health in terms of relationships with other people. These concepts, too, suggest a need to consider whether sociopolitical environments promote the opportunity for maximizing these effects, as shown in Figure 4-1.

If we consider Blaxter's research as providing some evidence of the occupation/health connection, then Kass's (1981) notion that health is best described as the well-working of the organism as a whole which reveals itself in activity is important. This is because it captures the essential concept that, through actively and regularly engaging in a well-balanced range of occupations, the organism as a whole works well, as shown in Figure 4-1. If we also consider the work of Fox (1999), who suggested that health is a process of becoming, relative to the enactment of human potential, we find ourselves close to some of the concerns of early occupational therapists.

THE PROMISE AND POTENTIAL OF HEALTH PROMOTION IN OCCUPATIONAL THERAPY: A HISTORICAL PERSPECTIVE

Occupational therapists began espousing "optimum health" through occupation about the same time that health promotion was emerging as a popular concept (West, 1967, p. 312). West encouraged occupational therapists to envisage a future in which occupational therapists would function as health agents with responsibility toward enabling normal growth and development (1969). She suggested that occupational therapists should consider more fully the socioeconomic and cultural causes as well as the biological causes of disease and dysfunction. She put forward the idea that occupational therapists could offer programs within primary prevention to stop disability or illness from occurring at all, and within secondary prevention to effect early detection and retardation of progression to more serious conditions. At the Fifth International Congress of the World Federation of Occupational Therapists (WFOT), West proposed a health model for occupational therapy practice that centred on a balanced regime of age-appropriate, work-play activities for people prior to the advent of disease or disability (West, 1969). She also called for a client/community-centred practice to enrich development of people's physical, mental, emotional, social, and vocational abilities.

At the same WFOT Congress, Florence Cromwell (1970) argued for occupational therapists to move into "well care." She addressed what she described as a global trend toward preventive as opposed to curative programs, and appreciated that the profession's concern about patients' behavior in ordinary environments was not only of great importance to health but would also fit well into preventive programs. Geraldine Finn (1977) expressed similar ideas and proposed that a model of occupational therapy practice should be developed based on the notion of how occupation can keep people in a state of health. That, she suggested, would enable occupational therapists to make a unique contribution by providing a greater understanding of a basic structural element of society which, to that date, had not been sufficiently considered. Those forward thinking occupational therapists were in tune with health promoting initiatives of the times, for it was then that the WHO (1978) made the famous plea, "Health for all by the year 2000" at a conference held at Alma Ata in the United Soviet Socialist Republic. There, the need for health professions to re-orient toward preventive programs integrated with those of a curative, rehabilitative, and environmental nature was stressed. The optimal health through occupation approach for all described by West, Cromwell, and Finn linked all people, sick or well, as occupational beings, and alluded to the impact of environments on the promotion of health.

Adding to the environmental notion of occupational therapists having something of value to add to the health promotion debate was Elnora Gilfoyle. In her Slagle lecture, *Transformation of a Profession*, (1984) she advocated for occupational therapists to increase their awareness of social, economic, and political environments so that they could move toward new understandings about the value of occupation and occupational processes within the promotion of self-health. At that time interest in health promotion was gaining ground. However, although some therapists have continued to espouse and establish practice within health promotion spheres, it remains a largely unacknowledged role both within the profession itself, and by the health promotion fraternity.

With present hindsight it is obvious that health for all was not achieved by the year 2000. Viewed retrospectively, the profession of occupational therapy also failed in the ensuing decades to respond to those early pressures regarding a health promotion stance.

Considerably simplifying the number and types of environmental pressures that contributed to that loss, it has been suggested many times that one of the most important failures was when occupational therapy lost its focus on occupation in its increasing alignment with biomedicine and the rehabilitation movement. In light of the importance of contributing to those, it was deemed unnecessary to delve further into the relationship between health and occupation.

THE GOODNESS OF FIT FOR HEALTH PROMOTION IN OCCUPATIONAL THERAPY PRACTICE

The first key concept and value identified in the emergent health promotion discourse was a recognition of the limits of biomedicine, and its attendant influence on the health policies of nations. Second was the identification of the role of the physical social, cultural, political, and economic environments in contributing to an individual's or community's state of health. This centralization of the role of environment is exemplified in the 1997 Jakarta Declaration on Leading Health Promotion, which reiterates the earlier Ottawa Charter's identification of prerequisites for health, namely: peace, shelter, education, social relations, food, income, sustainable resource use, social justice, and respect for human rights and equity (cited in Wass, 2000). Third, and of particular importance to occupational therapists, was an acknowledgment of the relationship between activity, productivity, and health. Finally, valuing the process of empowerment in addressing health inequities has been central in the development of health promotion. How these concepts were framed with respect to desirable outcomes (i.e., improving the health of populations) is perhaps best described through an examination of what is agreed upon by most authors as the single most important document in the history of health promotion, The Ottawa Charter (Breslow, 2000; Carey, 2000; Gorin, 1998; Jones, 1997; Wilcock, 1998).

The Charter, which was the outcome of the first WHO international conference on health promotion, has been described as a "landmark document, laying out a clear statement of action that continues to have resonance for health workers around the world" (Wass, 2000, p. 17). It named three essential issues for all those involved in health promotion. These were caring, holism, and ecology. Additionally, the Charter identified five directions necessary for the achievement of better health for communities globally, these being: the development of personal skills, the strengthening of community action, the creation of supportive environments, the building of healthy public policy, and the reorientation of health services toward positive health (WHO, 1986). Given the relevance of those to occupational therapy's occupational perspective of health promotion and its concern with the multilayered environments in which people engage in occupations (e.g., physical, cultural, socioeconomic, political), each direction is briefly and separately discussed. Woven into this discussion are current examples of how occupationally-focused occupational therapy practice and research initiatives exemplify the values described in each of the five areas (Figure 4-2). A further illustration of a health promotion initiative is shared in Chapter 5 of this text.

Developing Personal Skills

As occupational therapists at present tend to be more involved in the provision of services to individuals than to intervention at a community or political level, the development

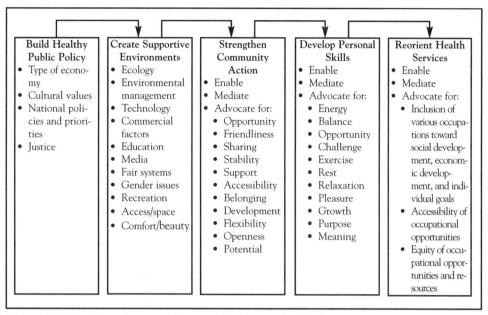

Build Healthy Public Policy	Create Supportive Environments	Strengthen Community Action	Develop Personal Skills	Reorient Health Services

• Type of economy
• Cultural values
• National policies and priorities
• Justice
	• Ecology
	• Environmental management
	• Technology
	• Commercial factors
	• Education
	• Media
	• Fair systems
	• Gender issues
	• Recreation
	• Access/space
	• Comfort/beauty
		• Enable
		• Mediate
		• Advocate for:
		  • Opportunity
		  • Friendliness
		  • Sharing
		  • Stability
		  • Support
		  • Accessibility
		  • Belonging
		  • Development
		  • Flexibility
		  • Openness
		  • Potential
			• Enable
			• Mediate
			• Advocate for:
			  • Energy
			  • Balance
			  • Opportunity
			  • Challenge
			  • Exercise
			  • Rest
			  • Relaxation
			  • Pleasure
			  • Growth
			  • Purpose
			  • Meaning
				• Enable
				• Mediate
				• Advocate for:
				  • Inclusion of various occupations toward social development, economic development, and individual goals
				  • Accessibility of occupational opportunities
				  • Equity of occupational opportunities and resources

Figure 4-2. Five directions to take to achieve better health for communities (WHO, 1986).

of personal skills is the obvious starting place for discussion. The development of personal (and adaptive) skills through enabling occupation is a current focus within occupational therapy. In large part that is concerned with the practicalities of life in the real world of domestic, school, work, and play environments. The usual recipients are those who have or are still experiencing illness or disability of some kind, although the principles are also applicable to others who are experiencing social, mental, or physical challenges outside the context of acute medical care.

Involvement in personal skill development could be viewed as a reflection of the wellness movement that has, to a greater or lesser extent, been incorporated into occupational therapy knowledge and practice in the last decades (Reitz, 1992). It is, however, of much earlier origins within occupational therapy thinking, but that has been disregarded in health promotion circles in some ways because the language used differs. Terminology is often the key to whether occupational therapists and those with whom they work regard what they do as health promoting or an extension of medical care. Failure to promulgate the value of occupational therapy approaches generally is a more worrying reason for lack of awareness of the profession's health promotion potential. In the minds of many it remains associated with the medical model and earlier paradigms of practice.

"Enabling" is one of the key terms within the Charter, and its current centrality in occupational therapy is indicative of questions that also abound within health promotion. Numerous commentators have cautioned that education and skills development within a health promotional framework may potentially be characterized by professional dominance, may be culturally inappropriate, and may further reinforce power differentials (Carey, 2000; Jones, 1997; Peterson & Lupton, 1996). Health-promoting enablers need to be highly reflective of their practice in this arena, for not only is health promotion opposite to biomedicine in mechanistic versus holistic conceptualization of health, it "requires a belief in the strengths and capabilities of the client, who defines and determines health care decisions. The provider remains a resource and facilitator of this process. The client is the expert, and the provider offers useful and meaningful information and skills" (Gorin, 1998, p. 74).

Although statements in the Charter appear to fit closely with occupational therapy values, it is important to ask questions, such as "How is this best done? What is the evidence to support it?" One section in the Charter stresses the importance of enabling "people to learn throughout life, to prepare themselves for all of its stages and to cope with chronic illness and injuries" (WHO, 1986, p. 3). An example of an occupational therapist's attempt to do that is demonstrated in a study undertaken by Pols (2001). She employed qualitative research methods to marry a community-driven, health-promoting approach with personal development interests. Within a selected population in South Australia, Pols considered the question of what is best practice (health-promoting) for residents with dementia in hostels providing low dependency care. In-depth interviews were conducted with a group of people, from a range of disciplines, who were acknowledged by their peers as experts in the field. Analysis of their reported ideas exposed three principles of best practice that related to residents, staff, and management. These suggest health-promoting strategies, some of which are already integral to occupational therapy thinking. The first was that residents' needs should drive responses from staff and management, in spite of the progressive decline of their cognitive and functional capacities. The second principle held that an "enabling" approach is both a feature and a strategy of best practice because it encourages those involved to meet their own needs for autonomy and well-being. The third principle, which underpinned the others, recognized all three sectors as occupational beings in relationship with each other. That is, best practice in this community depended upon meeting the occupational needs of all those within the hostel—residents, staff, and management. This suggests that the health promotion needs of whole communities require attention, and that it is not enough to develop approaches focused on individual skill development alone. Similarly, it can be contended that it is insufficient within larger populations as a whole to develop personal skills, if the community of which individuals are a part does not have the same opportunities and supportive environments.

Strengthening Community Action

"Finding the balance between working with communities through community development and working for communities is an ongoing challenge for health professional workers. Finding the appropriate balance between these requires critical reflection" (Wass, 2000 p. 174). To develop flexible systems for strengthening public participation and direction of health matters, Wass suggests that the most important strategies that a health professional can adopt in doing this include supporting localism, planning action, providing resources, and supporting community members.

Quite a number of occupational therapists have taken community-based rehabilitation (CBR) as a model or "strategy within community development for the rehabilitation, equalization of opportunities, and social integration of all people with disabilities" (McColl, 1997, p. 513). True to a notion raised earlier, it is driven by them but also includes their families and communities, along with appropriate assistance from a range of services and organizations in health, education, vocational, and social services as required. McColl, in her discussion around this issue, was concerned about a lack of preparation of occupational therapists or, indeed, most rehabilitation professionals, for work within community development and health promotion. She suggests that possibly the most compelling advantage of CBR is how, through the processes of community development, communities "learn to listen to disadvantaged and marginalized groups, engage in collective problem-solving, and marshall resources which were often previously unknown or underused" (p. 514).

In order to achieve better health for all, the Charter (WHO, 1986) indicates that the need for communities to be self defining is requisite. Empowerment lies at the heart of the process, and steps toward this are achieved through community action that is both concrete and effective. Action includes the setting of priorities and making decisions as well as planning strategies and implementing them. To do that requires a supportive environment.

Creating Supportive Environments

In its discussion of supportive and rapidly changing environments, the Charter (WHO, 1986) recommends, particularly as part of technology, work, energy production, and urbanization, that systematic assessment of environments be carried out to ascertain their health impact. In this regard, the natural as well as built and socio-political environments are recognized. This should be followed by action toward promoting health using approaches of benefit to the population at large.

The protection and conservation of natural resources is seen as imperative in any health promotion strategy. In addition, care of the natural environment is linked with care of each other and our communities, in recognition of the complexity and interrelatedness of all aspects of life. It proposed the linkage as the overall guiding principle for communities, regions, nations, and the world alike because it is impossible to separate health from other goals. Indeed, it suggested a socioecological approach to health because of an inextricable link between people and their environment. Occupational therapists, while fully appreciating that health and environment are inseparable, might not immediately consider that their professional role is closely connected to wider ecological issues. It could well be an issue that requires debate within the profession.

In explanation of the social environment, the centrality of changing patterns of work and leisure and their impact on health is a point of focus, suggesting that both should be a source of health for people. Indeed, how work is organized by society should be aimed at healthy communities. Furthermore, health promotion should be generating safe, stimulating, satisfying, and enjoyable living and working conditions. This direction, which fits closely with the profession's philosophical foundation, points to the need to extend occupational therapy thinking and practice into the wider societal sphere of planning. In that case, intervention based on the interrelatedness of occupation and health would no longer be confined to corrective work. Issues important from the point of view of enabling the development of personal skills of people with disability could be extended to an advocacy role for wider population benefits. Supportive environments are, however, dependent on public policy that supports and enables their activity.

Building Healthy Public Policy

Public policy itself is as dependent on environmental factors as programs aimed specifically at promoting health by occupational or other means. The Charter proposes that health has to be "on the agenda of policy makers in all sectors and at all levels" in that they need "to be aware of the health consequences of their decisions and to accept their responsibilities for health"(WHO, 1986, p. 3). In the same way, if the occupational needs of all people are to be considered seriously, they too need to be on the agenda of policy makers in all sectors.

"Healthy public policy includes a good measure of health protection and prevention and builds on public health traditions. The main distinction is on getting the health sector to work with other sectors and agencies" (Jones, 1997, p. 72). As this statement suggests, a healthy population requires more than a good ill-health care system with keen, committed health professionals. It takes orchestrated effort within well-informed and appropriately constructed public policy. Legislation, fiscal measures, taxation and organizational change, for example, are not only economically important; all are relevant in some way to health promotion policy and outcomes. One of the major overall challenges, as Gorin (1998) notes, is that policy formulation and enactment requires "a delicate balancing of the state's power to act for the community's common good and the individual rights to liberty, autonomy, and privacy" (p. 40).

It has been argued that power is a central feature in defining justice (Young, 1990). Any community's system of justice is part of enabling or thwarting environments. In the terms of this subject, it can be asked whether the power brokers view positive health for all people as a matter of justice. In terms of occupational therapists' involvement in health promotion, it can also be asked whether the role of occupation in health is viewed as a matter of justice, economy, or serendipity. Refugees seeking political asylum provide an extreme example of occupation not being regarded as of particular health importance. Very limited opportunities for engagement in occupation exist within camps where they are confined, sometimes for long periods of time. Similarly, but less extreme, are the limited opportunities in many care facilities for the aged, and the set curricula of many education establishments—both of which fail to recognize individual or cultural difference in terms of choice according to potential capabilities or meaning. Although such scenarios may be much worse in some countries than in others, in all there is work to be done.

One of the problems of enabling, advocating, or mediating occupation as a health promoting mechanism is the issue of recognizing the "difference" between individuals in terms of justice. Distributive justice is a much more common concept. Young (1990) contrasts a justice of difference with distributive justice, with power as her central interest. Taking the view that biological and social differences are real, and that opportunity is a concept of *enablement*, Jones argues for a justice that enables all individuals to play, work, and live without exploitation or violence in the everyday world (Jones, as cited in Townsend & Wilcock, 2004). Some socio-cultural-political environments do not enable all people to experience and develop individually different occupational needs and capacities as a matter of health. Those environments only give support to occupational issues of an economic nature whether or not they appear to promote health. It can therefore be queried whether systems of social justice have sufficiently recognized a justice of occupational difference. If that is not the case, it suggests a role for occupational therapists in advocating that occupation, as a requirement for health, should be considered and addressed within socio-political institutions and future planning.

Social justice addresses concerns about equal distribution of resources and about social differences arising from individual or group characteristics associated with ability, age, culture, ethnicity, gender, race, and sexual orientation. An occupational justice (of difference) perspective would consider, in addition, what forms of enablement are needed to create individually meaningful occupation for the promotion of health for all people, with or without disabilities. The resources needed for different groups will not be equal; rather justice would be served only if each has the appropriate support to engage in occupation in a meaningful way (Townsend & Wilcock, 2004).

Raising the issue of what is considered or legislated as just in societies is central to the debate of what and where interventions are required for the WHO vision encompassed in the Ottawa Charter (WHO, 1986) to be realized. In the western world, founded on a belief in the equal worth of individuals regardless of difference (Daniels, Kennedy, & Kawachi, 1999), concepts of justice favor the empowerment of individual citizens (Dybward, 1989). McGary (1999) discusses how individuals, alone or collectively, are viewed as active agents who hold the power to allocate resources as means toward a visionary or *ideal* goal. Occupational therapists could become active agents toward changing plans or policy toward the allocation of resources, regulations, or understanding of the need for individually meaningful, health-promoting occupation.

What ideas could occupational therapists put forward to such debate? Most importantly the idea that humans are occupational beings, and that this is a matter of health, requires wider promulgation. Humans need and want to be engaged according to their capacities and interests. We know this to be true because every person seeks occupation of some sort—however affluent or restricted their circumstances are. Without it most people languish or become ill, as is clear from a wide range of unemployment studies. Those who are deprived of occupation through isolation or disability have been found not to flourish. Children or adults who are confined, just as elderly people who are institutionalized, all know that the lack of occupation undermines the human spirit as well as physical and mental capacities (Townsend & Wilcock, 2004).

However daunting a proposition toward environmental activism might be, it must be recalled that just as occupation shapes individual health, so it shapes that of societies, which reciprocally shape the opportunities for communities and individuals to engage in occupations. It is all people who design our communities and societies by action or by default. These either do or do not offer opportunities for people to become educated, exercise and play, relax and enjoy the arts, find employment that suits their capacities, and to alternate those with time to rest, socialize, and sleep. Belief in humans as occupational beings carries with it an understanding that the connection can be negative or positive in terms of health and well-being, according to societal values, institutions, and structures, and whether or not they meet the requirements of "difference."

It can be queried how much opportunities for meaningful occupation for all are based upon sound or informed inquiry. Occupational therapists and students of the profession need to see and think about the consequences of socio-political values and structures in relation to occupation within health promotion. They need to consider what regulations, injustices, or disadvantages occur that might restrict individuals and communities in the practical, everyday world of occupations—even when no social injustices appear to exist. It may be that it is not legislation, but cultural or spiritual values, habits, and routines that are the culprits in preventing people from engaging in what they want or need to do for their health's sake. Some of these are sensitive issues to address, taking both time and courage. The language of the Charter, which proposes "enabling," "mediating," and "advocating" are those worth bearing in mind particularly in situations of great sensitivity (WHO, 1986). It is apt to recall that "Rome was not built in a day," and that the Rome of today is far different to the city referred to in the phrase, and that change is inevitable in the longer term.

In addition to the notion of difference, the concept of an occupationally healthy world is founded on the notion of all humans being able to equitably access occupational resources, rights, and responsibilities that are valued for their health-giving potential rather than, or as well as, their economically productive role. It is also founded on equitable access to facilities or to personnel who can recognize and promote the diversity and differences of individual occupational needs and capacities. But first, there is a requirement to recognize groups whose occupationally determined health disadvantage has not been sufficiently recognized, such as individuals who are isolated, disenfranchised, marginalized, underemployed, overemployed, institutionalized, or socially "ill," the latter in ways not necessarily medically recognized, and to act in ways that will facilitate an enabling environment.

Reorientation of Health Services

"The role of the health sector must move increasingly in a health promotion direction, beyond its responsibility for providing clinical and curative services" (WHO, 1986, p. 4).

That statement does more than suggest that health providers such as occupational therapists need to play an expanded role in addressing the needs of individuals and communities for a healthier life. To do so requires open and explicit avenues of communication between health service providers and those working within the broader social, political, economic, and physical environment. The Charter stressed that "the responsibility for health promotion in health services is shared among individuals, community groups, health professionals, health service institutions, and governments" (WHO, 1986, p. 4) working together.

The Charter's directive that approaches taken to shape health-promoting environments are those of enabler, mediator, or advocate rather than direct policy advisors may hearten many occupational therapists who would not imagine or enjoy a role that was overtly political (WHO, 1986). Most, however, would be able to see themselves as acting to:

* Enable people to achieve their fullest health potential through ensuring, where possible, equity of occupational opportunities and resources, and recognizing individual difference.

* Mediate between differing interests in society toward the promotion of health. Such mediation would aim at enabling men and women to feel able to influence political, economic, social, cultural, environmental, behavioral, and biological factors that determine or support their occupations for health.

* Advocate for the just consideration and inclusion of different occupations seen as individually and communally important to health as a major resource for social and economic development and personal quality of life.

Such agendas can, and should, of course, be applied to all members of populations including those who have been the usual recipients of occupational therapy services as a special group of the greater mass.

Other issues need to be pursued to reduce inequities in accessibility of different opportunities or of obstacles to healthy choices of occupation for many people with handicap or disability of a physical or psychological nature. Taking on the role of advocates or mediators as well as individual enablers for the traditional client group is an obvious first step to having policy makers become aware of the occupational health needs of the population at large. One means of doing this may be for occupational therapists to adopt a "social entrepreneurship" approach (Botsman, 2001). In this approach, stakeholder groups, including business organizations, are galvanized into proactive measures that positively influence the health of populations and concomitantly shape public policy.

DISCUSSION

Occupation, health, and wellness are intimately connected. Doing too much, a common phenomenon in contemporary society for some, has grave attendant health risks. Doing too little, due to occupational dysfunction, alienation, or deprivation is also a major threat to the wellness of individuals and communities (Wilcock, 1998; Whiteford, 2000). Those who have been interested in the relationship between what people do and their health, such as occupational therapists, historically, were active in approaches to social issues and worked at the social environmental level to effect change for marginalized communities of people such as those living in poverty (Wilcock, 1998). In this capacity, they were important agents of change in society, and were characterized by a primary orientation to what would now be called social justice. This meant operating at all levels of environmental intervention, from individual and physical to sociocultural and political. Too close an alignment with the concerns and foci of biomedicine in the latter half of last century, however, meant that the broader occupational, environmental, and collectivist concerns of occupation for health receded. The impact of such an alignment has been in the delivery of instrumental, individually focused interventions, with a concomitant preoccupation with personal hygiene (Wilcock, 1998).

Occupational therapists have for some time been expressing their frustrations at how they are hampered by their difference to other biomedically based professions, by a generalized lack of understanding of their potential contribution, by their small numbers, and by political, social, and economic factors. As suggested in this chapter, a health promotion framework for practice represents the promise and potential for occupational therapy in the future. It signifies, more importantly, a return to the early occupation-focused, population approaches to health in which occupational therapists or their antecedents were champions of social justice. It is probable that, if the profession embraces this mandate, there will be significant occupational challenges to face on behalf of communities and people globally.

To meet the requirements of the sort of re-orientation described by the Charter, serious reflection on the part of occupational therapists is called for. It cannot be dismissed out of hand. The challenge of gearing research and practice toward promoting the health of all people has to be taken seriously, as it was by the WHO, which is a well-respected assembly representative of global concerns. It was addressed to all health professionals, and the spirit of holistic health with which it is imbued seems to have goodness of fit with many of the ideas and beliefs that form the foundation of occupational therapy. If the profession ignores the directive, or chooses not to take up the challenge, a major opportunity to prove the essential nature and worth of occupational therapy will be lost for a second time. Clearly, this is a big challenge but a pressing one. Suggestions and recommendations follow, along with some ideas as to how the profession could proceed to make this shift in the future.

CONCLUSION

The relationship between occupation and health promotion has been considered in the context of sociocultural and political environments. How constitutional health issues and holistic health promotion strategies are at the mercy of environmental factors to a surprisingly large extent has been explored and noted. Specifically, the chapter has addressed how occupational therapists can enable people to engage in health-giving occupations of personal, social, and cultural meaning within their communities as part of a framework founded on the Ottawa Charter for Health Promotion (WHO, 1986). Understanding the impact of environmental factors on what is possible, or what needs to be the focus of change, is essential to that process, necessitating the need for occupational therapists to become more socially responsive and politically aware and active. In closing, it is worth reflecting on one of the many significant statements in the Charter that reminds us of our collective responsibility and the importance of the environment. The promotion of health is dependent on: "...reciprocal maintenance... tak[ing] care of each other, our communities and our natural environment...[so] that the society one lives in creates conditions that allow the attainment of health by all its members" (WHO, 1986, p. 4).

Applications and Recommendations

Accepting the mandate to promote health through occupation for all individuals and communities requires that occupational therapists, in partnership with public policy makers and those they seek to assist, take a pro-active stand toward the attainment of occupationally healthy social, political, and economic environments. To do this, several fundamental steps are necessary:

1. To constantly keep in mind the profession's own mandate, which is concerned with how the health of people can be promoted and enhanced by engagement in meaningful occupation. That is often difficult to do when others, often in positions of power, impose their requirements ahead of those occupational therapists espouse.

2. To become part of health promotion, public health, and community development teams as well as medically-oriented health professional teams. Most importantly these include communities and the population itself, along with working relationships with politicians, social planners, research bodies, and the media.

3. To emphasize health promotion and environmental impacts in professional education and training.

4. To encourage, consume, and engage in research about the fundamental relationship between occupation and the promotion of health.

5. To seek jobs in the community sector that aim at health promotion.

Study Questions

1. List the five strategies of the Ottawa Charter for Health Promotion, and provide an occupational dimension of each.

2. What ideas have occupational therapists proposed that are central to the profession's participation in health promotion?

3. Consider how environmental factors impact the development of personal skills and community attitudes and growth. List three potential interventions to improve the occupational health of individuals and communities.

4. Pick a story out of a current newspaper that illustrates a case of occupational deprivation likely to lead to ill health in the future. Compose a letter to the editor for publication that clearly makes the link between occupation and health status with regard to the article. Write a further letter to a political leader in regard to the same case.

5. Prepare an argument describing how occupational therapists might address or mediate the underlying social and political determinants of health to advance occupationally focused and enabling practice. Write a proposal to the manager of your service that defines how your current program could be changed or expanded to take on a health-promoting role. Detail the advantages.

REFERENCES

Blaxter, M. (1990). *Health and lifestyles.* London: Tavistock/ Routledge.

Botsman, P. (2001). Australia's new health care challenge, enabling good health through social entrepreneurship. In P. Botsman & M. Latham (Eds.), *The enabling state.* Annandale, Australia: Pluto.

Breslow, D. (2000). The societal context of disease prevention and wellness promotion. In M. Jamner & D. Stokols (Eds.), *Promoting human wellness.* Berkley, CA: University of California.

Carey, P. (2000). Community health promotion and empowerment. In J. Kerr (Ed.), *Community health promotion: Challenges for practice.* London: Harcourt.

Cromwell, F. S. (1970). *Our challenges in the seventies. Occupational therapy today—tomorrow.* Proceedings of the 5th International WFOT Congress (pp. 232-238). Zurich: World Federation of Occupational Therapists.

Daniels, N., Kennedy, B. P., & Kawachi, I. (1999). Why justice is good for our health: The social determinants of health inequalities. *Daedalus, 128*(4), 215-251.

Dybward, G. (1989). Empowerment means power sharing. In B. E. Hanft (Ed.), *Family-centered care: An early intervention resource manual, Unit 2.* Rockville, MD: American Occupational Therapy Association.

Finn, G. L. (1977). Update of Eleanor Clarke Slagle Lecture: The occupational therapist in prevention programs. *Am J Occup Ther, 31,* 658-659.

Fox, N. J. (1999). *Beyond health: Postmodernism and embodiment.* London: Free Association Books.

Gilfoyle, E. M. (1984). Eleanor Clarke Slagle Lecture. Transformation of a profession. *Am J Occup Ther, 38,* 575-584.

Gorin, S. (1998). Contexts for health promotion. In S. Gorin & J. Arnold (Eds.), *Health promotion handbook.* St Louis, MO: Mosby-Year Book.

Green, L., & Kreuter, M.W. (1991). *Health promotion planning.* San Francisco: Mayfield.

Jones, L. (1997). The rise of health promotion. In J. Katz & A. Pederby (Eds.), *Promoting health: Knowledge and practice.* Hampshire, UK: MacMillan.

Kass, L. R. (1981). Regarding the end of medicine and the pursuit of health. In A. R. Caplan, H. T. Engelhart, & J. J. McCartney (Eds.), *Concepts of health and disease: Interdisciplinary perspectives.* Boston, MA: Addison Wesley.

Kelly, B., & Charlton, S. (1995). The modern and postmodern in health promotion. In R. Bunton, S. Nettleton, & K. Burrows (Eds.), *The sociology of health promotion.* London: Routledge.

McColl, M. A. (1997). Meeting the challenges of disability. In C. Christiansen & C. Baum (Eds.), *Occupational therapy: Enabling function and well-being* (2nd ed.). Thorofare, NJ: SLACK Incorporated.

McGary, H. (1999). Distrust, social justice, and health care. *M Sinai J Med, 66*(4), 236-240.

Minkler, M. (2000). Health promotion at the dawn of the 21st century: Challenges and dilemmas. In M. Jamner, & D. Stokols (Eds.), *Promoting human wellness.* Berkley, CA: University of California.

Peersman, G. (2001). *Promoting health: Principles of practice and evaluation.* Buckingham, UK: Open University.

Pols, V. (2001). *Experts views of what is best practice in dementia care for hostel residents.* Unpublished master's thesis, University of South Australia, Adelaide, South Australia.

Peterson, A., & Lupton, D. (1996). *The new public health: Health and self in the age of risk.* London: Sage.

Reitz, S. M. (1992). A historical review of occupational therapy's role in preventative health and wellness. *Am J Occup Ther, 46,* 50-55.

Stokols, D. (2000). The social ecological paradigm for wellness promotion. In M. Jamner, & D. Stokols (Eds.), *Promoting human wellness.* Berkley, CA: University of California.

Townsend, E., & Wilcock, A. A. (2004). Occupational justice. In C. Christiansen & E. Townsend (Eds.), *Introduction to occupation,* (pp. 243-273). Upper Saddle Hill, NJ: Pearson Education.

Wass, A. (2000). *Promoting health: The primary care approach.* Marrackville, NSW: Harcourt.

West, W. (1967). The occupational therapists changing responsibilities to the community. *Am J Occup Ther, 21,* 312.

West, W. (1969). The growing importance of prevention. *Am J Occup Ther, 23,* 223-231.

Whiteford, G., & Barns, M. (1999). The world turns. In W. Pentland, A. Harvey, M. Lawton, & M. McColl (Eds.), *Time use research in the social sciences.* New York: Kluwer.

Whiteford, G. (2000). Occupational deprivation: Global challenge in the new millennium. *British Journal of Occupational Therapy 63,* 200-204.

Wilcock, A. A. (1998). *An occupational perspective of health.* Thorofare, NJ: SLACK Incorporated.

Wilcock, A. A. (2001). *Occupation for health: Volume 1: A journey from self-health to prescription.* London: British College of Occupational Therapists.

World Health Organization (WHO). (1946). *Constitution of the World Health Organization.* International health conference. New York: Author.

World Health Organization (WHO). (1978). *Report of the international conference on primary health care.* Alma Ata, USSR. Geneva: Author.

World Health Organization (WHO). (1986). *Ottawa Charter for Health Promotion.* Geneva: Author.

Young, I. M. (1990). *Justice and the politics of difference.* Princeton, NJ: Princeton University.

Chapter

5

Enabling Citizen Participation of Older Adults Through Social and Political Environments

Lori Letts, PhD, OT Reg. (Ont.)

CHAPTER OBJECTIVES

* To place citizen participation in an occupational therapy framework.
* To describe a role for occupational therapists in modifying social and political structures.
* To encourage readers to consider the value of community organizing strategies in varied occupational therapy practice settings.

INTRODUCTION

Traditional occupational therapy practice in the community with older adults has taken many forms including home rehabilitation, prevention education, accessibility consultation, community centre/drop-in centres, stroke recovery groups, support groups, and family advocacy groups. Many of these roles have developed in response to such trends as an increase in the proportion of older adults in the population, decreasing lengths of hospital stays, aging in place, and increasing recognition of the need to support informal caregivers. In many of these roles, occupational therapists are involved in modifying environments. However, the emphasis tends to be on physical built environments, or social environments at the level of the family or group.

This chapter describes an expanding role for occupational therapists to use community-organizing strategies to modify social and political structures so that older adults have opportunities to participate in their communities in ways that are satisfying and meaningful for them.

Citizen Participation as a Focus in Occupational Therapy

Citizen participation is an area of occupational performance that is as valid for occupational therapists to focus on as dressing, perhaps even more meaningful for some clients. In democratic societies, citizens have a right, and in some people's view an obligation, to participate in the decisions made at all levels of government, from local to national. This is at a minimum demonstrated through voting during elections. However, many describe that in a civil society, citizens need to be engaged in ongoing debate about issues of importance to them. Definitions of citizen participation vary, but often include "participating as a voter, as a volunteer and as a member of an interdependent community" (Himel, 2000). Citizen participation has been measured in some studies and has included: informal social contact, social contact through activities in public places, participation in organized group hobbies or sports, participation in individual or group civic activities, and participation in community groups with social and civic purposes (Baum, et al., 2000).

Although occupational therapists' clients may not frequently set goals related to their ability to practice their rights to engage in civic life, they may identify goals to return to former roles in community organizations. Regardless, occupational therapists should have an understanding of the importance of citizen participation to the life of the community and should describe civic participation as an occupational role that they are willing to address in therapy. This idea can be tied to occupational therapy literature on social justice and community development in public health (Townsend, 1993; Wilcock, 1998; see also Chapter 4 of this text). Townsend and Wilcock both argue that occupational therapists need to be prepared to work at the level of the group, community, and society to ensure that people are given opportunities to engage in civic and community life.

Participating in community life can be an important area of occupational performance for older adults. Civic engagement is often described in terms of volunteering, and older adults have time available to contribute to their communities in this way. Older adults make significant contributions to civic life and policy development. In the United States for example, the Gray Panthers have been involved in advocacy initiatives. Such organizations as the American Association of Retired Persons (AARP) and the Canadian Association of Retired Persons (CARP) offer advice to older adults and governments alike. For the older adults involved, these kinds of roles add purpose and are often viewed as their productive roles when they are no longer in the paid workforce. Therefore, civic engagement is an area that occupational therapists should be prepared to address when working with older adults.

Environmental Influences on Citizen Participation

For older adults to be able to participate in their role as citizens and engage within their communities, structures need to be in place to enable them to do so. At the local level, there are many decisions that are made by city governments that influence the health of its older citizens. For example, transportation, housing, and public health policies can all have significant influence on the health and occupational performance of older adults.

Transportation is known to influence the health of older adults. Seniors themselves often identify getting from one place to another as an important contributor to their health and independence (Dunn, 1990; Twible, 1992). La Gory and Fitzpatrick (1992) found that depression was more common among older adults who reported living in neighborhoods with transportation problems. Some of the difficulties reported for older adults using public transportation systems include getting on and off vehicles, walking to the stop, lack of shelter at the stop, standing in vehicles while moving, and difficulty seeing from the bus (Goss Gilroy Inc., 1995; Rittner & Kirk, 1995). Chapter 15 provides an example of how occupational therapists can participate in planning processes to make public transportation more accessible.

Transportation is but one example of the kinds of services provided by local governments that influence the health and quality of life of older adults. It is one of the services that in turn influences older adults' abilities to engage in civic life (an area of occupational performance) and this in turn influences their health. How then can occupational therapists use that information to ensure that environments enable citizen participation? The following scenario describes a project that was undertaken in the city of Toronto, Canada, where an occupational therapist worked with a group of older adults to modify the social and political structure of the city to ensure that older adults were enabled to participate in civic life.

CASE SCENARIO: ORGANIZING THE TORONTO SENIORS COUNCIL

In 1998, Toronto was facing significant changes at the level of the municipal government. Six former municipalities and the metropolitan government were amalgamated into one large municipality at the beginning of that year, as mandated by legislation created by the provincial government (with significant opposition from citizens of the city). Amalgamation meant that rather than having six mayors and city councils and one metropolitan regional chair and council, there would be one mayor and council to oversee local governance in Toronto. At the same time, the city faced major restructuring of municipal and provincial responsibilities, with a number of new areas given to the city that influenced the health of older adults, including public housing, public health, and transportation. Prior to amalgamation, there were two main seniors' organizations within the former cities that had mandates related to monitoring policies at the level of the local governments, and both were involved in education and advocacy initiatives at the local level.

As amalgamation was initiated, many older adults in Toronto were concerned that the voices of older adults would not be heard in the newly amalgamated city. They felt that without a structure in place to organize seniors, seniors might not have a say in decisions made at the local level that could have significant impact on their health and well-being.

A participatory research project was undertaken with a group of Toronto seniors to explore whether or not seniors from across the city shared that concern and felt there was a need for a city-wide seniors' organization. (Participatory research in occupational therapy is discussed in greater detail in Letts, 2003, and Cockburn & Trentham, 2002.) I became involved in the project as part of my doctoral studies. I wanted to explore aging, environments, and health promotion in my studies, and I was interested in conducting my research in partnership with older adults (because I wanted my research to be based on values associated with client-centred practice). I spent the year prior to amalgamation developing a relationship with one group of older adults in the former municipality of North York. As we formed a partnership for research, concerns were expressed about the upcoming amalgamation and restructuring. Together we formulated a set of research questions to explore whether there was a need for an independent seniors' organization within the newly amalgamated Toronto, and what functions and structures such a group might need for it to be created and sustained.

The organizing process involved three phases. First, *reaching out to seniors* involved expanding the membership of the planning group to include older adults from across the city, identifying seniors' organizations across the city and informing them about the initiative, linking with other seniors and community organizations across the city, and applying for and receiving funding. Second, *bringing seniors together* involved planning and conducting a consensus conference. Over 100 people from across the city came together for 1.5 days in November 1999 to identify the key issues facing older adults in the newly amalgamated city of Toronto. This confirmed the need for a new, independent seniors' organization in Toronto. Following the conference, we needed to *plan further action*, including evaluating our efforts to date, clarifying an organizational structure, and setting in place action plans for the future.

These efforts resulted in the creation of a new seniors' organization in the city, known as the Toronto Seniors Council (TSC). The TSC's mission statement "is to be actively involved in working for an improved quality of life and well-being of all seniors in the city of Toronto, whatever their cultural, ethnic, religious background, or socioeconomic status" (TSC, 2000). As the TSC was forming, members of the group articulated a vision for an organization independent of City Hall, but focused on policy development at the local level, communicating regularly with older adults across the city, and working in partnership with other organizations and groups on issues of mutual interest to seniors.

Through the organizing process, we sought to ensure that structures were in place within the city of Toronto to ensure that older adults' voices were heard. In other words, we worked to create an infrastructure (i.e., a seniors' organization) that would be able to identify and improve the circumstances for older adults in Toronto. Participating in the work of the TSC offers older adults in Toronto opportunities for civic engagement. Although it is too early to be sure that the voice is in fact effective, the creation of the organization itself represents a modification to the social fabric of the city.

SOCIAL AND POLITICAL ENVIRONMENTS: AN IMPORTANT FOCUS FOR INTERVENTION

Occupational therapists often focus on physical environmental modifications. At times they concentrate on social environments when they focus on families and support groups. However, this work demonstrates that there is value in focusing on social and political infrastructure at the level of the community, to ensure that opportunities for older adults are created so that they can participate as active members in their communities.

This same need to ensure that social and political environments enable participation is important for other groups of citizens. For example, people with limited literacy may face particular challenges when trying to access information relevant to their eligibility for services. Children with developmental disabilities want to be able to engage in community programs at the local library or community centre. Participation is an essential component of citizenship, and social and political structures must be in place to ensure that people are enabled to be part of that process.

OCCUPATIONAL THERAPY SKILLS
BROUGHT TO COMMUNITY ORGANIZING

From this chapter, the importance of social and political environments to the health and well-being of older adults should be clear. The research project resulted in the development of a new organization designed to ensure that seniors' voices would be heard in the new Toronto. The creation of the organization itself changed the social and political structure of the city.

While recognizing the importance of the work to enable older adults to participate in their community, it is useful to consider how the skills and knowledge of an occupational therapist contributed to the organizing process. At times during the organizing, members of the group who had been professional social workers asked me if I thought this type of work was more appropriately facilitated by social workers. Although it might be more common for social workers to be engaged in community organizing initiatives, this is an area to which occupational therapists can also contribute. Occupational therapists have numerous skills that can be brought to the process of organizing to modify social and political environments. Some of these are highlighted here, based on my experiences in working with community members to organize the TSC.

Instrumental Support

For groups to be able to successfully organize, a number of resources are required, including developing a system to record and implement decisions. Throughout the TSC organizing process, part of my role was to provide instrumental support to the group, through minute-taking and gathering information about funding sources, etc. In many ways, I played the role of a staff member (albeit unpaid) to the group. Kegler, Steckler, Malek, and McLeroy (1998) caution that the least effective outcomes occur when staff members perceive themselves as being responsible for singlehandedly doing the work of the group, and effective outcomes occur when staff members perceive themselves as coaches, linking agents, or coordinators. My approach to the task as an occupational therapist reflected my values in enabling and client-centred practice (Canadian Association of Occupational Therapists [CAOT], 1997). It only made sense to me that I provide a supportive role to the group as we were organizing.

Small Group Facilitation

Since much of the work of community organizing occurs at a group level, it is not uncommon for groups to go through phases of group development. Bettencourt, Dillmann, and Wollman (1996, p. 170) state that "the success of ongoing grassroots organizations may have as much to do with the dynamics of the group as with the cause propelling the activism." Not unlike other small groups, the organizing group of the TSC experienced occasions when low attendance became an issue and other times when there were so many people at the table it was difficult for everyone to feel that their voices were heard. Facilitation from within the group was often needed, both by the older adults leading the process and by me as a participant in the process. Because of my role as a researcher and doctoral student working with the group, I was continually observing processes and documenting decisions. I was able to share my own reflections with the group, and facilitate decision-making by the group on a regular basis.

Writing Funding Proposals

Funding for organizing initiatives is often needed if groups are going to be able to undertake outreach and bring community members together. As the TSC was being organized, we identified a need to obtain funding to assist in organizing the consensus conference. Although writing proposals is not often a specific skill that occupational therapists are taught in their academic programs, they are often required to use writing skills to critically analyze situations and to explain occupational therapy roles in a variety of settings. These skills can be fairly easily translated to developing proposals for funding, which require that an initiative be clearly described, the need for the initiative justified, and with outcomes clearly described. In the case of the TSC, a collaborative approach was taken to developing proposals, so that members of the group (including me) further developed skills in this area.

When considering funding, it is also necessary to identify sources of funding that might be interested in the project being undertaken. In the case of the TSC, although the project was undertaken as a research project, it was funded by an organization (the Ontario Trillium Foundation) as a community organizing initiative. In this case, it was important to present the project to match the interests of the funders. Although the funder required an evaluation of the project, and the research findings contributed to that, the proposal itself emphasized the nature of the organizing initiative. It was useful to consider the project from a number of different perspectives, so that the project might be of interest to a variety of funding agencies. Further, it was helpful to tap into a variety of resources to identify potential funders. Potential funding agencies were identified through personal contacts of members, through organizations with whom the forming seniors' group had links, and through a Canadian directory of foundations that funded both research and community projects.

Evaluation

One of the benefits of conducting community organizing within the context of research is that there is an explicit goal to link the initiatives being undertaken and lessons gained from that to questions that were originally formed. When the TSC organizing process began, we framed our work as four research questions. The four research questions were: 1) Is there a need for an independent city-wide seniors' organization within Toronto? 2) What functions should this group serve? 3) What organizational structure meets the needs of these participants? 4) How can such an organization be created and sustained? Throughout the organizing process, we regularly returned to these questions. Because I was working with the group as part of my doctoral studies, it was part of my function to revisit those questions on a regular basis. Our reflections on our organizing process not only resulted in new learning about the outcomes of organizing, but also about the process. For example, in reaching out to the diverse groups of seniors in the city, we learned the importance of not only sharing information, but also seeking input from others. In terms of the process of organizing, we learned to be patient about the need to revisit past decisions as our membership changed. We have shared our learning at a number of conferences for older adults and health professionals (Letts et al., 2001).

This evaluation framework also proved useful when it came time to evaluate our consensus conference, and report on our progress to our funders. As an occupational therapist with some knowledge and experience in program evaluation and participatory evaluation, I was able to contribute to the process of evaluating our progress and outcomes.

Challenges of Occupational Therapy Involvement in Community Organizing

It is important to acknowledge that occupational therapists may face challenges in working to expand their roles to incorporate community organizing so that social and political environments can be modified. Although the argument has been made here that citizen participation is an important area of occupational performance that should be valued by occupational therapists and others, occupational therapists need to be able to explain that link. Without clearly articulating why social and political infrastructure influence occupational performance, few funders are going to be willing to support occupational therapists working in this way. In turn, if communities or groups do not see the value of involving an occupational therapist in the process, they are not likely to invite one to join their initiatives. Fraser, Letts, and Carswell (1993) described a cycle of education, demand, and opportunity for occupational therapists' engagement in health promotion that is also relevant here. Without education about community organizing, occupational therapists will not be prepared to seek or find opportunities. Without demand from the public, funders will not be willing to fund occupational therapy services at this level. And without education and demand, opportunities will not exist. The three components are all needed to ensure that occupational therapists are able to demonstrate their potential contributions in community organizing and modifying social and political environments.

Occupational therapists also need to be prepared to think about their skills and abilities in generic terms and apply those at the level of the community or group. For example, rather than conducting individual assessments, a needs assessment may be needed at the level of the community. Groups may benefit from facilitation skills brought to the process by occupational therapists.

CONCLUSION

Summary of Environmental Intervention

The focus of this work with members of the TSC was on the social and political infrastructure in place in one city that enabled older adults to participate in civic life. Although occupational therapists don't often think of themselves as being in a position to influence social and political infrastructure, our skills can in fact influence that level of the environment. Although increasing numbers of occupational therapists are working with well older adults, this project was unique in its emphasis on organizing within the context of the community, and working to ensure that social and political structures were in place to enable seniors to participate in their communities. Emphasis throughout this project was in collaboration with an expanding group of older adults. Participatory research was deliberately chosen to ensure that the questions, actions, and reflections were based on a partnership between the seniors and the occupational therapist working with them. To some extent, this project broadens or expands our traditional understandings of evidence-based practice. It is an innovative project that is grounded in research and learning in partnership. The life experiences of the older adults involved and the learning that we gained throughout the process are in themselves evidence. The reflections we shared throughout the process provided an opportunity to ensure that the group felt some consensus on the learning. In fact, members of the group have been involved in reviewing products from the research (such as this chapter) to ensure that their perspective is reflected in any information shared about our work together.

The research can also be linked to current occupational therapy practice and theory. Occupational therapists working in community and institutional settings can be involved in community organizing. Our practice skills of needs assessment, group facilitation, and client-centred practice with individuals and groups put us in a good position to assist with community organizing initiatives. It may, however, be necessary for us to justify a role in this arena. Our understandings of the link between engagement in meaningful occupations and health provide important explanations for why participation is an important goal of community groups.

Occupational therapy theory acknowledges broad definitions of environment; however, in practice we tend to spend less time on social structures or political environments. This work encourages occupational therapists to consider these as potential sources of environmental modification to promote optimal occupational performance.

Applications and Recommendations

Based on the experiences from this initiative, occupational therapists can work with community groups to ensure that citizens are given opportunities to participate in civic life. This would represent a new approach to working with communities, though it fits well with occupational therapists' emerging interest in health promotion processes. There are many groups with whom occupational therapists work and for whom the lessons from this work might be applied (e.g., working with families in childcare centres, working with older adults living in sheltered care settings, working in developing countries, working with people with physical disabilities, and working with people with chronic persistent mental illness). Although the issues addressed with these groups are variable, the need to consider social and political structures is consistent, as is the potential role of community organizing initiatives by occupational therapists.

Why should occupational therapists be involved in focusing on social and political environments when there is so much to be done working with people with disabilities, and at the level of the individual? Labonte (1994) clearly argues for the need to consider both roles as important:

> "Unless professionals think simultaneously in both personal and structural ways, they risk losing sight of the simultaneous reality of both. If they focus only on the individual, and only on crisis management or service delivery, they risk privatizing by rendering personal the social and economic underpinnings to poverty and powerlessness. If they only focus on the structural issues, they risk ignoring the immediate pains and personal woundings of the powerless and people in crisis (p. 259)."

Although this chapter has demonstrated the potential role that occupational therapists can play in working with community members to modify social and political environments, that is not to imply that we should not continue working with individual clients to enable their occupational performance. Rather, occupational therapists should consider an expanded role, where they can contribute to occupational performance and health at the level of the individual as well as the community. By doing so, they are more likely to optimize opportunities for people to engage in meaningful roles as citizens in their communities, thereby optimizing occupational performance.

Study Questions

1. Is citizen participation an area of occupational performance? Explain.

2. What contributions can occupational therapists make to community-organizing initiatives?

3. Think about yourself in your community. How do you participate in community life? What activities do you engage in that fit the following categories: informal social groups, formal social/recreational groups, individual civic activities (e.g., signing petitions), or group civic activities (participating in protests)? Are these activities important to you personally? How are they important to the community?

4. Think about your role as a student in the occupational therapy program (now or in the past). What are the social or political factors that influence your student experience? What resources/supports are available to you? What issues create problems for the student body? How do students work to change the social and political environment to their own advantage? How could those strategies be translated into working with clients in the community?

REFERENCES

Baum, F. E., Bush, R. A., Modra, C. C., Murray, C. J., Cox, E. M., Alexander, K. M., & Potter, R. C. (2000). Epidemiology of participation: An Australian community study. *J Epidemiol Community Health, 54*, 414-423.

Bettencourt, B. A., Dillmann, G., & Wollman, N. (1996). The intragroup dynamics of maintaining a successful grassroots organization: A case study. *Journal of Social Issues, 52*(1), 169-186.

Canadian Association of Occupational Therapists (CAOT). (1997). *Enabling occupation: An occupational therapy perspective.* Ottawa, ON: CAOT Publications ACE.

Cockburn, L., & Trentham, B. (2002). Participatory action research: Integrating community occupational therapy practice and research. *Can J Occup Ther, 69*, 20-30.

Dunn, P. A. (1990). The economic, social and environmental obstacles which seniors with disabilities confront in Canada. *Canadian Journal of Community Mental Health, 9*(2), 137-150.

Fraser, B., Letts, L., & Carswell, A. (1993). Health promotion issue paper—Effective change: The education, demand, opportunity equation. *The National (Canadian Association of Occupational Therapists' Newsletter), 10*(2), insert.

Goss Gilroy Inc. (1995). *Transportation and disability in Canada—An overview* (TP 12545E). Ottawa, ON: Author (prepared for Transportation Development Centre, Policy and Coordination, Transport Canada).

Himel, S. (2000). *Citizen engagement.* Ontario Health Promotion E-mail Bulletin, 158. Retrieved March 26, 2002 from: http://www.ohpe.ca/ebulletin/.

Kegler, M. C., Steckler, A., Malek, S. H., & McLeroy, K. (1998). A multiple case study of implementation in 10 local project ASSIST coalitions in North Carolina. *Health Education Research, 13*(2), 225-238.

La Gory, M., & Fitzpatrick, K. (1992). The effects of environmental context on elderly depression. *Journal of Aging and Health, 4*(4), 459-479.

Labonte, R. (1994). Health promotion and empowerment: Reflections on professional practice. *Health Education Quarterly, 21*, 253-268.

Letts, L. (2003). Occupational therapy and participatory research: A partnership worth pursuing. *Am J Occup Ther, 57*, 77-87.

Letts, L., Berger, B., Berger, B., Dix, D., Knowles, A., McClelland, B., & Wells, L. (2001). Seniors organizing in a city undergoing amalgamation: Lessons from a participatory research project. *Gerontology, International Journal of Experimental, Clinical and Behavioral Gerontology, 47*(suppl 1), 518.

Rittner, B., & Kirk, A. B. (1995). Health care and public transportation use by poor and frail elderly people. *Social Work, 40*, 365-373.

Toronto Seniors Council. (2000). *Working to improve the quality of life of seniors in Toronto* [Brochure]. Toronto, ON: Author.

Townsend, E. (1993). Occupational therapy's social vision. *Can J Occup Ther, 60*, 174-184.

Twible, R. L. (1992). Consumer participation in planning health promotion programmes: A case study using the nominal group technique. *Australian Occupational Therapy Journal, 39*(2), 13-18.

Wilcock, A. A. (1998). *An occupational perspective of health.* Thorofare, NJ: SLACK Incorporated.

Culture as Environment: Complexity, Sensitivity, and Challenge

Sue Baptiste, MHSc, OT Reg. (Ont.)

CHAPTER OBJECTIVES

* To enhance practitioners' existing awareness of the importance of culture as a seminal piece of occupational therapy practice.
* To position this awareness in a context of culture as an environmental influence.
* To explore culture's inherent complexities.
* To explore and describe opportunities to promote sensitivity to culture as an environmental influence in occupational therapy practice.

INTRODUCTION

"Culture is like a security blanket... [that] has great meaning to its owner."
(Spradley & McCurdy, 1987, p. 37)

Contextually, over the past 2 decades, occupational therapists have tended to introduce "culture" as an element in practice largely due to the increase of attention afforded to it within the literature; and also, because of the implications of practicing within multicultural environments be they one's own country or the global village, be they mosaics or melting pots. Many of us strive to become more aware of, and comfortable with, our own cultural "baggage" and assumptions. However, how many of us utilize the true power and richness of culture when assisting our clients toward a more occupationally engaged future?

From the time of my formative years in England, I knew that the world and its people fascinated me. I seemed to gravitate to those who looked different, spoke differently, or were attracted to different life styles. Once I entered the profession of occupational therapy, those feelings of wanting to know and understand others from other places only became stronger. It was specifically during my work with chronic pain patients that I became totally committed to the belief that culture as an element of environment is a critical piece of the occupational therapy picture; that without considering it and celebrating it, we miss a vastly important component of what it means to practice from a truly occupation-based perspective, within a client-centred context. It is with this in mind that this chapter was written.

This chapter explores the very essence of what culture is, relating this to the underpinnings of occupational therapy practice and providing opportunities for the reader to become intimately aware of the potential that exists in cultural interfaces with culture as environment. This chapter is formatted differently from other chapters in this book to encourage readers to reflect actively on the ideas presented in relation to themselves and their practice.

WHAT IS CULTURE?

"Culture: learned and shared human patterns or models for living; day-to-day living patterns... Culture is mankind's primary adaptive mechanism."

(Damen, 1987, page 367)

This is a complex and very challenging statement. In the literature, there is a multitude of definitions from various bodies of knowledge. It has been defined as "every aspect of life: know-how, technical knowledge, customs of food and dress, religion, mentality, values, language, symbols, socio-political and economic behavior, indigenous methods of taking decisions and exercising power, methods of production and economic relations and so on" (Verhelst, 1990, p. 39). Culture is "learned and shared... Human beings cope with their natural and social environment by means of their traditional knowledge" (Spradley & McCurdy, 1987, p. 4). Similarly, "culture is at once socially constituted (it is a product of present and past activity) and socially constitutive (it is part of the meaningful context in which activity takes place)" (Roseberry, 1989, p. 42). Within the occupational therapy literature, there are many descriptions of "culture" as a core construct for practice (Dyck, 1989; Kinebanian & Stomph, 1992; Paul, 1995). As Krefting states, culture is "a blueprint or organizing framework to guide daily behavior" (Krefting, 1991, p. 2).

We also find varying schools of thought concerning whether culture is learned or inherited; whether it is constituted by visible, tangible symbols and rituals or whether it is a living, dynamic process that changes as people cope with, and adapt to ever-emerging conditions and expectations (Banks & McGee, 1989; Damen, 1987; Hofstede, 1984; Kluckhorn, 1976).

A common element across many interpretations of "culture" is that it is learned. We all learn what it means to be a part of a family, a certain social group, a certain region, and a particular country or part of the world. We are not born with these behaviors and understandings, but we assimilate them as we grow by watching others around us and by mimicking behaviors, acquiring skills, and responding to a natural desire to participate as one of the crowd. As we become more independent and mature, we then recognize the need to act upon some personal choices. At this stage, we may elect to jump wholeheartedly into the cultural heritage from which we stem or we may choose the reverse. We may decide that a totally different lifestyle is what we crave. It is thus that culture often is a result of choice. Without exception, however, culture is shared. This almost seems overly simplistic—an obvious and therefore extraneous statement. However, I believe it bears mentioning, because it is from the knowledge that culture evolves from shared beliefs, values, norms, and expectations that new ideas are fostered and alternative models for living emerge. Following this line of thinking even further, it becomes apparent that culture is cumulative.

Culture also is not simply that which arises from shared ethnicity. All too often, the notion of being sensitive to cultural issues in professional practice focuses on where clients are from, the color of their skin, their facial features, and so on. This is only making assumptions from the superficial. While not wishing to patronize the reader, it seems necessary to restate a well-recognized truth—that generalizations have no place in health care practice, particularly in client-centred, evidence-based occupational therapy practice. Culture can also be construed as referring to any group that forms to support a lifestyle based upon a specified set of values, beliefs, rites, and behaviors (Banks & McGee, 1989). Consequently, it is critical to view certain lifestyle alternatives as essential cultural influences when working with clients and attempting to partner with them to create positive occupational change.

> *Anna was a very personable-looking, 30-year-old woman. She was attending the chronic pain clinic after experiencing a serious road accident that resulted in upper back and neck pain. She approached our initial interview with openness and a clear sense of what she hoped to gain from the relationship. Within the first half-hour of our conversation she advised me she was a leading member of a local Wiccan coven, that her faith and lifestyle were very important to her, and she hoped this would not be a problem for me.*

✳ What would be your response to this situation?

After many years of practice within a chronic pain clinic, and working predominantly with clients of different ethnic cultures, I prided myself on being interested and able to respond in a caring manner to whatever faced me; however, this was a first for me. Anna and I shared many conversations over the next few weeks, during which she enlightened me about the different approaches to the occult. This case example opens up another aspect of the cultural debate—that of the close relationship of culture to spirituality. In fact, I often wonder if one could view one's own spirituality as one's inner culture. This remains but an unformed idea, albeit one that certainly merits exploration. This leads us to the next point of inquiry—namely, how aware are we of the cultural nature of practice, on a day-to-day and client-by-client basis?

THE THERAPIST AND PERSONAL CULTURAL AWARENESS

To begin this discussion, it is essential to explore our sense of our own cultural selves, and to raise our awareness of personal bias, preference, and choice. As with any other area of practice, to enhance cultural awareness and sensitivity for the overall benefit of quality service, requires a declared comfort with our own "baggage."

> *Brian strolled into my office one late summer morning, complete with tattoos, heavy leather boots, and studs everywhere. The red bandanna around his head managed to contain his curly, long red hair. He had suffered a broken thumb and soft tissue injuries of his left hand in an accident on his motorcycle.*

✳ What are your first thoughts upon hearing about this client?

This was another challenge for me. Brian tested my beliefs about myself. Before the end of the first assessment interview, he had noticed a book on the coffee table in my office–*Zen and the Art of Motorcycle Maintenance* (Pirsig, 1984). He asked if he could borrow the book, and I willingly agreed, though I was wondering if he knew that the title was not truly a reflection of the content. Two weeks later, at our second appointment, Brian talked in a thoughtful and conceptual manner about the excellence of the book—how much he had appreciated what it had shown him. In essence, his articulation about that complex and existential work was one of the most profound I had heard. In fact, in his reflections he shared many insights that I had not understood for myself. I was both ashamed and shocked. I realized that I had made a set of assumptions because of his appearance.

So, how can we best reflect on our own cultural history in order to declare it, bring it into the open, and thereby minimize any negative impact on building therapist-client partnerships?

One of the most effective strategies is to reflect upon our own cultural development and consider the implications for effective client-centred practice. The following questions can be used to initiate a conscious process of self-examination.

* Where do my parents and grandparents come from?
* Where did I grow up? The city, the country, traveling around?
* What position do I hold in my nuclear family—only child, oldest, youngest, middle, twin?
* Have I traveled outside my own country?
* Who are my closest friends? Are they like me?
* When I think about my life, which people have been most influential?
* Do I enjoy different kinds of food?
* Do I speak more than one language? If not, would I travel in foreign countries without knowing the language?

CULTURE AND COMPLEXITY

One of the most difficult aspects of addressing "culture" and trying to understand it is that it means so many different things. As mentioned earlier, culture can refer to mainstream concepts such as race and ethnicity, or to emerging contexts such as sexual orientation, lifestyle choices of sports, substance use, choice of domicile, and so on. Similarly, "culture" can refer to the context in which a relationship is engaged at any given time; as such, "culture" becomes an environment of immediacy—it is the way in which we attempt to relate one to the other; it is the kind of setting we create for such interchange; it is the very nature of interaction through language, gesture, symbol, posture, and expression. Anthropology is the science of cultural study; in his excellent overview of key anthropological theorists, Moore (1997) offers many useful insights, not the least of which is the following "partial list" of insights into human nature gained through the first century of anthropological thought:

* Race does not account for variations in human behavior.
* Other cultures are not "fossilized" representatives of earlier stages in human evolution.
* There is a complex [relationship] between individual and culture in every society. Individuals are shaped by and shape the culture they experience.
* Culture is not a thing of "shreds and patches," but neither is it a smoothly integrated machine.
* Our knowledge of other people is shaped by our own cultural experience. "There is nothing simple about understanding another culture" (Moore, 1997, p. 274).

It is against this backdrop of concepts and ideas that I would like to embark upon an exploration of how we engage in the client-therapist contract. What are the elements of the different components of clinical interaction in which we participate together with our clients utilizing client-centred principles, seen through a "culture as environment" lens?

THE INITIAL INTERVIEW

Coming from a background of predominantly institution-based professional roles, I first mentally configure the idea of an initial interview as a meeting in an office, clinic, or an environment that is obviously one related to health care. I see it as a small room, often without a window, with a table or desk, several chairs, bookshelves, perhaps an x-ray viewing box, a weighing scale, and a side table containing boxes of dressings. I could go on. It is certainly not a warm, personal, or inviting environment. It gives off a message of distance, efficiency, and fiscal concerns (the furniture often tends to be shabby or sparse); this sense is further exaggerated by the expectations of clinicians to wear white coats or the sporting of other symbols of presumed expertise such as a stethoscope. In the community, the sites of initial client contacts tend to become much broader and are often situated within clients' homes or within widely varying workplaces. In this sense, the interview environments are often familiar to the clients, thus providing more of a sense of comfort for them, albeit not necessarily for the clinician.

So, what then is the culture of therapy, of health care?

* In which kinds of environments do I work?
* Do I feel comfortable in the interview space when I see a client for the first time?
* What kind of impression do I think the interview space gives to someone seeing it for the first time?
* How do I usually set up the space to participate in an initial interview?
* Do you think the interview space, and the way it is set up, is congruent with client-centred principles?
* If not, what could/should be changed?

Obviously, this discussion and reflection on the space in which we do what we do has relevance to the other stages of the therapeutic endeavor, so let's consider the process of assessment and intervention in a similar light.

ASSESSMENT AND INTERVENTION

Depending on the scope of our practice, the places in which we complete our assessments and interventions are many and varied. They range from the formal therapy department to a construction site, and all steps in between. Regardless of the specific environmental focus, each and every assessment location brings with it its own cultural considerations. In addition, we bring our own cultural expectations and assumptions as well as those of the client in question. There emerges this intensely complex matrix of interdependent and interactive elements that tends to fall into the background of the operational agenda of both assessment and intervention.

Establishing a client-centred context for assessment is, in essence, the beginning of intervention. Despite their different foci, assessment and intervention often interweave and the boundaries, of necessity, must be fluid. Similarly, the idea of utilizing the most culturally appropriate environment in which to complete the assessment, and, later, to continue the intervention, must be central to the overall enterprise.

> *Fred was referred by his insurance company for an occupational therapy evaluation of his current status relative to returning to work. Fred is a 60-year-old pig farmer who has a small acreage outside a large town in Ontario, Canada. He was injured in his barn, when he slipped on a very muddy, wet day. He fell heavily on his side, injuring his back, and had to stay away from his farming chores for over 6 weeks. A neighborhood youth has done the heavy work during that time. Fred has four children, none of whom have been interested in continuing working on the family farm. They are all away either at school or working in another province. His wife, Sonja, also 60 years old, is most supportive despite her encroaching osteoarthritis. The couple came to Canada from Germany approximately 40 years ago; both had been raised on farms in Bavaria. Their command of the English language is adequate for day-to-day conversation. They both have agreed to the occupational therapy interview, assessment, and evaluation being completed on their property, although they have stated that they are concerned about what the occupational therapist will do when she comes and what kind of questions she will ask.*

As the occupational therapist in question, I drove out to Fred and Sonja's farm on a very snowy day in January. The property appeared quite old, but immaculately kept. The sheds where the pigs were kept looked newly painted; despite the usual odor associated with pigs, it was not overwhelming. The farmhouse, although quite small, was a charming century-old building in gray stone. I was invited into the kitchen where the table was laid with mugs and dishes, and a wonderful smell was emanating from the large wood-burning stove. Fred presented as a quiet man while Sonja bustled around supporting herself by holding on to the furniture, using her walker only as much as was absolutely necessary.

* What cultures are interfacing here?
* What potential cultural issues can be seen to be emerging?
* Are there any foreseeable problems that are emerging?

If we consider the overlapping "cultures" at play in this scenario, there are several.

Therapist: I am a first generation immigrant from a rural, small town, English background; I have been educated to a graduate level in my profession. At that time, my role was a mixed one of educator, clinical practitioner, and consultant. As a professional, I am aware of the expectations placed upon me by my professional association and the regulatory body, the provincial college. I know that I am expected not to accept gifts, and that I must behave in an ethical, professional manner. I also know that this farmhouse reminds me very much of experiences from my youth, and that I feel immediately at home with Sonja and Fred.

Clients: They too are first generation immigrants; they have brought their values and cultural priorities from the place of their birth to their new adopted country. They have striven over the years to create a firm foundation from which they could raise their family, while pursuing farming as their livelihood. Both Sonja and Fred are from backgrounds that value strength, independence, and who possess a very profound work ethic. They experienced a serious sense of loss when their children decided not to follow in the family business. Therefore, it became even more imperative that they both strive to maintain their world on their farm for as long as possible. They are private and proud people who do not wish to reject offers of help, but find it very difficult to accept them. They see themselves in a current situation of helplessness and dependence, not the least of which is dependence upon external financial support, albeit minimal.

Insurance Company: In this scenario, there exist many interrelating cultural entities, not the least of which is that of the third party payer—the insurance company. There has developed a definitive culture of disability management that incorporates the payers, the workers, the assessors, the treatment providers, and the employers. As with any independent cultural entity, this one demands knowledge and the confidence to determine a special interpretation and understanding in order to negotiate the rules and guidelines and recognize potential pitfalls.

Cultural Interactions: The therapist and clients share much in common; their backgrounds in their countries of origin are not dissimilar; their search for new opportunities brought them all to Canada. Their appreciation of rural lifestyles is also a common element. It is when the interfaces with the third party payer are examined that the areas of core differences are observed. Both the clients and the therapist are coming from cultural contexts to which the context of the insurance industry is foreign and unclear. Of particular importance in this scenario are the differences between the professional philosophy and ethics of the occupational therapist, and the expectations and demands of the third party. In most cases one can expect an initial parrying between these players to determine the scope of assessment and the desired intervention. In a few situations, there will remain a serious difference of opinion and belief concerning the role that the occupational therapist believes he or she should fulfill, and that which is expected by the payer.

* Review your responses to the previous list of questions.
* Think about how you would respond now.
* Reflect on the differences between these two responses.
* How do they differ?
* What are the key areas of learning for you in this scenario?
* How does understanding these elements better impact on your practice?

Intervention at the Institutional Level

There are clearly different skill sets that are more applicable in some settings than others. The skill of practicing from a cultural perspective is one of the core skills of occupational therapy: the customizing of this skill set is based upon the sensitivity of the practitioner to determine which cultural elements are at play in each situation.

Cecil has recently moved into The Colonnade, a modern long-term care facility, situated on the outskirts of a major city. His wife of 55 years died 3 months ago, leaving Cecil unable to take care of himself. He had a stroke last year and has been left with a moderate degree of disability, particularly in his mobility. He also suffers from severe pain in his back, a legacy from many years of working as an auto mechanic. Cecil has three children, all of whom live within driving distance of The Colonnade; however, all have very busy lives, and felt it best for everyone if "Dad" went into a facility where he could be well cared for. Cecil has decorated his room with bright wall hangings; a steel drum hangs from the wall above his bed; there are many photos of sunny beaches, palm trees, and smiling people in sparkling costumes. Cecil came from the West Indies about 30 years ago and met his wife in Canada. She was from Guyana. They enjoyed a very active social life, traveling widely, attending gatherings with their friends every weekend, and going to dances and "fetes" whenever they could. The couple was very proud of their seven grandchildren, and spent as much time with them as possible.

Cecil presents a very familiar picture for those of us who reside within multicultural communities. Bearing in mind the discussion process utilized in the last scenario with Sonja and Fred, think about Cecil and his life situation using a "culture as environment" perspective.

* Which cultures are interfacing in this situation with Cecil?
  • Individual
  • Institutional
  • Other
* What potential cultural issues are beginning to emerge?
  • Individual
  • Institutional
  • Other
* Are there any potential problems emerging within this scenario?

Cecil was referred to occupational therapy service at The Colonnade after one of his daughters voiced her concern regarding his mood change. She stated that her father was always so involved with the world around him, exhibited an active sense of humor, and was continually "on the go." Now, since he moved into his room in The Colonnade, he was keeping himself very much to himself, had stopped calling her and her siblings on the phone, and did not participate in any of the activities within the residential setting. She said that she was finding herself getting irritated with her father and she did not like that at all. After having this conversation, the charge nurse on the wing where Cecil had his room took these comments to a team meeting and there was a decision to refer Cecil to occupational therapy, as well as to include him in the therapeutic recreation programs that were ongoing throughout the week at different times of the day.

On first meeting, Cecil talked to the occupational therapist in a very cordial manner, although he appeared rather quiet and waited to answer questions rather than offering any information of his own. Nevertheless, after completion of the COPM (Law et al., 1998), it was apparent that Cecil saw his main problems stemming from his sense of total disconnection from his family and the life he knew. The key areas of concern for him were his lack of any productive roles and his loss of familiar and cherished leisure pursuits. He wanted to feel useful by still spending time with his grandchildren and taking care of them; similarly, he missed being able to socialize with his friends and enjoy the music of the islands. After a relatively brief time period, the occupational therapist succeeded in establishing a good working relationship with Cecil, sharing her experiences of visits to the West Indies, and looking at photographs of his home and family members. With his approval, she contacted his daughter, through whom there were plans made for the grandchildren to visit Cecil and for Cecil to go to their houses for short visits on weekends and some evenings. The occupational therapist spoke with the recreationist, who agreed to include calypso, steel band, and brass band music in her entertainment afternoons. Cecil was encouraged to join these sessions from time to time, and he made the acquaintance of several residents who were eager to share their life experiences, particularly around traveling and family. Throughout these negotiations and the development of the intervention plan, members of the administration of the facility were helpful and interested in achieving a positive outcome. This was most fortunate, because there could have been a very different result if the institution did not embrace a client-centred perspective. Overall, Cecil became more interactive with his environment, in the facility as well as within the homes of his children.

Again, let us explore the overlapping "cultures" in this particular scenario.

Therapist: The occupational therapist in this particular scenario was very open to experiencing new cultures and had traveled widely over most of her adult life. She was married to someone from a totally different background, and enjoyed the challenge of helping to make Cecil's day-to-day life in his new residence one that was as congruent with his previous lifestyle as possible. She had a good relationship with the therapeutic recreationist, therefore providing a positive, caring context against which the plans could be made. She functioned from a strong occupation-based practice framework, embracing client-centred principles easily in her daily practice.

The therapeutic recreationist was someone who was committed deeply to the philosophy of her discipline, valuing the importance of facilitating an environment that supported the values, beliefs, and preferred occupational patterns of the residents.

Clients: In this case, both Cecil and his daughter were clients. Cecil brought the rich heritage of the West Indies to these interactions, as well as a strong investment and belief in the importance of family. The daughter, while accepting the need for her father to move into a protected environment, was not comfortable seeing him so changed and having lost his joy in living. She, in turn, cherished the value of family relationships and connections, and was delighted to see the arrangements made for her father to become more engaged in his world.

Long-Term Care Facility: The philosophy of care within The Colonnade is one of normalization; when new residents move in, there is an expectation that they will bring items of personal furnishing as well as memorabilia, pictures, ornaments, and other possessions that are happy reminders of their rich lives. Each room is decorated slightly different from the next. Colored linens are used, and there are many plants and comfortable chairs and couches available throughout the common areas. The staff members wear uniforms that do not look like uniforms. This is an environment that radiates warmth and caring.

Cultural Interactions: Cecil's children may have done their homework well when selecting suitable living arrangements for their father. This is an institution that does not think like one; therefore, there is an openness and willingness to engage with the residents in creating a harmonious living environment. This environment provides for individual preferences and a link with the past, thus facilitating from the very first day of residence the creation of an acceptable bridge from one chapter of life to the next. Cecil's cultural needs are well respected, the systems in place for recreational events are flexible, and the clinical staff are interested in the residents and committed to doing their job in a professional and caring way. The only concern remaining is that these adjustments and individualized plans were not forthcoming without Cecil's daughter's concerns being voiced. This would suggest that, while the institutional culture is welcoming and flexible, not all staff members perhaps buy into this.

INTERVENTION IN A SYSTEM

There have been several occasions during my career when I have had the chance to work with broader systems to bring about opportunities for emerging roles for occupational therapists. One in particular provides an excellent example of converging cultures from an individual to a system level.

Trinity Cottage is a large house on the outskirts of the urban centre of a medium-sized city. It is one of three similar buildings that house a network of residences for individuals who are recovering from addictions of varying kinds. Trinity House is the main residence for women who have experienced problems with addictions to alcohol, drugs, gambling, and self-destructive lifestyle patterns. The administration members of the network have identified that they see a definite need for occupational therapy services; however, despite several attempts to gain additional funding, they feel it is a hopeless cause. The occupational therapy faculty at the neighboring university were approached by occupational therapy practitioners within the community and a meeting was arranged between the faculty and the administration of the addiction network. The focus of this meeting was to find out whether there would be any chance of working with occupational therapy students and faculty to pilot a role in Trinity House with potential for consulting to the other residences. The results of a pilot project of this nature would provide excellent evidence upon which to base future funding requests.

This scenario is reflective of others that have become much more common over the past few years. Occupational therapy practice has been moving slowly away from the structured, contained world of institutional practice. This shift provides a wide variety of emerging opportunities for innovative practice settings to showcase the unique contributions that can be provided by occupational therapists. Moving our thoughts away from the "person" level of health care delivery, think about the different cultural connections that are represented within this latest case study.

Reflect upon the cultural interfaces that we have explored so far:

* Which cultures are interfacing in the case of Trinity House?
  • Individual
  • Institutional
  • Social
  • Political
* What potential cultural issues are emerging?
  • Individual
  • Institutional
  • Social
  • Political
* Are there any potential problems emerging here?

The professional practice coordinator met with the administrator of Trinity House and began negotiations around the kind of relationship that could be possible. Very soon, other occupational therapy faculty members were involved in the discussions at the university and it was agreed that there would be a placement offered at the site for the first time in the coming semester. One of the students had already shown an interest, because she was one of the individuals to identify the possible opportunity in this environment. Consequently, the first student placement at Trinity House began that spring with supervision being shared by the program administrator on site and one of the occupational therapy faculty members who would periodically visit. Discussions were still ongoing because there was a belief that the formal evaluation of the pilot placement could be part of the research internship that was a required part of the curriculum. The same student requested to complete her research internship at Trinity House, and was subsequently approved by the curriculum committee.

During the total placement time of 12 weeks across two terms, the occupational therapy role began to come together. Initially, the student recognized that she had to move very slowly, trying to become an accepted member of the household in as unobtrusive a manner as possible. Together with other house staff, she began forming relationships with residents as they became more comfortable with her presence. Intervention strategies tended to be more individualized than group-based. Interventions identified by the residents varied from learning how to dress appropriately for job interviews to finding out how to enroll in evening classes; one resident wanted to become capable of having her son visit, requiring many hours of coaching in parenting skills and the use of strategies for increasing her self-esteem. A small group stated their desire to learn how to create a budget and then keep to it. The student participated in the daily routines at Trinity House, becoming recognized as an important member of the small community. Parallel to her direct service role, the student was completing the program evaluation in order to make a presentation at a regional meeting addressing the needs and future directions of the addiction network. While dollars are yet to be forthcoming, there have been two other students placed within the setting with equal success.

In this scenario, the cultural interfaces are somewhat different than those that have been discussed so far.

Therapists: There were two distinctly different perspectives inherent within the therapists who interacted with the role development at Trinity House. The first was a student occupational therapist. She was a mature student who had changed career paths when entering the occupational therapy program, married, from strong Italian roots. Despite being nearly 30 years old, she embraced her student role with great energy. Consequently, she exhibited great enthusiasm for the tasks with which she was presented. Her political persuasion was toward the left, with a clear sense of social justice and appreciation of the needs of vulnerable populations. The second therapist component is the faculty members involved—the professional practicum coordinator and the program chair. The practicum coordinator is a woman in her early forties, of Chinese descent, who has committed a major part of her career to the role she now holds. The program chair is this author, about whom the reader knows a great deal already; in essence, a woman in her mid-fifties who emigrated from England and has been involved in the service sector in senior positions as well as academia. These two faculty members bring very different world views to this innovation, one more conservative than the next.

Clients: The women of Trinity House came from across Canada, from cities, small towns and rural areas. All had struggled, or were still struggling, with trying to get out of the nightmare of addiction. Most were dealing with drug withdrawal; many with alcohol dependence. Some had had to extricate themselves from abusive relationships, of which several were associated with prostitution and negative relationships with pimps. Others had exhibited self-mutilating behaviors and the majority had attempted suicide at least once. Their ethnic backgrounds resembled a microcosm of the world in the color of their skin, their religion, and their country of origin. Predominant were Caucasian and native North American women. Together with this broad ethnic base came the complex culture of the drug world and the frightening profile of the world of prostitution.

Trinity House: Built at the turn of the century, Trinity House provided a sense of old world charm and affluence. This is a large house with many bedrooms, high ceilings, and renovated plumbing and electricity. There is a small garden at the front, and another at the side and the back that links with the next residence in the network. The furnishings are comfortable and inviting; the interior design was provided by local businesses, so it is tastefully decorated without being ostentatious. Each resident has her own room, and there are three bathrooms, one on each floor. The kitchen is friendly and chaotic; it is where everyone lends a hand, where many informal "group sessions" take place over a good meal and where changes of mood are handled, recognition of problems is given credence, and new friendships come to light.

The Addiction Network: The network is, in essence, a service agency with a difference. It is funded through governmental monies, following a lengthy advocacy process at the local and provincial levels. Champions of the cause were identified within the ranks of local politicians, philanthropists, and business people. The three houses are on the edge of property owned by the local general hospital and are rented for a nominal sum. The residences provide services to both men and women, although only one house is designated for the treatment of women at this time. Running these houses requires a strong sense of political know-how, together with sound business sense and solid clinical knowledge and skills. The staff is comprised of nurses, addiction counselors, social workers, a psychologist, a chaplain, and an administrator. Consequently, this rich collection of expertise brings with it a very complex set of mixed professional mandates, philosophies, and missions. However, all come together with a common purpose: to help those who are fighting addictions.

Cultural Interactions: Throughout this description of the individual cultures, there have been indications of some of the cultural interactions that exist within this complicated environment. The therapists bring a mixture of naivety and experience; conservatism and radicalism; and youth and age. The clients bring such a complex collective of so many cultures that it becomes difficult to describe: the subcultures of drugs and the sex trade, country upbringing and survival on city streets, immigrants and indigenous people, educated and unskilled. All of these attributes are represented in this community of women. The network exists with thanks to wealthy donors and government subsidy. The staff members come from multiple professional preparations, with mixed experiences but a common focus. In order to function within such a matrix of cultures, it is critical that one has or develops an extremely culturally sensitive style for communicating, interpreting, planning, understanding, and supporting.

A Personal Practice Perspective

We have shared a journey through the complexities of culture and environment, exploring these constructs from various views of occupational therapy practice. However, you have been a captive reader for the author to describe her experiences and try to extrapolate from them kernels of value to any occupational therapy clinician. It is now your turn, reader. The process that will follow could be done alone or with a group of colleagues and friends. It is intended to provide an opportunity to review personal practice using culture as environment lenses. By approaching this task openly, it is hoped that you will be able to gain fresh insights into areas of your own clinical work where growth towards a more culturally sensitive practice style is either possible and/or desirable.

1. Select a recent practice scenario of which you have been a part and where there are potential cultural interactions to explore.

   * Was this:
     * An initial interview
     * Intervention session
     * Follow-up session
     * After discharge
     * Long-term follow-up

   * Why is this an issue? What would be the different implications of the cultural elements of the client/therapist interaction at these different points in time?

2. Following the format used throughout this chapter, analyze this scenario using:

   * Environmental elements that provide cultural cues (space, furniture, location, clothing of clinician(s), "tools of the trade"); relevance to client-centred principles

   * Individual elements (therapist, client(s), other clinicians)

3. Which elements of which cultures are interfacing in your scenario?
    * Individual
    * Institutional
    * Political
    * Social
    * Other
4. Which potential cultural issues are beginning to emerge?
    * Individual
    * Institutional
    * Political
    * Social
    * Other

Having looked at your chosen scenario through a cultural microscope, it now becomes clearer where the potential changes, shifts, and enhancements in practice style may be appropriate. Undoubtedly, following this process with one solitary client or client system will not provide a particularly rich set of data from which to determine personal change and growth. Therefore, it would seem to be useful to apply this model to a few other client situations that you feel may yield some powerful information.

Once we start to think in a different manner, however small the difference may be, we will find that the questions become natural to ask at any time when a new client-therapist relationship is being forged.

CONCLUSION

In many ways, this is the wrong heading; this is not a conclusion, but a commencement of the potential application of new learning with direct relevance to professional practice. Perhaps it will be helpful to take a few steps back from the "busy-ness" of daily professional life, and allow time to reflect on some of the less tangible elements of culture and culture as environment.

Anthropology as a science has experienced a checkered and unsettled evolution, as any true science should. From Taylor, who argued for the non-biological basis of social difference (Moore, 1997, p. 27) to Boas, who posited that "culture could only be explained in reference to specific cultural patterns... that one can only understand a specific society's practices within its specific cultural context" (Moore, 1997, p. 65), the early founders opened up a world of rich and exciting opportunities to view human kind from a different set of principles. However, as the science emerged, warnings came with the newfound insights. One has particular relevance to health care practice and that supports the whole notion of client-centredness and recognition of the unique nature of everyone with whom one relates; this comes from the work of Ruth Benedict, whose study focus is synthesized in the following quote: "On the one hand, culture is an expression of core values, which most people learn and absorb; on the other hand, there are individual personalities that lie outside the particular segment of the arc of possibilities that define culture. Therefore, not only are cultural values relative, but the very definition of deviance as well" (Moore, 1997, p. 85).

Perhaps one of the most powerful yet simple sentences to support the importance of cultural sensitivity from a personal level comes from describing the work of Edward Sapir. As Moore states, "He (Sapir) believed that broad generalizations about society are misplaced and that 'There are as many cultures as there are individuals in a population' [quoted by Benedict 1939, p. 407]" (Moore, 1997, p. 92). Another powerful quote from Sapir provides a wonderful backdrop to the whole notion of cultural interactivity, although his context for making the statement was not one of considering the broader definitions of the term: "A healthy [national] culture is never a passively accepted heritage from the past, but implies the creative participation of the members of the community… it is just as true, however, that the individual is helpless without a cultural heritage to work on [Sapir, 1968, p. 321]" (as quoted in Moore, 1997, p. 94). It would seem that this gives us an invitation to view our work, our impact upon it and our clients, and the impact of our clients upon us, in a manner that celebrates diversity and uniqueness.

As the occupational therapy profession forges ever forward, this is a time of immense change and, with it, fascinating challenges to our practice. "Culture" and its importance, hand-in-glove with the partner construct of "environment," may well prove to be the next central ideas that influence our discipline for the betterment of our clients and our society. They will enhance the importance of client-centred thinking and planning; they will enrich the manner in which the client-therapist relationships are formed. It behooves us to pay attention, to embrace these ideas with enthusiasm, and to shape our practice with them at the core.

Evidence related to working with cultural sensitivity relative to environment is sparse to say the least. Leavitt (1999) provides clear guidelines and suggestions for considering environmental elements through a cultural lens, and on an international scale (pp. 173-300). Much of what is reported and described addresses particular circumstances and reinforces the critical need for practitioners to consider these examples and extrapolate from them the key learning for other environments. Perhaps the key issue for occupational therapy practitioners is to consider culture naturally as an essential underpinning to the context of daily professional practice, whether it is from a personal, occupational, or environmental perspective.

Study Questions

1. What are the key elements of culture that make it such a critical part of contemporary occupational therapy practice?

2. How broadly can we, and should we, define culture in the context of contemporary occupational therapy practice?

3. Describe and outline a template for ensuring cultural content in occupational therapy assessment and intervention.

4. Reflect on the exercises within this chapter and formulate a personal study guide based on the learning you have achieved by reading this chapter.

REFERENCES

Banks, J. A., & McGee, C. A. (1989). *Multicultural education.* Needham Heights, MA: Allyn & Bacon.

Damen, L. (1987). *Culture learning: The fifth dimension on the language classroom.* Reading, MA: Addison-Wesley.

Dyck, I. (1989). The immigrant client: Issues in developing culturally sensitive practice. *Can J Occup Ther, 56,* 248-255.

Hofstede, G. (1984). National cultures and corporate cultures. In L. A. Samovar, & R. E. Porter (Eds.), *Communication between cultures.* Belmont, CA: Wadsworth.

Kinebanian, A., & Stomph, M. (1992). Cross-cultural occupational therapy: A critical reflection. *Am J Occup Ther, 46,* 751-756.

Kluckhorn, F. R. (1976). Dominant and variant value orientations: In P. J. Brink (Ed.), *Transcultural nursing* (pp. 63-81). Prospect Heights, IL: Waveland.

Krefting, L. H. (1991). The culture concept in the everyday practice of occupational and physical therapy. *Phys Occup Ther Pediatr, 11*(4), 1-16.

Law, M., Baptiste, S., Carswell, A., McColl, M. A., Polatajko, H., & Pollock, N. (1998). *The Canadian occupational performance measure* (3rd ed.). Ottawa, ON: CAOT Publications ACE.

Leavitt, R. L. (1999). *Cross-cultural rehabilitation: An international perspective.* London: W. B. Saunders.

Moore, J. D. (1997). *Visions of culture: An introduction to anthropological theories and theorists.* Walnut Creek, CA: Altamira.

Pirsig, R. (1984). *Zen and the art of motor cycle maintenance: An inquiry into values.* New York: Bantam Books.

Paul, S. (1995). Culture and its influence on occupational therapy evaluation. *Can J Occup Ther, 62,* 154-161.

Roseberry, W. (1989). *Anthropologies and histories.* New Brunswick, NJ: Rutgers University.

Spradley, P., & McCurdy, D. W. (1987). *Conformity and conflict: Readings in cultural anthropology.* Boston, MA: Little Brown and Company.

Verhelst, T. (1990). *No life without roots.* London: Zed Books.

Chapter
7

Universal Design of the Built Environment to Enable Occupational Performance

Laurie Ringaert, MSc, BMR-OT, BSc

CHAPTER OBJECTIVES

* To provide occupational therapists with an understanding of universal design.
* To describe occupational therapists' traditional role in accessibility and promote an expanding role in universal design.
* To provide examples of universal design education and research projects and how they apply to occupational therapy.
* To encourage occupational therapists to work in the universal design field.

INTRODUCTION

Universal design is an emerging field of which occupational therapists are beginning to become aware. Occupational therapists are well suited to have major roles in this field. The purpose of this chapter is to acquaint occupational therapists with universal design, to discuss occupational therapy competencies that will benefit the field, and to provide examples of how occupational therapists can, and do, work in this area. Universal design represents a broad focus considering a large range of abilities (hearing and visual disabilities, people of short stature, for example) and ages in a broad range of contexts including public buildings such as schools, museums, hotels, parks, and public rights of way in urban areas. One needs to have an extensive knowledge of theories and principles of universal design, human rights laws, standards, and codes, functional requirements of a variety of users, as well as involvement of user groups in the process. Rather than working at the individual level, this chapter will demonstrate the need for occupational therapy practice in a larger contextual approach.

WHAT IS UNIVERSAL DESIGN?

Universal design is a way to create products and environments that are more usable by everyone, regardless of age or ability (Mace, 1985). The intent of universal design is to simplify life for everyone by making products, communications, and the built environment more usable by as many people as possible at little or no extra cost (North Carolina State University, 1997). The needs of persons with ranges of cognitive, visual, hearing, mobility, and agility functions are taken into account, as well as the needs of persons of various heights, widths, and ages. Conventional design caters to the "average" person, while universal design recognizes that people have a range of capabilities and design needs to include this range. Universal design is important to the design of environments, such as houses, office buildings, hotels, restaurants, parks, streetscapes, urban planning, swimming pools, trails, historic attractions, museums, exhibits, auditoria, Web sites, communication, product design, services, and policies.

> *"Universal design is an approach to design that acknowledges the changes by everyone during his or her lifetime. It considers children, the elderly, people who are tall or short and those with various disabilities. It addresses the lifespan of human beings beyond the mythical average person...Universal design respects human diversity and promotes inclusion of all people in all activities of life. It is not an identifiable architectural or design style. Rather it is a way of thinking about the design process and how it can provide comfort and convenience for all people"* (Assistive Technology Network, 2001, p. 1).

Other terms have been used in the past such as accessible design, adaptable design, and barrier-free design; however, these terms tend to refer to the needs of a specialized population rather than designing for the general population. For instance, barrier-free design is an older term that often connotes catering to persons who are manual wheelchair users and adding a ramp on to a building after it has been built (as an after thought) rather than designing a level access from the beginning. This process occurred in the 1970's and 80's as advocates worked to change existing inaccessible environments. Adaptable design refers to adapting an environment for a particular individual. An example is traditional home modification recommendations made by occupational therapists. Universal design, on the other hand, implies designing features that address the needs of a range of human abilities into the product or environment or service from the design development stage.

The important characteristic of universal design solutions is that they do not segregate users and do not become less usable by persons who are not disabled. Caplan (1992) states that universal design is a good term although in a rational world you wouldn't have to use it because that's what design itself would be, just as food would be to health food.

By thinking of the environmental needs of a range of human functioning, a better environment is created for everyone. For example:

* A level entrance is easier to use for a person pushing a baby carriage or someone moving boxes or furniture into a building.

* Wider hallways, doors, and washrooms make it easier for cleaning and moving personnel to do their work. These same features provide a more aesthetically pleasing environment than a cramped style.

* Color contrasts between walls and floors and at stairways and entrances provide visual interest and also make wayfinding easier.

* Wayfinding is also improved through use of easily read (including tactile and Braille) and understood signs at appropriate heights located at decision-making points

throughout the environment. Talking signs also provide wayfinding information especially to distinguish buildings and exhibits in museums and outdoor trails.

* Lever door handles on doors are easier to use by people with wet or cold hands or who have their hands and arms filled with packages.

The construct of universal design encompasses many areas including (but not limited to) universal design theory, current trends in universal design, building codes, standards and guidelines, human functioning, human development and aging theory, human rights issues, guidelines and laws, assistive technologies, design theory, anthropometrics, ergonomics, building construction, historical properties, lighting, and sound design. The construct of universal design is currently in a developmental stage with leaders, researchers, and educators debating what makes and does not make up the construct. For instance some leaders argue that gender, culture, and sustainability issues should be included in the construct and this area is only beginning to be developed (Kanes Weisman, 2001; Sandu, 2001).

WHY UNIVERSAL DESIGN HAS BECOME A SIGNIFICANT WORLDWIDE ISSUE

Universal design has become important for a variety of reasons including the demographic changes toward an aging society leading to more persons with disabilities (Steinfeld, 1994), the rise of the Independent Living Movement (ILM) (Crewe & Zola 1984; DeJong 1979; Enns, 1986), human rights laws, changes in building codes and standards, and the general desire for better, more functional designs by informed consumers.

Independent Living Movement and Human Rights Legislation

Within the ILM, problems are defined in terms of barriers in the environment including economic, architectural, or support systems rather than in terms of the consumer's physical and/or mental disabilities (DeJong, 1979). Harlan Hahn, a well-known American disability rights advocate, has described the "minority-group" model (1988). This model emphasizes that environmental barriers have more impact than biological or psychological forces in shaping major life experiences of persons with disabilities. He argues that having to live with the shared constraints of inaccessible physical, social, and communicative environments and being denied equal access to education, employment, transportation, and housing while contending with negative stereotypes and minimal political power has constructed a distinct minority-group experience that includes the segregation, discrimination, and exploitation of people with disabilities (Hahn, 1988). Hahn's minority perspective is congruent with the principles of the ILM, which place emphasis on the environment rather than on the individual.

The primary meaning of independence to the ILM is the ability to have control over one's life in the community (Crewe & Zola, 1984). Persons with disabilities gain control of their lives in the community though their day-to-day experience with facilitators and barriers in the built environment. They learn which situations have to be avoided, what works, and what has to be advocated for future betterment. Through this day-to-day interaction over time, they become experts in maneuvering through the built environment. This expertise provides a sense of control.

Recognizing persons with disabilities as user-experts in the process of universal design rather than as "patients" in a medical model with minimal input is in congruence with independent living principles. The ILM was a major factor in changes that have occurred with respect to human rights regarding accessibility. Human rights declarations provide broader mandates for inclusion than do building codes and standards. An important development has been the United Nations' adoption of the 1982 World Programme of Action (United Nations, 1982) and the 1992 Standard Rules on the Equalization of Opportunities for People with Disabilities (United Nations, 1992). Human rights legislation regarding accessibility has occurred in several countries including England, Australia, Canada, and the United States. The latter two will be briefly discussed here.

Canadian Human Rights Legislation

In Canada, the Charter of Canadian Rights and Freedoms prohibits discrimination on the basis of disability. For example, Canada has declared that "individuals with disabilities shall be assured access to fundamental elements of daily life available in the community. Wherever possible the effects of an impairment or disability on an individual's life shall not be determined by environmental factors" (Principle 5 from the Declaration on the Decade of Disabled Persons, 1983). Human rights legislation federally and provincially addresses private sector entities.

Americans With Disabilities Act

The Americans With Disabilities Act (ADA) (US Department of Justice, 1990) is a civil rights law to address discrimination against people with disabilities. The ADA directed the US Architectural and Transportation Barriers Compliance Board (Access Board) to develop technical criteria for accessibility compliance and the US Department of Justice (DOJ) to write the implementing regulations and adopt specific standards for enforcement (Salmen, 2001). The ADA Accessibility Guidelines (ADAAG) (US Architectural Barriers and Compliance Board, 1998) were developed by the Access Board. The guidelines provide technical requirements for accessibility. There are continual updates and new guidelines being developed. For example, in addition to the original ADAAG, there are guidelines for public rights of way, outdoor recreation facilities, and other areas.

There are important differences between civil rights laws and building codes that are noted by Salmen (2001) and Mazumdar & Geis (2001). For instance, lawsuits regarding accessibility are usually based upon human/civil rights legislation. An architect could technically build according to the building code, however if the building still excluded a user (for example, lack of lines of sight for wheelchair users in an auditorium), the user could sue because human rights legislation would indicate that a person cannot be discriminated against by the built environment. Human rights legislation overrides the building code. This example demonstrates that occupational therapists working in the area of universal design need to be aware of both legislation and building codes in their country.

User-Expert Involvement in Universal Design

In order to design effectively using universal design concepts, designers must know and consult with the various users. As Salmen (2001) states, "the reality demands collaborative efforts between designers, environmental decision makers, and users as the diversity and complexity of our global society increases" (p. 12.8). Users should include anyone using and working in a space. This must include the range of staff from maintenance personnel to administration and the range of ages, human functionality, and assistive technologies. One of the most important members of the user consultation group is the user-expert, defined by Ostroff (1997) as:

> "Anyone who has developed natural experience in dealing with the challenges of our built environment. User-experts include parents managing with toddlers, older people with changing vision or stamina, people of short stature, limited grasp, or who use wheelchairs. These diverse people have developed strategies for coping with barriers and hazards they encounter everyday. The experience of the user-expert is usually in strong contrast to the life experience of most designers and is invaluable in evaluating both existing products and places as well as new design development" (pp. 33-35).

It cannot be assumed that the designer's concepts will meet the needs and wishes of the residents and staff, or that nursing home administrators will be aware of resident needs. Duffy, Su, Beck, and Barker (1986) conducted a study to investigate the differences that existed among the preferences of nursing home residents, nursing home administrators, and designers on a series of design alternatives. Results indicated that a significant difference existed in the preference of the lounge design, dining arrangement, and private room furniture between residents and administrators/designers. Nursing home residents consistently chose design/patterns that facilitated privacy whereas administrators and designers chose design/patterns that encouraged social interaction. It is necessary to emphasize the need for user research in the design process as it may be a more accurate portrayal of residents' needs (Duffy et al., 1986).

In recent years there has been more recognition of user-experts providing input to design projects. Very little has actually been documented in the literature about this process. Anecdotal evidence includes calls for advice from designers and building contractors to disability specific groups (such as the local paraplegic association), and disability generic groups (such as independent living organizations). Calls are also made to individuals known to the designers or contractors. The problem with this approach is the premise that anyone with a disability will have expertise in access or universal design issues, which is an incorrect assumption. The specific study of universal design is a complex subject that develops over a period of years. Also, for the person seeking advice, they often have to call a number of disability organizations to put together a complete picture. Importantly, most representatives are not paid for their consultation. Monetary recognition must be considered for the important contribution the user-experts make to the design.

Another approach to recognition of disability consumers to the process is through committees. Committees have been struck at municipal, provincial, and state levels that include representatives from a variety of user groups. Representatives also have been asked to provide input to national building codes, standards, and guidelines. In Canada, the Canadian Standards Association (CSA) includes a variety of consumer representatives on its standards development committees. In most cases these people represent a unilateral dimension to universal design. To address these issues a training program (Ringaert, 2001) was developed for persons with disabilities in the universal design process at the Universal Design Institute in Winnipeg, Canada. The issue of participant involvement in the design process is discussed further in Ringaert (2001) and in the post-occupancy evaluation (POE) study by Ringaert, Rapson, and Lagman (2002) outlined in the case study.

BUILDING CODES, STANDARDS, AND GUIDELINES

Building codes and standards provide regulations to create the minimum level of accessibility. Universal design per se is not a set of technical standards as is found in codes, and standards. As Salmen (2001) writes, while "*accessibility* is a function of compliance with regulations or criteria that establish a minimum level of design necessary to accommodate people with disabilities, *universal design* is the art and practice of design to accommodate the widest variety and number of people throughout their lifespans" (p. 12.1).

However, it is necessary to understand and know these requirements in order to design universally or go beyond these standards. The Canadian National Building Code (NBC) Section 3.8 (Institute of Research in Construction, 1995) and the CSA's Barrier-Free Standard B651-95 (CSA, 1995) have codes and standards for the planning, design, and construction of the built environment for persons with disabilities. The NBC is a model code that Canadian provinces can either adopt in its entirety, partially, or not at all, in each province. It is considered the minimum basic code requirement. In the United States, the ADAAG (US Architectural Barriers and Compliance Board, 1998) are used to direct accessibility of buildings and outdoor environments. State and local governments as well as private entities must comply with the ADAAG. Federally owned or funded entities must comply with the Uniform Federal Accessibility Standards (UFAS) (US Access Board, 1984).

Current building codes and standards are generally not termed "universal design." They tend to be called "barrier-free" or access codes. Current codes and guidelines dealing with barrier-free access tend to concentrate on the needs of persons with mobility impairments, particularly manual wheelchair users. This has led to inadequacies in design requirements and in some cases created problems for persons with other types of disabilities. For example, some current designs of curb ramps are not suitable for persons with visual disabilities, as there is little tactile design requirements (no change in the texture of the pavement close to a curb). Another example is the design of roll-in showers in current CSA standard and the ADA guidelines. There is no requirement for a fold-up seat in a roll-in shower (J. Sanford, personal communication, May 30, 2001), therefore many users (particularly seniors) could not use this type of shower. By adding a fold-up seat, the design would be more usable by a wider range of people—thereby making it more universally designed. This is particularly important in commercial applications, such as hotels, that can then use the same guestroom with a wider range of guests.

A flaw of current codes and standards is that many of the measurements have been based upon the reach ranges of young able-bodied army personnel with little consideration of any functional differences. The focus on manual wheelchairs also leaves out the significant population of people using power wheelchairs, scooters, and sports wheelchairs (Ringaert, Rapson, Qiu, Cooper, & Shwedyk, 2001). On the positive side, there have been some recent changes due to consumer involvement in the standards and code development process in North America and other parts of the world. Many jurisdictions have begun to incorporate the needs of more user groups in their building codes.

SCOPE FOR OCCUPATIONAL THERAPY

How Universal Design Affects Occupational Performance

If people with disabilities and seniors are unable to use products, services, or environments, their occupational performance is affected in any of their interactions with society. For instance, a child is unable to get into a school because it is inaccessible. This will affect his or her occupational performance. If persons who are blind are unable to obtain documents from their local government on disk, then their ability to participate equally in society is affected. A young man with a spinal cord injury is rehabilitated but finds that there are no accessible arenas to view hockey and very few accessible restaurants in his town. He cannot visit the homes of friends because none of their homes have an accessible entrance or bathroom. Even though the rehabilitation has been "successful," the individual now has to live in an inaccessible environment in society.

Universal design relates to the areas of participation and environmental factors in the World Health Organization's (WHO) International Classification of Functioning, Disability, and Health (ICF) (WHO, 2001). Universal design consultants tend to operate more at the societal, policy development level rather than at the individual level of functioning. Often occupational therapists are working with a client at the individual level. An example of why it is important for them to understand the larger picture is illustrated in the following scenario. Occupational therapists often prescribe powered mobility devices as part of the rehabilitation process. Occupational therapists who are working at the individual level may be unaware that the built environment is actually designed by law to be inaccessible to persons using scooters and power wheelchairs because the codes and standards were designed with manual wheelchair users' requirements in mind. Thus their client may be unable to enter public restrooms or use public ramps with his or her power mobility device. Occupational therapists who focus on the societal level of functioning will develop knowledge about codes and standards, and universal design principles, as they impact on society as a whole, which will in turn benefit their clients. The remainder of this chapter will discuss the knowledge, skills, and abilities that occupational therapists need to develop to work in the field of universal design.

What Competencies Do Occupational Therapists Have That Make Them Suited to Work in Universal Design?

Occupational therapists have a constellation of knowledge, skills, and abilities that make them valuable team members in the universal design process, including:

* Knowledge of human functioning
* Knowledge of disability
* Knowledge of occupational performance
* Knowledge of the person-environment interaction
* Knowledge of assistive technologies
* Skills in occupational/task analysis and environmental adaptation/modification

Designers, architects, and builders do not possess these same skills, thus the occupational therapist becomes an important asset to the team. Occupational therapy competencies can be analyzed in terms of the seven principles of universal design. The definition of uni-

Table 7-1

Analysis of Occupational Therapists' Skills in Relationship to the Seven Principles of Universal Design

Principle of Universal Design	Description	Occupational Therapy Skill Set
Equitable Use	The design is useful and marketable to any group of users.	OT knowledge of human rights legislation, policies, and laws not well developed.
Flexibility in Use	The design accommodates a wide range of individual preferences and abilities.	Knowledge of human functioning.
Simple and Intuitive Use	Use of the design is easy to understand, regardless of the user's experience, knowledge, language skills, or current concentration level.	Knowledge of cognitive functioning; knowledge of activity analysis.
Perceptible Information	The design communicates necessary information effectively to the user, regardless of ambient conditions or the user's sensory abilities.	Knowledge in terms of needs of persons with visual and hearing disabilities not well developed; knowledge of cognitive disabilities.
Tolerance for Error	The design minimizes hazards and the adverse consequences of accidental or unintended actions.	Knowledge of human functioning, disability, and occupational analysis.
Low Physical Effort	The design can be used efficiently and comfortably and with a minimum of fatigue.	Knowledge of energy conservation and work simplification principles.
Size and Space for Approach and Use	Appropriate size and space is provided for approach, reach, manipulation, and use regardless of user's body size, posture, or mobility.	Knowledge of activity analysis, human functioning, assistive technology, and human-environment interactions.

versal design has been elaborated upon into seven principles by the Centre for Universal Design, (North Carolina State University, 1997), and is widely accepted, both in the United States and internationally. The seven principles clarify the universal design criteria requirements, in order to improve the quality of life for everyone. Table 7-1 presents the author's analysis of occupational therapy skills in relation to these principles. Some areas are well developed while others require more development.

What Areas Require Development in Occupational Therapy Clinicians?

Occupational therapists do bring a multitude of competencies to the universal design table; however, there are a number of areas that require further development in order to work in this area. These include:

* Knowledge of universal design theory
* Knowledge of codes, standards, and guidelines
* Skills to interact with designers, architects, and builders
* Knowledge of best practices in universal design
* Knowledge of environments beyond housing (streetscapes, parks, commercial buildings)
* Ability to review architectural drawings

It should be noted that this analysis is only a first attempt by the author and further dialogue among occupational therapy colleagues will be interesting.

Knowledge about a new area of practice traditionally develops through educational programs and courses. To date, very few occupational therapy programs include courses in universal design. More courses in the occupational therapy curriculum as well as continuing education courses will enhance the knowledge and skills of occupational therapists. The Council of Europe Committee of Ministers feels so strongly about the educational issue that they developed a resolution regarding training of occupations working on the built environment. They resolved that universal design and accessibility play a key role in the promotion of human rights and fundamental freedoms and should therefore be incorporated into all levels of the education and training programs of all occupations working on the built environment (Council of Europe, 2001).

The Clinician Versus the Universal Design Expert

The development of new knowledge is the first step for a clinician who is interested in becoming a universal design expert. Shamberg (2001) distinguishes between the expectations and competency requirements of the clinician ("generalist") and the expert in this area:

> "Generalists advise consumers about relatively simple adaptations to the home or job site; provide off-the-shelf adaptive equipment, such as reachers, grab bars, and tub benches; instruct in work simplification strategies... [Expert] clinicians typically make recommendations about design, adaptive equipment, auxiliary aids, policy changes, reasonable accommodations (alternative methods of providing equivalent services) and environmental adaptations to remove barriers." (Shamberg, 2001, pp. 787).

This speaks to the need for occupational therapy clinicians to focus their learning and skill development on the societal level of functioning, participation, and environmental factors if they wish to become experts in the area of universal design. There are many opportunities for occupational therapists to develop competencies in this area, and the next section highlights some strategies and examples from occupational therapists who are working in universal design.

New Frontiers for Occupational Therapists

There are several areas in which occupational therapists are already contributing to the field of universal design, and there is great potential for the profession to continue to expand skills and knowledge. Contributions of occupational therapists working in this area include:

* Providing consultation with the universal design team on building projects
* Providing education to designers, architects, contractors
* Serving on accessibility committees and building codes and standards committees
* Developing new assessment tools
* Working as researchers with a universal design team

Many individual occupational therapists have continued to work in the area of home modification but at a different level compared to traditional occupational therapy practice. Other therapists have expanded their roles into new areas of universal design. Examples of innovative occupational therapy practice are described below to encourage other occupational therapists to begin working in this field.

PROVIDING CONSULTATION WITH THE UNIVERSAL DESIGN TEAM ON BUILDING PROJECTS

Occupational therapists can play an important consultation role on the universal design team. It is not enough for designers and contractors to know the dimensions of the space they are designing; they must understand the "why" of the design. Occupational therapists provide the "why" due to their knowledge of human functioning, disability, and the human-environment interaction. Some occupational therapists have become involved in this type of consultation by starting up their own consulting businesses. One example is Susan Mack from Murrietta, California, who won "The Best of Seniors' Housing Awards 2001" from the National Association of Home Builders/National Council On Seniors' Housing in the category of "Aging in Place—Universal Design" (S. Mack, personal communication, September, 2002). Figure 7-1 is an example of her work in designing a kitchen using universal design principles.

Occupational therapists can also encourage persons with disabilities to become user-experts in the universal design process. There are emerging models of occupational therapists who work with user-experts and designers to form powerful universal design teams. This is a new and emerging role for occupational therapists.

PROVIDING EDUCATION TO DESIGNERS, ARCHITECTS, CONTRACTORS, AND OCCUPATIONAL THERAPISTS

Occupational therapists can make an important contribution to teaching aspects of human functioning and disability to designers, architects, and contractors so that they understand why they should design in certain ways. Several occupational therapists now teach designers or co-teach with designers (A. Morris, personal communication, January 2002), and are involved in the development of internet-based mentoring program on accessibility for rehabilitation professionals (S. Shamberg, personal communication, January, 2002).

As occupational therapists develop expertise in universal design, they are sharing their knowledge through the writing and publication of books and other documents. Several occupational therapists have written articles about universal design issues in journals for the occupational therapy profession as well as other related professions (Ringaert, 2000; Shamberg, 1993, 1995, 1996). Some examples of books to which occupational therapists have contributed are listed at the end of this chapter under Related Readings.

Fig 7-1. Kitchen designed by Susan Mack illustrating multiple work surface heights.

SERVING ON ACCESSIBILITY COMMITTEES AND BUILDING CODES AND STANDARDS COMMITTEES

Occupational therapists can be involved with the universal standards and codes development process and can contribute to access committees in their communities. In this way, they can provide input at the societal (policy development) level.

There are many national, provincial/state, and municipal opportunities for this type of involvement. This is a particularly important area as occupational therapists provide the functional knowledge that code inspectors, representatives of manufacturers, and designers on these committees do not have. This author currently sits on the NBC of Canada Section 3.0 (Fire, Safety, and Occupancy) and is chair of the working group on the barrier-free section (3.8). She also serves on the CSA Barrier-Free (B651) Committee and chairs the B651-1 Automated Teller Machine. An example of the importance of having an occupational therapist in this area of universal design occurred at one committee meeting. Currently in the NBC there is a requirement for a shelf that only appears in the accessible toilet stall. There is no requirement for this shelf in any other stall. The committee wondered why it was there and wanted to eliminate it until the author asked them if they wanted an explanation of how a person with a spinal cord injury used the toilet. They promptly said "no." The requirement will remain in the code to the relief of many people with disabilities who must place various medical appliances on the shelf while in the stall. Another example was the justification for the length of a level space in front of a door, which the author explained by discussing how a wheelchair user opens a door. Occupational therapists have a wealth of knowledge to bring to codes and standards committees. At the same time occupational therapists can increase their knowledge and understanding of this area by serving on these committees.

DEVELOPING NEW ASSESSMENT TOOLS

Several occupational therapists have been involved in the development of assessment tools and protocols for universal design. One example is Margaret Christenson, who founded Lifease Inc. in the United States. She has developed computerized person/home assessments such as Ease 3.1 (Lifease, New Brighton, MN) (Christenson, 2001). The Ease 3.1 software features the ability of producing reports of home with assistive technology products and idea solutions based on functional capabilities and problem areas in the home. Occupational therapists are encouraged to offer their unique expertise and knowledge to the development and use of assessment measures that incorporate universal design principles.

WORKING AS A RESEARCHER WITH A UNIVERSAL DESIGN TEAM

Due to the relative newness of this field, there are many areas in universal design that have never been researched. Occupational therapists can serve as important members of the research team due to their knowledge and skill set. Some examples of the type of research the author has been involved in are:

* Working as a Beta test site coordinator to test the new International Classification of Impairments, Disabilities and Handicaps ICIDH (i.e., ICF) for the WHO.
* Developing an accessibility audit methodology to assess large urban spaces.
* Determining dimensions in the environment to accommodate scooters and power wheelchairs.
* Studying accessibility codes and standards internationally to determine best practices.
* Post-occupancy evaluation of a long-term care facility using universal design principles.
* Examination of the tourism industry to determine their level of readiness for the market of seniors and persons with disabilities.
* Examining the banking industry policies on physical access and developing a model code of accessibility.

Two examples of research areas that are of particular interest to occupational therapy are environmental simulation and POE. These areas are elaborated upon here to show the natural fit between them and occupational therapy.

Environmental Simulation

Environmental simulation has been used to test the relationship between the environment and human functioning. Sanford and Megrew (1999) define it as "a means of matching the design of the specific environmental features to human performance capabilities" (p. 183). The purpose of the simulation is for researchers to observe people moving through and manipulating the environment while doing real life tasks. This in turn assists the observer to determine problems encountered with use and to identify limitations of the design directly, rather than having to rely on subjective evaluations of the user (Steinfeld, 1988). Environmental simulation research has been used previously in accessibility research (Bostrom, Malassigné, & Sanford, 1984; McLelland, 1972; Ownsworth, Galer & Feeney, 1974; Sanford & Megrew, 1999; Steinfeld, Schroeder, & Bishop, 1979; Walter, 1971). Some studies have used fixed construction with moveable dimensions while others used movable partitions to vary the measurements. Many of these studies have focused on bathroom simulations. Use of space in bathrooms has been studied by McLelland (1972); Ownsworth, et al. (1974); and Steinfeld et al. (1979).

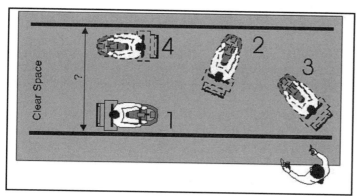

Fig 7-2. Testing for width requirements of a path of travel for a three-point turn. (Reprinted with permission from Ringaert, Rapson, Qui, Cooper, & Schwedyk [2001].)

User space required for opening and closing of doors has been studied by Brattgard (1974); Nichols, Morgan, & Goble (1966); Ownsworth et al., (1974); Steinfeld et al., (1979); and Walter (1971). Voordt (1999) found that the dimensions recommended in these studies were widely divergent. His review also shows that the scope of the studies varied widely. For instance, the direction of the approach to the door and the rotation of the door would affect the outcomes.

Figure 7-2 illustrates one of the environmental simulations undertaken in a study conducted by the author (Ringaert et al., 2001), which is an example of environmental simulation research.

Post-Occupancy Evaluation Studies

Post-occupancy evaluation (POE) is a term that has been used since the late 1960's and a wide variety of techniques and methods have been developed since that time (Hughes & Bowler, 1996). POE is a method of evaluating buildings after they have been built and occupied for some time. POE's focus on the building's occupants and their needs. It provides insight into building performance and consequences of the past design decisions. This knowledge provides the opportunity for the creation of better buildings (Prieser, Rabinowtiz, & White, 1988).

Environment and behavior studies emerged over 25 years ago when social scientists, planners, and designers came together with the common goal of improving the built environment through research and design in order to maximize the degree of fit between the designed environment, human needs, and aspirations (Verderber & Refuerzo, 1999). Research-based design is the generation and application of environment and behavior research findings in the actual design and construction of buildings. It is characterized by various streams that then can be subdivided into specialty areas, for example, user needs analysis; programming theory and technique; the scope, content, and format of design guidelines; and POE theory and technique (Verderber & Refuerzo, 1999).

Post-occupancy evaluations in architecture have been concerned with social and behavioral issues as opposed to aesthetic issues. They compare actual building performance with explicitly stated human performance needs. Variables such as task performance, privacy, communication, safety, and thermal comfort may be considered (Cooper, Ahrentzen, & Hasselkus, 1991).

Cooper et al. (1991) reported on three POEs conducted in the health care sector. The POEs were conducted on a hospital, nursing home, and independent living apartment units. An example of a POE study is also discussed in this chapter's case scenario.

Issues Facing Occupational Therapists Involved in Universal Design

The previous section describes numerous roles and functions in which occupational therapists are developing expertise. Occupational therapists can develop their knowledge and expertise through self-study; attending conferences and workshops; serving on access, codes, and standards committees; networking; collaborating with people in other fields; and keeping up-to-date on new developments in the field. There currently is no master's program in universal design in any discipline, and therapists are encouraged to participate in further study in related fields such as architecture and design.

Occupational therapists who are interested in working in the area of universal design need to network with each other to develop a collective expertise. There are some challenges in doing this, however. For example, the annual registration forms for many occupational therapy associations do not include universal design as an area of practice or expertise that therapists can select. Occupational therapists can connect with each other through list serves, especially those that promote multi-disciplinary discussion. For example, a list serve for occupational therapists involved in home modifications was initiated in 1999 at a universal design conference in Rhode Island (M. Carter, personal communication, January, 2002). Other multi-disciplinary list serves that focus on specific areas of design, such as home modifications, are in operation for occupational therapists to join (see Related Readings and Resources at the end of this chapter). This type of resource helps occupational therapists to network and learn more about universal design, and also to share their knowledge with others.

In summary, there are many issues facing occupational therapists who want to work in the area of universal design, as it is a new frontier. Role-emerging opportunities often mean that occupational therapists move into positions that are not called "occupational therapy." This does not, however, mean that an occupational therapist is not using the skills and knowledge of the profession. Furthermore, in many situations, the occupational therapist is working at the policy-societal environmental level rather than at the individual level. The author believes that more occupational therapists should be involved at this level because they have many unique skills to offer.

CASE SCENARIO: AN OCCUPATIONAL THERAPY EXPERT IN UNIVERSAL DESIGN

Working in this area is very rewarding and exciting. As the director of the Universal Design Institute in Winnipeg, Manitoba for the past 6 years, I have had the opportunity to consult in different areas of universal design including public buildings, public rights of way, parks, houses, historical properties, tourism facilities, homes, and long-term care facilities. I have been able to teach universal design principles to university students, and I have participated in research studies.

The following briefly describes one research project using POE that utilized my occupational therapy skills and expertise in universal design. It is one example of how research in this area, with a focus on the environment, can enable occupational performance.

Toward Development of Universal Design Guidelines for Long-Term Care Facilities: Post-Occupancy Evaluation Case Study

A post-occupancy study using the principles of universal design was conducted at a long-term care facility (Ringaert, Rapson, & Lagman, 2002). At the time of the study, many residents had lived within the new facility for approximately 4 years and most staff had at least 2 years working in the facility. This provided ample opportunity to test and use the new centre. This was an opportunity to conduct a POE and explore with the residents and staff what was working and what was not working and ultimately to determine design of future universally designed long-term care facilities.

Findings: There were several aspects of the design that were appealing to staff and residents. These included the layout and décor in the common areas, such as the cafeteria, hallways, and garden areas. Residents and staff appreciated the heated sidewalks leading from the parking area to the main door. However, there were several problem areas identified, including the size and layout of resident rooms and bathrooms, the size and layout of the unit kitchens and common rooms, lack of adequate readable signage, accessible parking too far away from main entrance, a large revolving door at the main entrance, and lack of walkways and sitting areas in the grounds. Upon analysis, many of the principles of universal design were violated by the design. Suggestions were made by the focus group participants and also by the researchers for changes that could be considered by this particular long-term care facility as well as for development of future long-term care building guidelines. One of the lessons to be learned from this case study for occupational therapists is that they can provide valuable input into designs at the early design stage to ensure that the space is adequate for the functional needs of their clients. They can also ensure that the voices of persons with disabilities are heard during the design process. With the combined input of the occupational therapist and user-experts, a more functional and usable space will be created for more users.

CONCLUSION

Universal design is a field that needs the expertise of occupational therapists. Occupational therapists must also recognize this field as an important area to which they can contribute. Occupational therapy practice and theory link to universal design theory via the profession's understanding of human functioning, disability, and the person-environment interaction. There are basic principles of universal design that appear to dovetail with occupational therapy competencies. However, there are gaps in the knowledge and skill set of occupational therapists, and there is a need to increase the knowledge of occupational therapists through continuing education courses as well as in occupational therapy education programs.

The environmental focus of universal design is at the social and political level. The purpose of this chapter was to demonstrate to occupational therapists how they can become involved at this larger contextual level of environmental intervention. This field is client-centred as it views persons with disabilities as user-experts who need to be involved in the process and who are important contributors to the team. Occupational therapists who are involved in universal design policy development directly affect clients by providing policies that influence their participation at the societal level.

There have been relatively few studies done in universal design; however, this chapter provides some examples of studies that would be of particular interest to occupational therapists. There are many ways that occupational therapists can study the outcomes of their work as they begin to incorporate universal design principles into their practice. Some suggestions are:

* Adding a new environmental assessment tool that uses universal design principles to the occupational therapy assessment protocol, and keeping track of the results.

* Asking clients to be user-experts and identify what they need in terms of universal design and environmental modifications.

* Developing a "best practice protocol" for universal design in an area of practice, and evaluating the protocol over a specific time period.

* Setting up a program evaluation of new services that use universal design principles, and including client questionnaires or interviews as part of the evaluation.

Practicing in universal design is innovative as few occupational therapists work in this area at this time. Universal design itself is just developing as a recognized field for many disciplines. As a result, there are many opportunities for occupational therapists to capitalize on unfilled roles. Occupational therapists can take on the challenge and consider becoming consultants to such places as museums, parks, and housing developments, and to designers, architects, landscape architects, planners, and contractors. The field is ripe for occupational therapists to become involved and make their mark in universal design.

Study Questions

1. What is the difference between building codes and universal design?

2. What are the seven principles of universal design?

3. Who benefits from universal design?

4. What accessibility committees exist in your community, state/province, or country and how can you serve on them?

REFERENCES

Assistive Technology Network. (2001). *Issue brief on assistive technology and universal design*. Retrieved September 30, 2002 from: www.atnet.org.

Bostrom, J. A., Malassigné, P. M., & Sanford, J. A. (1984). Development and evaluation of four shower prototypes. *Proceedings of the Second International Conference on Rehabilitation Engineering, 4,* 91. Arlington, VA: RESNA Press.

Brattgard, S. (1974). *Maneuvering space for indoor wheelchairs*. Goteborg, Sweden: University of Goteberg, Department of Handicap Research.

Canadian Standards Association (CSA). (1995). *CCAN/CSA-B651-65 Barrier-free design B A national standard of Canada*. Etobicoke, ON: Canadian Standards Association.

Caplan, R. (1992, August). Disabled by design. *Interior Design,* 88-91.

Christenson, M. (2001). *Ease® 3.1 software*. New Brighton, MN: Lifease.

Cooper, B., Ahrentzen, S., & Hasselkus, B. R. (1991). Post-occupancy evaluation: An environment-behaviour technique for assessing the built environment. *Can J Occup Ther, 58,* 181-188.

Council of Europe. (2001). *Resolution ResAP(2001)1: On the introduction of the principles of universal design into the curricula of all occupations working on the built environment*. Adopted by the Committee of Ministers on 15 February 2001, at the 742nd meeting of the Ministers Deputies. Retrieved September 30, 2002 from: http://cm.coe.int/ta/res/resAP/2001/2001xp1.htm.

Crewe, N., & Zola, I. (1984). *Independent living for physically disabled people.* San Francisco: Jossey-Bass.

DeJong, G. (1979). Independent living: from social movement to analytic paradigm. *Arch Phys Med Rehabil, 60,* 435-446.

Duffy, M., Su, B., Beck, B., & Barker, D. G. (1986). Preferences in nursing home design: A comparison of residents, administrators, and designers. *Environment and Behavior, 18*(2), 246-257.

Enns, H. (1986). An excerpt from the historical development of attitudes towards the handicapped: a framework for change. In A. D'Aubin. (Ed.), *Defining the parameters of independent living* (pp. 94-97). Winnipeg, MB: Council of Canadians with Disabilities.

Government of Canada. (1983). Declaration on the Decade of Disabled Persons. Ottawa, ON: Author. Retrieved March 31, 2003 from: http://www.schoolnet.ca/aboriginal/disable8/index-e.html.

Hahn, H. (1988). The politics of physical differences: Disability and discrimination. *Journal of Social Issues, 44,* 39.

Hughes, T., & Bowler, B. (1996). Calling occupants. *Hospital Development, 27*(9), 27-32.

Institute of Research in Construction. (1995). *The national building code.* Ottawa, ON: National Research Council.

Kanes Weisman, L. (2001). Creating the universally designed city: Prospects for the new century. In W. Preiser. & E. Ostroff (Eds.), *The universal design handbook.* New York: McGraw-Hill.

Mace, R. (1985). Universal design: Barrier-free environments for everyone. *Designer West, 3,* 147-152.

Mazumdar, S., & Geis, G. (2001). Interpreting accessibility standards: Experiences in the U.S. courts. In W. Preiser, & E. Ostroff (Eds.), *The universal design handbook* (pp. 18.1-18.20). New York: McGraw-Hill.

McLelland, I. (1972). *Bath aids for the disabled.* Loughborough, UK: University of Technology Foundation.

Nichols, P. J. R., Morgan, R. W., & Goble, R. E. A. (1966). Wheelchair-users. A study in variation of disability. *Ergonomics, 9,* 131-139.

North Carolina State University. (1997). *The center for universal design.* Retrieved October 20, 2002 from: www.design.ncsu.edu/cud.

Ostroff, E. (1997). Mining our natural resources: The user as expert. *Innovation, the Quarterly Journal of the Industrial Designers Society of America, 16,* 33-35.

Ownsworth, A., Galer, M., & Feeney, R. J. (1974). *Housing for the disabled. Part 2: An ergonomic study of the space requirements of wheelchair users for bathrooms.* Loughborough, UK: University of Technology, Institute of Consumer Ergonomics.

Preiser, W. F. E., Rabinowitz, H. Z., & White, E. T. (1988). *Post-occupancy evaluation.* New York: Van Nostrand Reinhold.

Ringaert, L. (2000, Fall). Teaching city centres to care. *Exchange: design on the edge.* Toronto, ON: Design Exchange.

Ringaert, L. (2001). User/expert involvement in universal design. In W. Preiser. & E. Ostroff (Eds.), *The universal design handbook* (pp. 6.1-6.14). New York: McGraw-Hill.

Ringaert, L., Rapson, D., & Lagman, R. (2002). *Towards development of universal design guidelines for long-term care facilities: Post-occupancy evaluation case study.* Winnipeg, MB: Universal Design Institute, Faculty of Architecture, University of Manitoba.

Ringaert, L., Rapson, D., Qui, J., Cooper, J., & Shwedyk, E. (2001). *Determination of new dimensions for universal design codes and standards with consideration of power wheelchair and scooter users.* Winnipeg, MB: Universal Design Institute, Faculty of Architecture, University of Manitoba.

Salmen, J. (2001). US accessibility codes and standards: Challenges for universal design. In W. Preiser. & E. Ostroff (Eds.), *The universal design handbook* (pp. 12.1-12.8). New York: McGraw-Hill.

Sandu, J. (2001). An integrated approach to universal design: Toward the inclusion of all ages, cultures, and diversity. In W. Preiser, & E. Ostroff (Eds.), *The universal design handbook* (pp. 3.3-3.14). New York: McGraw-Hill.

Sanford, J., & Megrew, M. B. (1999). Using environmental simulation to test the validity of code requirements. In E. Steinfeld, & G. S. Danford (Eds.), *Enabling environments: Measuring the impact of environment on disability and rehabilitation* (pp. 183-206). New York: Kluwer Academic/Plenum.

Shamberg, S. (1993). The accessibility consultant: A new role for occupational therapists under the Americans with Disabilities Act. *Occupational Therapy Practice, 4,* 14-23.

Shamberg, S. (1995). Opening doors: Universal barrier-free environments make independence possible. *Health Care Dealer/Supplier, 7,* 100-5.

Shamberg, S. (1996). Strategy for occupational therapy accessibility, assessment and consultation. In R. Null (Ed.), *Universal design: Creative solutions for ADA compliance* (pp. 95-99). Belmont, CA: Professional.

Shamberg, S. (2001). Optimizing access to home, community, and work environments. In C. Trombly, & M. Vining Radomski (Eds.), *Occupational therapy for physical dysfunction* (5th ed., pp. 783-800). Philadelphia, PA: Lippincott Williams & Wilkins.

Steinfeld, E. (1994). *The concept of universal design*. Buffalo, NY: Adaptive Environments Laboratory, State University of New York at Buffalo.

Steinfeld, E. (1988). *Full scale modeling in architecture*. Presentation at the Architectural Research Centers Consortium 1988 Conference. Champaign, IL: ARCC.

Steinfeld, E., Schroeder, S., & Bishop, M. (1979). *Accessible buildings for people with walking and reaching limitations*. Washington, DC: U.S. Department of Housing and Urban Development.

United Nations. (1982). *World programme of action concerning disabled persons*. Adopted by the General Assembly by its resolution 37/52 of 3 December 1982. Retrieved September 17, 2002 from: http://www.un.org/esa/socdev/enable/dissre01.htm

United Nations (1992). *Standard rules on the equalization of opportunities for people with disabilities*. Adopted by the General Assembly by its resolution 32/2 of 20 February 1991. Retrieved September 17, 2002 from: http://www.un.org/esa/socdev/enable/dissre01.htm..

US Access Board. (1984). *Uniform federal accessibility standards* (UFAS). Retrieved October 20, 2002 from: htpp://www.access-board.gov/ufas/ufas-html/ufas.htm.

US Architectural Barriers and Compliance Board. (1998). *Americans with Disabilities Act: Accessibility guidelines for buildings and facilities (ADAAG)*. Washington, DC: US Architectural Barriers and Compliance Board (Access Board).

US Department of Justice. (1990). *Americans with Disabilities Act*. Washington, DC: Author.

Verderber, S., & Refuerzo, B. J. (1999). On the construction of research-based design: A community health center. *Journal of Architecture and Planning Research, 16*(3), 225-241.

Voordt, D. J. M. (1999). Space requirements for accessibility: Cross-cultural comparisons. In E. Steinfeld & G. S. Danford (Eds.), *Enabling environments: Measuring the impact of environment on disability and rehabilitation* (pp. 59-88). New York: Kluwer Academic/Plenum.

Walter, E. (1971) *Four architectural movement studies for the wheelchair and ambulant disabled*. London: Disabled Living Foundation.

World Health Organization (WHO). (2001). *International Classification of Functioning, Disability, and Health*. Geneva: Author.

RELATED READINGS AND RESOURCES

Christenson, M. (1990). *Aging in the designed environment*. Binghamton, NY: Haworth.

Christophersen, J. (Ed.). (2002). *Universal design: 17 ways of thinking and teaching*. Norway: Husbanken.

Cooper, B., Cohen, U., & Hasselkus, B. R. (1991). Barrier-free design: A review and critique of the occupational therapy perspective. *Am J Occup Ther, 45*, 344-350.

Cooper, B., & Hasselkus, B. R. (1992). Independent living and the physical environment: Aspects that matter to residents. *Can J Occup Ther, 59*, 6-15.

Finkel, G. (Ed.). (2001). *Access: A guide to accessible design for designers, builders, facility owners and managers*. Winnipeg, MB: Universal Design Institute, Faculty of Architecture, University of Manitoba.

Frye, J., Frue, K., & Sandilands, R. (2000). *Accex. The universal design expert software system*. Winnepeg, MB: Universal Design Institute, Faculty of Architecture, University of Manitoba.

Kornblau, B., Shamberg, S., & Klein, R. (2000). Americans with Disabilities Act position paper: Occupational therapy and the Americans with Disabilities Act. *Am J Occup Ther, 54*, 622-3.

Null, R. (Ed.). (1996). *Universal Design: Creative solutions for ADA compliance*. Belmont, CA: Professional.

Preiser, W., & Ostroff, E. (Eds.). (2001). *The universal design handbook*. New York: McGraw-Hill.

Ringaert, L. (2002). A universally designed universal design course. In J. Christophersen (Ed.), *Universal design: 17 ways of thinking and teaching* (pp. 269-285). Norway: Husbanken.

Ringaert L., Knutson, B., & Rapson, D. (2001). *Is your business open to all?* Winnipeg, MB: Universal Design Institute, Faculty of Architecture, University of Manitoba.

Susan Mack, OTR/L, CAPS. Homes for Easy Living. Universal Design Consultants. 25060 Hancock Avenue, Suite 103-186 Murrieta, CA 92562. Phone: 760 409-7565. www.homesforeasyliving.com. Email: smack@homes-foreasyliving.com.

Steinfeld E., & Danford, G. S. (Eds.). (1999). *Enabling environments: Measuring the impact of environment on disability and rehabilitation.* New York: Kluwer Academic/Plenum.

Web-Based Resources

Adaptive Environments (Massachusetts): http://www.adaptenv.org/index.php

Canada Mortgage and Housing Corporation: www.cmhc-schl.gc.ca

Center for Universal Design (North Carolina): http://www.design.ncsu.edu/cud/

IDEA Center & the RERC on Universal Design (Buffalo): http://www.ap.buffalo.edu/idea/

Lifease Inc.: http://www.lifease.com/lifease-home.html

National Resource Center on Supportive Housing and Home Modification (California). www.homemods.org

UDC @ Sheridan (Toronto, Canada): http://www.sheridanc.on.ca/udc/content/overview.html

UN Standard Rules: http://www.un.org/esa/socdev/enable/dissre00.htm

Universal Design International Consulting (Winnipeg, Canada): Universal Design Institute (Manitoba, Canada): http://www.arch.umanitoba.ca/CIBFD

U.S Access Board: http://www.access-board.gov/

SECTION THREE

Using Environments to Enable Occupational Performance with Communities, Groups, and Individuals

Section Three of this book presents a wide range of applications of environment and person-environment theories and models to the daily practice of occupational therapists. The focus of this section is on innovative ways that occupational therapists are using environments to work with social groups and individuals to enable their occupational performance. The term "communities" in this section refers to groups of people with common interests and beliefs. Some are schools, nursing homes, childcare centres, and public transportation. These applications fit well with the chapters that focus on using environments with groups of individuals, such as people with mental illness and youth with physical disabilities. Other chapters describe specific applications of environmental theories and models to different clients with whom occupational therapists work on a daily basis. These applications challenge all occupational therapists to think about ways that they can use environments to enable clients to experience satisfaction and well-being in their daily roles and occupations.

Chapter

8

An Interdisciplinary Approach to Environmental Intervention: Ecology of Human Performance

Melisa Rempfer PhD; Wendy Hildenbrand, MPH, OTR;
Kathy Parker, MS, OTR; and Catana Brown, PhD, OTR, FAOTA

CHAPTER OBJECTIVES

* To describe applications of the ecology of human performance (EHP) framework in designing interventions that address environmental issues.
* To demonstrate how occupational therapy professionals can work with colleagues from other disciplines in environmentally focused interventions.

INTRODUCTION

Faculty from the Occupational Therapy Education Department at the University of Kansas Medical Center developed the EHP framework (Dunn, Brown, & McGuigan, 1994) to guide our teaching, service, and research activities. The EHP framework addresses the interaction of the person, context, and task for performance; however, the main goal of the developers of the framework was to promote greater attention to context. Following a brief introduction to the EHP framework, this chapter illustrates how the framework has been used in the design and implementation of three very different programs. In the first scenario, an intervention to improve the grocery shopping performance of people with psychiatric disabilities is presented. The second scenario describes a model for making accommodations in adult basic education (ABE), while the last scenario demonstrates how occupational therapy students applied EHP principles to a service learning project at an assisted living facility. Each of the programs will be described in terms of the EHP constructs of person, context, and task and the intervention strategies: establish/restore, adapt/modify, alter, create, and prevent.

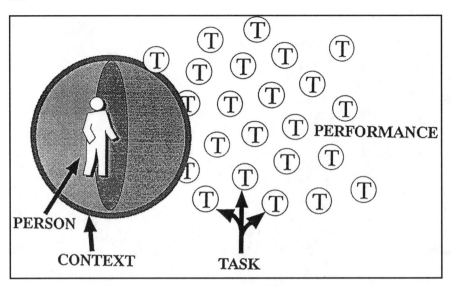

Figure 8-1. Schemata for the ecology of human performance framework. (From Dunn, Brown, & McGuigan. (1994) The ecology of human performance: A framework for considering the effect of context. *Am J Occup Ther*. Reprinted with permission from the American Occupational Therapy Association.)

Introduction to the Framework

In the EHP framework, the person is pictured as embedded in the context, indicating that the person cannot be seen or understood without his or her surrounding context (Figure 8-1). The person brings skills, abilities, and experiences to the situation. This includes performance components, such as motor or cognitive skills, but it also encompasses abilities, such as how to operate a computer, and experiences (e.g., moving to a new home). The context includes the social, cultural, physical, and temporal environment. The social environment is made of individuals, such as family, friends, and coworkers, as well as social institutions and systems, such as clubs, churches, schools, health care systems, economic systems, and political systems. The cultural environment involves those groups that contribute to the person's beliefs and values. The physical environment is the non-human aspects of context with a tangible presence such as objects, tools, buildings, and the natural world. The temporal environment addresses time-oriented issues associated with the person such as chronological age, developmental stage, and life cycle stage, and includes time conditions associated with task performance (e.g., when the task is performed, how often, and for how long).

Tasks are another important construct in the EHP framework. The whole universe of tasks is depicted as surrounding the person and the context. Whether or not a task is available to a particular person depends on what both the person and the context bring to the situation. Those tasks that are available are described as falling within the performance range. A task may be outside of the performance range due to person and/or context factors. For example, a person may lack the cognitive skills or training/experience to work as an air traffic controller or may live in a rural area where there are no airports, making this task unavailable.

The selection of the term *task* (as opposed to occupation) was intentional in the development of the EHP framework. The faculty in our department and most of our research and service activities involve an interdisciplinary team. We designed the framework so that it would be easily accessible to other disciplines and so that other disciplines could adopt the framework in their own work. In fact, the programs described in this chapter involve collaboration with other agencies and people with a wide range of professional backgrounds. The manner in which task is used in the EHP is compatible with the construct of occupation. In the occupational therapy literature, occupation is often described in terms of tasks that take on meaning for the individual (Christiansen & Baum, 1997; Clark, Parham, Carlson, Frank, Jackson, Pierce, Wolfe, & Zemke, 1991; Nelson, 1996). Although not explicit in the EHP, tasks take on meaning for the individual through the person's experiences and interaction with the context.

The EHP framework describes five intervention approaches: establish/restore, adapt/modify, alter, create, and prevent. A primary motivation for distinguishing the intervention strategies is to bring greater attention to the use of environmental approaches. The first intervention strategy, *establish/restore*, is the only approach that is focused on changing the person. In establish/restore, the intervention involves establishing new skills and abilities in the person or restoring skills and abilities that were lost. Learning how to use a computer is an example of an "establish" intervention, whereas increasing range of motion after a stroke is a "restore" intervention. The context is still a major consideration in establish/restore interventions. For example, there is literature indicating that people acquire skills more successfully when intervention involves engagement in everyday tasks that use natural objects and environments (Holm, Santangelo, Fromuth, Brown, & Walter, 2000; Ma, Trombly, & Robinson-Podoloski, 1999; Trombly & Wu, 1999).

In *adapt/modify*, the intervention involves making changes to the environment and task. Occupational therapists commonly use adapt strategies when applying assistive technology. However, the EHP aims to broaden the concept of environment so that adapt strategies are not exclusively aimed at the physical environment, but consider the social, cultural, and temporal contexts as well. For example, creating a support network uses an adapt strategy to modify the social environment, while reorganizing daily routines changes the temporal environment.

The *alter* strategy does not change the person or the environment, but is an approach that matches the person with the best fitting environment. The environmental match may involve a large context such as moving from a two-story home to a single-level home for someone who has trouble managing stairs or making the best match of roommates in a long-term care facility. Other alter strategies may make smaller changes such as the selection of elasticized pants versus those with zippers and buttons. An often-neglected consideration in matching people and environments is culture. However, the beliefs and values of a particular work setting, neighborhood, or club can have a major impact on an individual's ability to successfully participate in those settings.

The create and prevent strategies can use any of the approaches described above; establish/restore, adapt/modify, or alter. The difference is the goal of the intervention. In *prevent* strategies the goal is to avert a potential problem. The prevent strategy may use approaches targeting the person, such as teaching lifting techniques, or the strategies may target the environment, such as cushions to prevent decubitus ulcers. In *create* strategies the goal is to enhance performance without the assumption that a problem exists. For example providing new parents with information on child development (person-focused approach) and buying larger-handled cooking tools (environment-focused approach) are examples of create strategies. In both of these examples it is assumed that the individuals can manage the given tasks successfully, but their performance will be enhanced or more satisfying with these interventions.

The next sections of this chapter will provide three examples of how the EHP has been used in three programs: a grocery shopping intervention for people with psychiatric disabilities, accommodations for adult basic education, and a student service learning project at an assisted living facility. For each example, the person, environment, and task features will be described along with the specific intervention strategies employed.

Case Scenario 1: An Environmental Approach to Teaching Independent Living Skills to Persons With Serious Mental Illness

Psychiatric rehabilitation programs typically emphasize interventions that address impairments rather than environments (Heinssen, 1996). However, exceptions include the consideration of social environments in social skills training and clubhouse models (Harding & Keller, 1998). Because social functioning is frequently difficult for people with serious mental illness, these approaches often recognize the role of interpersonal skill as relevant to the social environments. Although many daily life activities require significant interaction with a social environment, others such as grocery shopping, housekeeping, and transportation are more dependent upon successful negotiation of the physical environment.

Although the physical environment is well recognized as a potential barrier to individuals with physical disabilities, it is virtually ignored in interventions for individuals with psychiatric disabilities. There is, at best, a small amount of literature in psychiatry regarding the potential benefits of addressing environmental accessibility from a physical perspective. For instance, some hospital-based programs for individuals with psychiatric disabilities have developed specific adaptations for the cognitive difficulties associated with psychiatric disabilities, using labels, signs, and other cognitive adaptations (Bopp, Ribble, Cassidy, & Markoff, 1996; Velligan, Mahurin, True, Lefton, & Flores, 1996). Although these research studies have supported the notion of using environmental interventions within the hospital setting, inpatient environments are different from those experienced by individuals with severe mental illness who live in the community. Therefore, there is a need to develop environmental interventions that are aimed at assisting individuals with serious mental illness to live successfully in community settings.

The purpose of this project is to develop and evaluate an independent living intervention designed to enhance accessibility of the physical environment for individuals with psychiatric disabilities. The independent living skill of grocery shopping was chosen as the target skill due to the essential nature of the activity and the highly complex physical environment in which the activity takes place. Using grocery shopping as an archetypal skill, this project emphasizes the interaction between a person and his or her physical environment. The intervention program was designed with an intentional interdisciplinary focus; members of our core research team include a psychologist (M. Rempfer, PhD), an occupational therapist (C. Brown, PhD, OTR) and a psychiatric nurse practitioner (E. Hamera, PhD, ANP). Development of our intervention was informed by theory and research in each of these disciplines. The person factor critical to the present study is cognitive functioning. The task factor in the present study is grocery shopping as an essential independent living activity. The environment or contextual factor is the grocery store.

Grocery Shopping as the Task

The task of grocery shopping was chosen because it is an important independent living activity and it is an excellent illustration of the complexity of person-environment interaction in daily life. The process of finding the desired items in a grocery store is a cognitively complex task that takes place in a physical environment that presents an array of information designed to capture attention and prompt impulsive purchases. Grocery shopping, as the task exemplar, presents workable opportunities to reduce environmental barriers by taking advantage of environmental supports.

COGNITIVE FUNCTIONING AS THE PERSON FACTOR

The cognitive impairments often associated with serious mental illness are well documented, and have been found in several studies to be associated with daily life functioning (Green, Kern, Braff, & Mintz, 2000). In particular, the grocery shopping skills program was intended to compensate for memory difficulties as well as the higher-order cognition involved in executive functioning, such as planning, organizing, and performing meaningful behavior.

THE GROCERY STORE AS THE CONTEXT

According to the EHP framework, the environment encompasses physical, social, cultural, and temporal aspects of context. While all aspects of context influence grocery shopping performance, the physical environment is most conspicuous. The physical environment of the grocery store includes the building, the aisles, the carts, checkout stations, and the numerous grocery items. This environment, brimming with physical information, presents a challenge in finding particular items. Moreover, the physical environment of the grocery store is organized to encourage the purchase of items the store wants to sell, sometimes obscuring items for which a customer is looking. At the same time, however, there are naturally occurring cues in the grocery store's physical environment that support effective performance. These cues include overhead signs and consistency in store layout.

The social environment of the grocery store includes other shoppers and grocery store personnel. This aspect of the context can facilitate grocery shopping performance when a grocery store clerk helps someone locate an item. On the other hand it can also be a barrier to performance when the store is crowded with shoppers. Cultural expectations related to choice and efficiency also affect the grocery shopping experience. Increasing consumer demands for a wide range of choices make the grocery store a busier environment. Cultural influences indicate that shoppers are not to interfere with the progress of others and that certain behaviors are expected in terms of shopping etiquette.

Finally, the temporal environment influences grocery shopping. The fact that shopping in most North American cultures is usually done less than daily means that the person has fewer opportunities for practice. In addition, the temporal environment is also a factor when time of day is considered; clearly some times of day are more supportive of grocery shopping than other times.

DESIGNING AN INTERVENTION TO ADDRESS CONTEXT

Our intervention used the above analysis of the grocery store environment to identify barriers and supports for the task of grocery shopping. According to the EHP framework, understanding of the person-environment transaction is the basis for designing interventions that support engagement in chosen tasks. In this case the person-environment interaction involves the task of grocery shopping for individuals with known cognitive difficulties in an information-rich environment, the grocery store.

Our intervention was designed with two primary assumptions: 1) It would be difficult to make significant environmental changes to reduce the amount of information in a grocery store; and 2) Given limits on current rehabilitation technologies it would be inefficient to attempt the amelioration of cognitive dysfunction directly (Bellack, Gold, & Buchanan,

1999). Therefore, the focus of this program is to compensate for cognitive dysfunction by providing instruction on helpful environmental cues. The intervention increases the accessibility of the grocery store by teaching individuals how to make sense of the shopping process and the environment of the grocery store.

This independent living skills intervention uses several techniques to compensate for cognitive impairments in an environment that poses barriers to performance. Some techniques are non-specific and reflect common therapeutic factors in psychiatric rehabilitation programs, such as the use of reinforcement and other behavioral learning strategies. In addition, several techniques were included specifically to address the environmental aspects of this intervention, such as utilizing naturally occurring cues including overhead signs and assistance from store personnel.

The instruction consists of nine group sessions that are organized around three questions: 1) Where is it? 2) Is this what I want? and 3) Is this the lowest price? Each question is used to assist group members in becoming familiar with the physical environment of the grocery store and to teach specific strategies to discern the salient environmental cues that help guide performance of the shopping task. For instance, in teaching group members how to find "what I am looking for," the group leader will teach three environmental strategies: using overhead signs, understanding the organization/layout of the store, and asking other people for help. Each of these three options is a naturally occurring cue that can be taken advantage of while shopping. Therefore, when experiencing difficulty locating a particular item while shopping, a group member is encouraged to use one of these compensatory strategies (which are common strategies for any shopper, in fact). Additional intervention techniques are illustrated in Table 8-1.

This skills training program has been examined in preliminary research and has been found to improve the grocery shopping accuracy and efficiency of persons with schizophrenia (Brown, Rempfer, & Hamera, 2002). The program has been incorporated into several community mental health agencies as part of their psychosocial rehabilitation programming and has been successfully implemented by staff from a variety of disciplines including social work, nursing, psychology, and occupational therapy. Additional research is underway to document the effectiveness of this environmentally focused treatment approach.

Case Scenario 2: Accommodating Adults With Disabilities in Adult Basic Education Programs

Adult educators face many challenges as they serve individuals in their community programs. Many students in adult education programs have unique learning needs. As professionals have become familiar with the specific needs of these adults the challenge to accommodate those needs takes on new features. While adult educators have been accommodating learning needs throughout their programs, there is an increasing need to know how to organize these strategies so they can be applied at the right time and for the right needs (Dunn, Gilbert, & Parker, 1997).

The need for occupational therapists to intervene with students with disabilities in public schools is not only well respected, it is also mandated (Individuals With Disabilities Education Act [IDEA] Public Law 105-17). However, adults with similar needs in education settings find themselves without support services and with instructors with little educational preparation serving their needs.

Recognizing the need to support adult education teachers in their task of educating adults who had significant barriers to learning, the Department of Education funded the Division of Adult Studies of the University of Kansas Center for Research on Learning to develop staff training materials for instructors around issues of persons with disabilities in

Table 8-1

Grocery Shopping

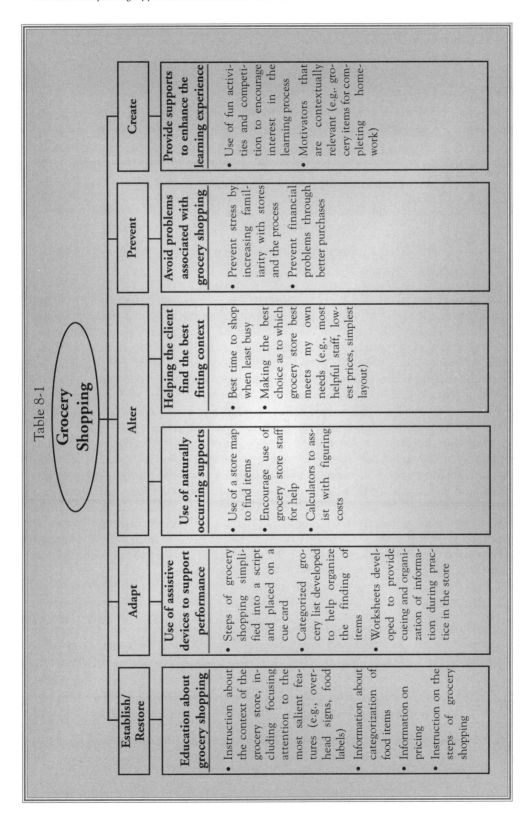

Establish/Restore	Adapt	Alter		Prevent	Create
Education about grocery shopping	**Use of assistive devices to support performance**	**Use of naturally occurring supports**	**Helping the client find the best fitting context**	**Avoid problems associated with grocery shopping**	**Provide supports to enhance the learning experience**
• Instruction about the context of the grocery store, including focusing attention to the most salient features (e.g., overhead signs, food labels)	• Steps of grocery shopping simplified into a script and placed on a cue card	• Use of a store map to find items	• Best time to shop when least busy	• Prevent stress by increasing familiarity with stores and the process	• Use of fun activities and competition to encourage interest in the learning process
• Information about categorization of food items	• Categorized grocery list developed to help organize the finding of items	• Encourage use of grocery store staff for help	• Making the best choice as to which grocery store best meets my own needs (e.g., most helpful staff, lowest prices, simplest layout)	• Prevent financial problems through better purchases	• Motivators that are contextually relevant (e.g., grocery items for completing homework)
• Information on pricing	• Worksheets developed to provide cueing and organization of information during practice in the store	• Calculators to assist with figuring of costs			
• Instruction on the steps of grocery shopping					

adult education programs. The EHP framework was used to develop the appropriate strategies and materials for this project.

Adult Education as the Task

The goals of adult basic education include: providing adults with basic academic skills that will help them become more productive members of the community; helping learners meet personal goals, such as developing job readiness skills, finding employment, advancing on the job, becoming a better parent, developing skills for interpersonal relationships, or entering adult secondary education classes; increasing students' self-respect and sense of self-worth; providing a non-threatening program environment to assist students in reaching their goals; and assessing students' skill levels (Work Investment Act, 2000). Typically adult basic education is accomplished one-on-one with a teacher or volunteer and the student using a prepared curriculum that consists of worksheets that the student is expected to complete independently.

The Context of Adult Education

Adult education takes place in institutional settings (school district, community college, corrections), community programs, or in the workplace. Adult education is most often state funded and includes ABE, English as a Second Language (ESL), and adult secondary education. Preparation of instructors in ABE is considered one of the greatest needs in ABE. Many ABE instructors receive little or no training, either in subject matter content or in the process of teaching adults (Kutner, 1992). Instructors in adult basic education tend to be part-time personnel (39%) or volunteers (48%) and only 19 states require a bachelor's degree (US Department of Education, Office of Adult and Vocational Education, 1999). Many states have instituted a certification program for adult educators.

Adult education classrooms are unique. Most consist of one large open space with tables surrounded by chairs. Because students are not obligated to attend (as in K through 12 education) there is more fluidity among the students, with students entering the program at different times and continuing for varying lengths of time depending on their goals. Some students attend daily while most attend only 1 to 3 days per week. Social relationships often develop between students who attend regularly. As a result of the non-obligatory nature of the process, students who do not feel welcome do not participate in the program. Project staff observed that many students with disabilities did not feel included and therefore did not return after their first day.

Adult Education Students: The Person Factors

In addition to being undertrained for their roles, adult basic educators receive no support services such as occupational therapy, special education, psychology, etc. Most of the students they instruct are individuals between the ages of 18 and 44 who did poorly in school, have psychiatric conditions, low cognitive abilities, or learning disabilities, and could also be unemployed, welfare recipients, or incarcerated (US Department of Education, Division of Adult Education and Literacy, 1999). It was estimated at the time that we began our project that 6 to 30% of enrollees in adult education had some kind of disability (US Department of Education, 1992). An awareness of this problem became the catalyst for our project.

Designing an Intervention to Address the Person, Task, and Context Interface

Daryl Mellard, PhD, a school psychologist and researcher with a background in adult education, was the principle investigator of this project. He and his colleagues at the University of Kansas chose the EHP framework as a model for identifying needs and designing strategies to help adult basic educators make their teaching approach systematic (Mellard, Hall, & Parker, 1999). The EHP provides a mechanism for making decisions about a person's goals and skills, tasks they wish to perform, and for considering the environmental (contextual) supports and barriers to successful performance. The EHP also enables adult educators to organize their knowledge and expertise in order to make decisions about which accommodation strategies would be the best match for the person (Dunn et al., 1997).

Continuing Education Materials for Adult Educators

The project staff developed training materials and a training package for adult educators around issues of disability. Brochures were developed describing the rights of adults with disabilities in adult education settings; handbooks were written for both learners and instructors describing a process for identifying, verifying, and accommodating a disability. Pivotal to the accommodations process was a functional needs interview for instructors to use to help adult students articulate where needs might exist and a format for a formal discussion of what strategies had both worked and not worked in the past.

Before the project began, a national study was conducted of adult educators to assess their knowledge and use of accommodations in their practice (Hall, 1997; White, 1998). Armed with these results and the EHP, occupational therapists from the University of Kansas Medical Center assessed the occupational performance needs of students in adult education and made suggestions for accommodations within classrooms based on the EHP intervention strategies. Table 8-2 provides a sample for one task. A chart was made for each of the "essential functions" identified by the occupational therapists. Each chart lists recommendations for interventions based on the various EHP approaches outlined at the beginning of this chapter.

From an activity analysis of the student role in adult education, occupational therapists identified the necessary performance skills. Some of the performance skills identified include: reading, accessing information with low vision, handwriting, solving math problems, remembering, paying attention to oral directions/spoken words, attention to task, getting started, staying on track, staying organized, dealing with changes, and frustration.

Once these performance skill areas were identified, therapists identified interventions based on the EHP intervention strategies. Instructors were taught to use the functional needs interview to identify needs and strategies that worked for the person in the past. At this point instructors could decide if a person, task, or environmental modification would be best to serve the person's needs. Some of the environmental modifications addressed included: modifying the social context by providing a study buddy or providing written materials along with oral instructions for individuals who had trouble paying attention, peer coaches for an individual with difficulty getting started. For individuals who had difficulty taking tests a reader or recorder were effective strategies. Modifications to the physical environment included providing a study carrel or quiet corner for a learner who was having difficulty sitting and working quietly; using visual cues on the chalkboard or overhead for individuals who had difficulty paying attention to oral directions or to the spoken work; and recommending that students with difficulty hearing were placed closer to the teacher and away from distracting noises (Mellard, Gilbert, Parker, 1998; White & Polson, 1999).

Table 8-2

Sitting for Long Periods:
causes increased agitation and decreased attention to task

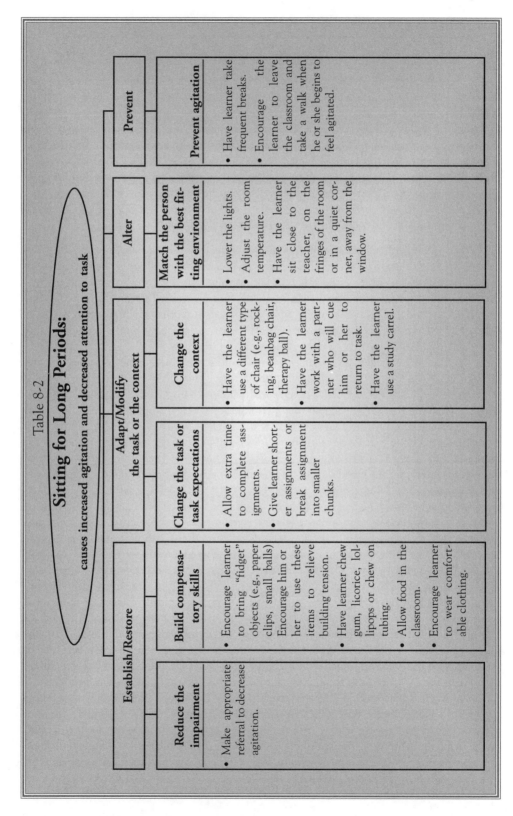

Establish/Restore		Adapt/Modify the task or the context		Alter	Prevent
Reduce the impairment	**Build compensatory skills**	**Change the task or task expectations**	**Change the context**	**Match the person with the best fitting environment**	**Prevent agitation**
• Make appropriate referral to decrease agitation.	• Encourage learner to bring "fidget" objects (e.g., paper clips, small balls). Encourage him or her to use these items to relieve building tension. • Have learner chew gum, licorice, lollipops or chew on tubing. • Allow food in the classroom. • Encourage learner to wear comfortable clothing.	• Allow extra time to complete assignments. • Give learner shorter assignments or break assignment into smaller chunks.	• Have the learner use a different type of chair (e.g., rocking, beanbag chair, therapy ball). • Have the learner work with a partner who will cue him or her to return to task. • Have the learner use a study carrel.	• Lower the lights. • Adjust the room temperature. • Have the learner sit close to the teacher, on the fringes of the room or in a quiet corner, away from the window.	• Have learner take frequent breaks. • Encourage the learner to leave the classroom and take a walk when he or she begins to feel agitated.

Our project staff provided staff in-service to adult education sites in nine states and all adult education programs in the state of Kansas. Occupational therapists have unique ways of looking at people and their tasks and environments that make them well suited for interdisciplinary projects like this one that serve the needs of many. We have applied the knowledge gained in this project to several other projects including Welfare to Work (Division of Adult Studies, 2000).

Case Scenario 3: Experiential Learning in a Contextually-Based Service Curriculum

Service learning, based in experiential learning, has been defined by the National Society for Experiential Education as "any carefully monitored service experience in which a student has intentional learning goals and reflects actively on what he or she is learning throughout the experience while the community organization is benefiting from this experience" (Sinay, 2000). As an educational model, key elements of effective service-learning include:

1. Service activities that are designed to meet community needs.
2. Structured and connected educational components that challenge participants to think critically about and learn from their experience.
3. Student reflection that educates those involved in the service learning activity.

Based on these defining points, it is important to remember that student participation in a community-based service organization alone does not constitute "service learning." Experiential learning does not necessarily result in a valued service to an organization. Likewise, involvement in service does not necessarily support student learning. The combined constructs of service and learning require collaboration among academic and community partners to determine the meaning of service for the agency and the meaning of learning for the students. Applied within the occupational therapy curriculum at the University of Kansas Medical Center, service learning provides a model of future interdisciplinary collaboration and an opportunity to establish professional competencies, while broadening the contextual understanding of occupational therapy.

A Context for Learning

Creating Health Options in Community Environments (CHOICE), an occupational therapy department service-learning project, provides community-based service experiences that require students to apply occupational therapy knowledge and skills. The project is completed during the first year of the Master of Occupational Therapy curriculum and is designed to allow exploration of community-based practice environments as well as the impact of context on a person's ability to participate in desired daily living tasks. Through direct and indirect service delivery, students acquire and demonstrate skill in assessment, program development, intervention, clinical reasoning, and collaboration. As recipients of these services, community partners that are often underfunded and understaffed are able to address additional service and programmatic needs. In addition, these sites are exposed to occupational therapy practice concepts, while also contributing to the learning experiences of the students. Examples of existing and previous community partners include supportive housing programs for people who are homeless, youth development organizations, child and family service agencies, and community support programs for stroke survivors.

Using the EHP as a conceptual lens, the project requires students to select and initiate contact with a health, education, or social service agency/facility that does not provide occupational therapy services. Student activities are guided by a thorough analysis of person, task, and context features. Throughout the course of the project, students receive support from occupational therapy instructors, the project coordinator, and community mentors who assist students in incorporating occupational therapy concepts in agency activity and defining opportunities for future occupational therapy practice.

A Demonstration Project

A specific example of a CHOICE community partner, established by a student trio for their service learning experience, is an area assisted living facility with the goal of supporting independence of older adults. Initial attraction to this site centred on newness of the facility, funding availability for projects/programs, interest in the older adult population, and student convenience. Interestingly, these same factors posed some barriers for the students, who reported "it was hard to come up with something to do at a place that has everything." Surveys, observations, interviews, and policy/program review helped students to identify program strengths and develop strategies for further serving the facility and its residents. The completed needs assessment determined that there was need for an accessible gardening community and structured gardening program, which would enhance programming and support independent engagement in a desired life task.

The Task of Gardening

Gardening was chosen as an intervention focus because of the multidimensional nature of this instrumental daily living task and the expressed interest of the facility population. Gardening provides an opportunity for relaxation, creativity, productivity, socialization, and community contribution. The participation requirements of this activity meet personal needs for physical activity (endurance, gross or fine motor activity), cognitive stimulation (problem-solving, sequencing), sensory processing (tactile, olfactory), meaning, and social belonging. It was observed that some residents were already engaging in personal gardening activities. However, this task was not systematically available to all interested residents. Having noted the discrepancy between expressed interest and actual involvement in gardening, the residents, students, and facility management determined that an accessible gardening program would be beneficial to student learning, resident satisfaction, and facility programming and appearance.

The Gardeners (Person Variables)

Many residents have come to live in the assisted living community as a result of environmental access needs, social changes in their circle of family and friends, and real or anticipated changes in personal performance. Certainly there are numerous physical, cognitive, and socio-emotional changes that occur as people age. In addition, the reality of increased susceptibility to illness as well as the likelihood of living with a chronic condition(s) is a significant concern and potential restriction for many aging people (Gormly, 1997). It is important to be mindful that the degree and pattern of natural change and health-related change varies greatly. Perhaps even more variable is the extent to which any of the changes interfere with a person's ability to engage satisfactorily in meaningful relationships, leisure activities, and productive lives. The people talk of past gardening experiences, remember gardening as a shared family activity, reminisce about Garden Club teas, and offer insights and hopes for future gardening activities. For this population, maintaining social connectedness, personal autonomy, and individual strengths are common goals that can be addressed within the context of an accessible gardening community.

THE "GARDEN" (UNDERSTANDING CONTEXT)

For this service learning experience to be effective, the understanding of context had to occur on two levels—the garden itself and the overall context of the facility. With regard to the garden itself, several contextual issues required attention. While there was a designated community garden area, the areas were located at ground level and access routes were ill-defined and rough. User-friendly gardening supplies were lacking. These physical barriers interfered with mobility for individuals dependent on ambulatory aids, and restricted contributions of those with difficulties with manipulation. The social environment of the garden is typically made up of other residents with occasional health care professionals or family and friends. The interaction of individuals in the garden facilitates socialization, supports nurturing needs, and encourages independence in task engagement. One barrier to sustained garden involvement was the lack of an organized social structure. The existing gardening situation had centred on gardening as an individual activity, rather than a community benefit or a program development opportunity.

Temporally, outdoor gardening is somewhat bound by the seasonal nature of the activity. The students quickly discovered that seasonal parameters also factored into the garden construction process as fall temperatures and weather events affected project timelines. Additional temporal considerations include the life stage of the facility residents. They are indeed representative of an aging population and/or people with disabilities associated with chronic conditions that may experience limitations in life participation. However, another temporal point is the need for these individuals to recreate routines and habits in their day (temporal reconstruction) that will support their collective interests, individual needs, and desire for community ownership through personal investment. Gardening provides a context for this engagement and contribution.

This analysis of gardening illustrates its contextual complexity; similarly, the overall context of the assisted living facility also warrants consideration. The facility is a newly established assisted living environment designed to support elder independence through safe accommodations, health care and other service professionals, and population-focused programming. Physically, the facility is bright and home-like, has individual living quarters as well as community kitchens (popular for "happy hours") and common areas, and has a range of environmental supports that ensure daily living in a least restrictive environment. Socially, the facility is located in a very financially sound suburban community known for its commitment to family, schools, and a healthful living environment. The cultural expectation that this suburban community will care for its elder citizens is evidenced by invested family units, accessible social supports, and family-focused community planning and policy making.

TARGETING A CONTEXT FOR INTERVENTION

Bringing gardening to the facility would require significant investment on the part of the establishment, the community, and the residents, not to mention incredible commitment by the student group that chose to embrace this project. Specific areas of intervention focus included creating an accessible community garden and establishing and institutionalizing a gardening program that will support gardening activities. Through these two avenues, the project would address physical, social, and cultural context components internal to the facility while also incorporating community resources as change agents in the project process and outcome. Table 8-3 demonstrates specific intervention strategies that were involved in this program development.

Table 8-3

Community Gardening at an Assisted Living Facility

Establish/Restore	Adapt	Alter	Prevent		Create
Education about community gardening/garden maintenance	Use of assistive technologies to support performance	Use of naturally occurring community supports	Prevent inactivity due to restricted leisure/productive work options	Prevent physical, cognitive, and psychosocial decline associated with aging	Provide supports to enhance the living experience
• Information and instruction about the task of gardening and garden maintenance (e.g., plant/flower knowledge, gardening techniques). • Education about energy conservation/work simplification and body mechanics when working in the garden. • Use of adapted gardening tools.	• Use of adaptive devices/gardening equipment (e.g., gloves, tools with straps, larger or longer handles, and lightweight or weighted tools). • Constructed raised gardens and work surfaces created to accommodate wheelchair users or those needing to sit while working. • Rough, uneven paths/surfaces smoothed through asphalt application o allow safe mobility	• Involved community partners through on-site labor (landscaping company) and in-kind donations (home improvement centre). • Area high school students joined project as labor through their own required community service involvement. • Garden Club incorporated as program structure. • Neighbor involved in sustaining Garden Club activities.	• Community garden provides activity option for leisure, productive work, and/or socialization. • Accessible garden environment supports health-promoting activity for residents with limitations due to disability.	• Prevent physical injury by use of adaptive devices and compensatory strategies when working in the garden. • Prevent cognitive decline by garden planning and problem-solving. • Prevent social isolation and psychological distress by involvement with peers through the Garden Club or garden work.	• Established the Garden Club to support resident gardening enthusiasts. • Collaborative planning and construction process that ensured project "buy-in" from residents. • Provided an activity option for residents, families, and friends. • Contributed to the aesthetic quality of the social/common ground that is shared by facility residents.

The community garden changed the physical face of context, created an arena for socialization, and incorporated community supports. The initial assessment of the existing garden highlighted many problems with accessibility. Mobility barriers (rugged pathway, ground height gardening platforms), obstacles to task participation (heavy and awkward tools/handles, lack of education/expert gardener support), and lack of resident investment in maintaining the garden (no structure for socialization, activity engagement) all impeded the development of a thriving community garden. Collaborative discussions/planning led the students to resources outside of the facility to gather expert guidance and additional labor support. An area "master gardener" was consulted to identify material/supply requirements and construct design plans appropriate for needs of residents and the facility. A local landscaping company installed a sprinkler system to help sustain the garden. A nationally known home improvement centre donated significant supplies to build the garden structure and to support safe work in the garden through tools, gloves, etc. Asphalt for the garden path/floor was donated by a local source. These aesthetic and structural modifications allowed the garden to be favorably viewed by the residents and facility staff.

The social community that came together to adapt the existing garden provides a model for incorporating social context in problem solving as well as problem recognition. In addition to professional and business support, an area high school with service designated in their graduation requirements was able to participate in the experience.

The garden club, comprised of facility residents, was organized to advise the direction of the garden construction and to structure activities to support ongoing socialization, leisure, and productive work based on gardening as an activity. The creation of the club introduced an additional social structure into the facility program and provided a social option for residents, thus enriching the social context. Because overall project activity and student leadership prompted the club, there was concern about the sustainability of the Club. This concern is being addressed through the volunteer involvement of a neighborhood resident who watched the construction progress and observed the renewed interest in community gardening. Again, this unforeseen social support emphasizes the value of context as an avenue for performance analysis, need assessment, and program development.

CONCLUSION

The EHP is one framework that can be used in program development to encourage the application of environmental interventions. As illustrated in the scenarios provided, this framework is especially appealing for interdisciplinary practice. In this chapter, different contexts were analyzed: the grocery store, ABE classrooms, and an assisted living facility. After the context was analyzed and considered in relation to relevant person and task factors, specific interventions were developed. In all three scenarios, interventions were used that enhanced skills in the person (establish/restore), made changes to the context or task (adapt/modify), made the best fit between the person and context (alter), enhanced performance (create), and avoided potential problems (prevent).

Summary of Environmental Interventions

Each of the three scenarios described in this chapter utilized a common intervention approach: examining person, task, and context factors to achieve multiple intervention options. The strategies described in all three projects aimed to be environmentally focused and innovative. For instance, each of the approaches involved entering into the actual environments of interest; the described projects have taken place at the grocery store, within the educational setting, and at the assisted living facility. By conducting the interventions in the natural environments, we were able to take advantage of several existing supports to facilitate the intervention process, such as utilizing grocery store signs to help individuals shop more efficiently, or using assistance from a neighbor to create and sustain the Garden Club.

These projects were also characterized by an interdisciplinary philosophy. Occupational therapists (and occupational therapy students) were core members of the project teams, but worked together with colleagues in other disciplines in developing the intervention models and strategies. Further, in implementing the projects within community settings, it was natural for the occupational therapy staff to partner with those professionals working in the contexts of interest. This was perhaps most salient in the adult education project in which occupational therapy professionals were entering into a relatively novel arena and consulted with the disciplines that traditionally work in this area, with educators, and with school psychologists.

Finally, each of the intervention approaches described are consistent with evidence-based practice. Two of the projects (adult education and grocery shopping skills training) were part of grant-funded research activities and all of the intervention strategies described were informed by existing evidence in disciplines such as occupational therapy, education, psychology, and nursing. Further, we are engaged in an ongoing research project to establish the effectiveness of our context-based intervention approach for skills training, as described in the first scenario.

Applications and Recommendations

We have described several specific intervention approaches in this chapter that we hope will be applied to other settings, populations, or tasks. The service learning project has been successfully applied to a wide variety of community settings, including child care centres, homelessness programs, and community programs for older adults. We have found that the model of using the EHP as an intervention approach has been attractive to community-based agencies that are unfamiliar with occupational therapy services.

The other projects have broader applications as well. We have recently applied our environmental accommodations from the adult education project in the welfare to work arena, where many individuals have occupational performance issues that keep them from effectively participating in the workplace. We have also developed learning modules for community college instructors and disabled student services personnel using our accommodation models. The skills training program we described has been pilot tested in a community program for adults with traumatic head injuries. Our preliminary work indicates that the strategies may need further adaptation for this population (e.g., more attention to the role of memory impairments in determining effective interaction with the shopping environment), but that the approach shows promise for extension into new areas.

The approaches described in this chapter show promise by providing a framework for environmentally focused interventions, particularly for community-based settings. The projects show preliminary effectiveness, both in terms of empirical evidence as well as in program success. However, further research is needed to determine how specific person-task-context interventions can be matched with specific populations or settings to promote effective occupational functioning within the natural environment.

Study Questions

1. How are the constructs of context and task described and related in the EHP framework?

2. How does the EHP framework promote interdisciplinary work between occupational therapists and colleagues in other disciplines?

3. Identify and describe the five intervention approaches in the EHP framework. How do these relate to the constructs of environment and task?

4. Select a context that an occupational therapist typically experiences in your practice area, and apply the five intervention approaches of the EHP framework.

REFERENCES

Bellack, A. S., Gold, J. M., & Buchanan, R. W. (1999). Cognitive rehabilitation for schizophrenia: Problems, prospects and strategies. *Schizophr Bull, 25*, 257-274.

Bopp, J. H., Ribble, D. J., Cassidy, J. J., & Markoff, R. A. (1996). Re-engineering the state hospital to promote rehabilitation and recovery. *Psychiatric Services, 47*, 697-701.

Brown, C., Rempfer, M., & Hamera, E. (2002). Teaching grocery shopping skills to people with schizophrenia. *Occupational Therapy Journal of Research, 22 (suppl. 1)*, 90-91.

Christiansen, C., & Baum, C. (1997). Understanding occupation: Definitions and concepts. In C. Christiansen & C. Baum (Eds.), *Occupational therapy: Enabling function and well being* (pp. 2-25). Thorofare, NJ: SLACK Incorporated.

Clark, R., Parham, D., Carlson, M., Frank, G., Jackson, J., Pierce, D., et al. (1991). Occupational science: Academic innovation in the service of occupational therapy's future. *Am J Occup Ther, 45*, 300-310.

Division of Adult Studies. (2000). *Opening doors: Empowering people with disabilities.* Lawrence, KS: University of Kansas, Center for Research on Learning.

Dunn, W., Brown, C., & McGuigan, A. (1994). The ecology of human performance: A framework for considering the effect of context. *Am J Occup Ther, 48*, 595-607.

Dunn, W., Gilbert, M. P., & Parker, K. (1997). The ecology of human performance framework: A model for identifying and designing appropriate accommodations for adult learners. In D. Mellard (Ed.), *Proceedings of the symposium on accommodating adults with disabilities in adult education* (pp. 25-33). Lawrence, KS: University of Kansas.

Gormly, A. V. (1997). *Lifespan human development* (6th ed.). Fort Worth, TX: Harcourt Brace.

Green, M. F., Kern, R. S., Braff, D. L., & Mintz, J. (2000). Neurocognitive deficits and functional outcome in schizophrenia: Are we measuring the right "stuff?" *Schizophr Bull, 26*, 119-136.

Hall, J. (1997). Results of a national survey and state interviews. In D. Mellard (Ed.), *Proceedings of the symposium on accommodating adults with disability in adult education* (pp. 14-21). Lawrence, KS: Kansas State University.

Harding, C. M., & Keller, A. B. (1998). Long-term outcome of social functioning. In K. T. Mueser & N. Tarrier (Eds.), *Handbook of social functioning in schizophrenia* (pp. 134-148). Boston, MA: Allyn & Bacon.

Heinssen, R. K. (1996). The cognitive exoskeleton: Environmental interventions in cognitive rehabilitation. In P. W. Corrigan & S. C. Yudofsky (Eds.), *Cognitive rehabilitation for neuropsychiatric disorders* (pp. 395-423). Washington, DC: American Psychiatric Association.

Holm, M. B., Santangelo, M. A., Fromuth, D. J., Brown, S. O., & Walter, H. (2000). Effectiveness of everyday occupations for changing client behaviors in a community living arrangement. *Am J Occup Ther, 54*, 361-371.

Individuals with Disabilities Education Act Amendments of 1997, Pub. L. No. 105-17, 34 C.F.R.

Kutner, M. (1992). *Staff development for ABE and ESL teachers and volunteers*. ERIC Digest. Washington, DC: Center for Applied Linguistics.

Ma, H., Trombly, C. A., & Robinson-Podolski, C. (1999). The effect of context on skill acquisition and transfer. *Am J Occup Ther, 53*, 138-144.

Mellard, D., Gilbert, M. P, & Parker, K. (1998). Procedural guide to accommodating adults with disabilities. In University of Kansas Institute for Adult Studies, *Accommodating adults with disabilities in adult education programs* (pp. 95-187). Lawrence, KS: University of Kansas Center for Research on Learning.

Mellard, D., Hall, J., & Parker, K. (1999). Assisting adult educators in preparing individuals with disabilities for employment. *American Rehabilitation, 25*(3), 24-31.

Nelson, D. L. (1996). Therapeutic occupation: A definition. *Am J Occup Ther, 50*, 775-782.

Sinay, T. (2000). Service learning in the undergraduate health administration curriculum: Theory and practice. *The Journal of Health Administration Education, 18*(4), 357-373.

Trombly, C. A., & Wu, C. Y. (1999). Effect of rehabilitation tasks on organization of movement after stroke. *Am J Occup Ther, 53*, 333-344.

US Department of Education, Division of Adult Education and Literacy. (1992). *Fact Sheet #9: Adult basic education programs for adults with disabilities*. Washington, DC: Author.

US Department of Education, Office of Adult and Vocational Education. (1999). *Adult education: Human investment impact 1994-1998*. Washington, DC: Author.

Velligan, D. I., Mahurin, R. K., True, J. E., Lefton, R. S., & Flores, C. V. (1996). Preliminary evaluation of cognitive adaptation training to compensate for cognitive deficits in schizophrenia. *Psychiatric Services, 47*, 415-417.

White, W. J. (1998). *Research report on the use and effectiveness of accommodations for adults with disabilities in adult education centers*. Manhattan, KS: Kansas State University, Department of Special Education.

White, W. J., & Polson, C. (1999). Adults with disabilities in adult basic education centers. *Adult Basic Education, 9*(1), 36-45.

Workforce Investment Act, Title II Adult Education and Family Literacy Act. (August 2000). *Model program standards for adult basic education: The California state plan*. Retrieved August 20, 2002, from www.otan.dni.us/webfarm/stateplan/2004revisedstateplan/ch03.html.

Creating Supportive Work Environments for People With Mental Illness

Susan Strong, MSc, BSc, OT Reg. (Ont.); and
Karen Rebeiro, MScOT, BScOT, OT Reg. (Ont.)

CHAPTER OBJECTIVES

* To encourage readers to consider environmental interventions for clients with mental illness who have goals related to work.
* To identify key physical, cultural, and institutional aspects of workplace environments that facilitate or hinder clients' participation in work.
* To suggest strategies to create supportive work environments.
* To offer a broad conceptualization of work and alternative paths to meaningful work.
* To frame occupational performance for clients with mental illness within the recovery process.

INTRODUCTION

Traditional approaches to assist individuals with mental illness to succeed in work roles have tended to focus exclusively upon interventions directed at the person with the illness, and have not been supported by research or the consumer literature. Despite intensive training and support of workers in community placements, less than 15% of people with serious mental illness successfully obtain and maintain traditional paid employment (Anthony, Cohen, & Farkas, 1990). In fact, "mental illness is associated with the lowest employment outcomes among various disabilities" (Davidson et al., 2001, p. 376). Thus, the potential for consumers of mental health services to enjoy the benefits of full citizenship and to escape the cycle of poverty, which significantly defines their existence, remains intangible (Church,

1997). Work not only provides the means for practical survival, but is also the basis for feelings of self-worth and belonging, a venue to contribute to the community (Clutterbuck & McKay, 1984; Rebeiro, Day, Semeniuk, O'Brien, & Wilson, 2001), and a strategy to manage mental illness (Strong, 1998). Therefore, there is a need for new innovative approaches to assist individuals with mental illness to achieve work-related goals.

This chapter describes changes to traditional vocational practice, what has been learned through both experience and research, and an expanding role for occupational therapists in community mental health practice, namely, that of creating and establishing a range of supportive work environments. New ideas that utilize the environment as a key therapeutic variable and intervention strategy for persons with serious mental illness are offered and discussed. Evidence from research studies is used to illustrate the benefits of creating supportive work environments to enable occupational performance for persons with a mental illness.

BACKGROUND

There has been a shift in the conceptualization of mental illness from a medical orientation of disability that focuses on an individual's functional impairments toward viewing disability as an interaction between the individual and environment (Hahn, 1984; Kirsh, 2000). Empirical studies have shown that vocational outcome is not related to severity of psychiatric illness or diagnosis or psychological tests, and there is little correlation between symptomatology and function (Anthony & Jensen, 1984; Kirsh, 2000). Kirsh's (2000) study of factors associated with return to work illustrates the impact and significance of the work environment and "the importance of looking beyond the person and pathology to include individual and environmental factors" (p. 19). It is recognized that a combination of a number of different client and program attributes must be considered to determine who does best in different locations (Cochrane, Goering, & Rogers, 1991).

Increasing evidence that social, cultural, and institutional environments contribute to unemployment supports the need for innovative approaches to focus on environmental interventions. High unemployment has been attributed to the lack of creative and flexible employment programs for the psychiatrically disabled, society's negative stereotypes of the mentally ill, and competition in today's job market (Hartl, 1992; Trainor & Tremblay, 1992; Trainor, Pomeroy, & Pape, 1993). Persons with mental illness feel that society's prejudicial attitudes and the resulting stigma of mental illness are the central barriers to working (Leete, 1989; Rebeiro 1999). This includes negative attitudes held by professionals such as believing a person with a mental illness will never be able to work and denying access to supportive employment programs (Corring, 1996; Davidson, 1991). Deegan (1992) describes how a central attitudinal barrier drives a "cycle of disempowerment and despair" by transferring control over consumers' lives to the health care system and encouraging consumers to behave as helpless patients (Figure 9-1). Economic barriers arising from disincentives built into the disability support payment system have also been a focus of criticism (Church, 1997; Rebeiro, 1999; Strong, 1995; Trainor & Tremblay, 1992).

New government policies recognize the value of self-help and supportive communities. Individuals and systems now share responsibility for health, and the new focus is to work with communities versus individual clients to create supportive, caring communities and to enable community action (National Health and Welfare Canada, 1988). Canadian government programs have been undergoing mental health reform as articulated in *A New*

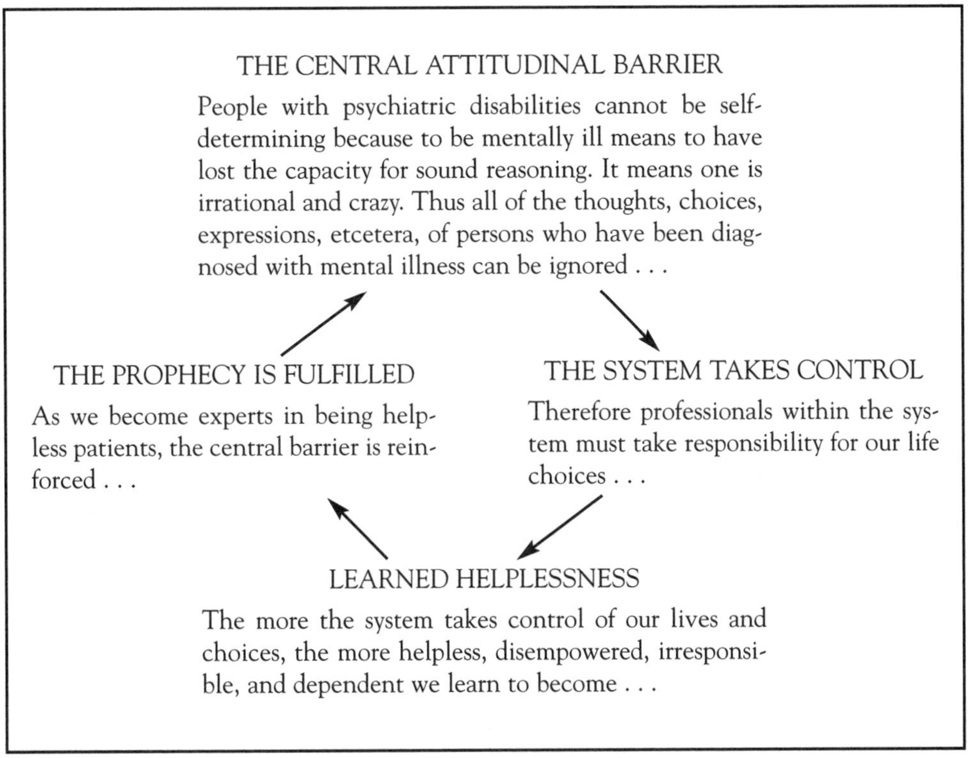

THE CENTRAL ATTITUDINAL BARRIER

People with psychiatric disabilities cannot be self-determining because to be mentally ill means to have lost the capacity for sound reasoning. It means one is irrational and crazy. Thus all of the thoughts, choices, expressions, etcetera, of persons who have been diagnosed with mental illness can be ignored . . .

THE PROPHECY IS FULFILLED

As we become experts in being helpless patients, the central barrier is reinforced . . .

THE SYSTEM TAKES CONTROL

Therefore professionals within the system must take responsibility for our life choices . . .

LEARNED HELPLESSNESS

The more the system takes control of our lives and choices, the more helpless, disempowered, irresponsible, and dependent we learn to become . . .

Figure 9-1. The Cycle Of Disempowerment and Despair (Strong, 1995 adapted from Deegan, 1992).

Framework For Support For People With Serious Mental Health Problems (Trainor et al., 1993), and a series of government documents: *Mental Health For Canadians: Striking A Balance* (National Health and Welfare Canada, 1988), *Putting People First* (Ontario Ministry of Health [OMOH], 1993), *Making it Happen* (OMOH, 1999), and *Making it Work* (OMOH, 2001). Through this process, mental health programs are shifting from a "service paradigm" to a "community process paradigm." As depicted in Figure 9-2 (Trainor et al., 1993, p. 1), the person with a mental illness is conceptualized in the centre of a community resource base, drawing resources from four sectors involving informal and formal community supports, and having access to the fundamental needs of housing, work, education, and income. Mental health consumers are becoming more organized and cognizant of their civil rights, and are increasingly seeking their rightful and equal access to work, in actions similar to those taken by individuals in the disability movement in the United States (Diasio, 1971).

In the rehabilitation arena, the widespread adoption of psychosocial rehabilitation philosophy by many mental health care providers has promoted services to be provided in a more flexible manner; driven by client-defined needs and client choices. The emphasis has shifted toward interventions that focus on aspects of the person's living, learning, and working environments. Client empowerment and client choice are emphasized. Similarly occupational therapists have begun incorporating into mental health practice the person-environment-occupation (PEO) model (Law et al., 1996; Strong et al., 1999), client-centred practice principles and the Canadian model of occupational performance (Canadian

Figure 9-2. Person at centre of community resource base. (Reprinted with permission from Trainor, Pomeroy, & Pape [1993]. *A new framework for support for people with serious mental health problems.* Toronto, ON: Canadian Mental Health Association.)

Association of Occupational Therapists, 1997). The occupational therapy profession views people, occupations, and the environments in which people engage in occupations as inseparable; the product of their relationship is referred to as occupational performance and is the focus of intervention (Law et al., 1996).

Each of these trends in the consumer, government, and rehabilitation arenas has shaped how therapists conduct vocational practice in mental health today. Client-centred practice requires therapists to assist consumers to engage and participate in the world of work, a regular aspect of community life. Because the goal is enabling occupational performance in environments of choice, therapists become informed of the many aspects of the community environment that pose as either barriers or enablers to this end. A variety of strategies can be utilized by therapists to address the identified barriers to participation, including, but not limited to, creating supportive work environments. For example, the therapist might intervene at the level of municipal or provincial policy to ensure that a client has access to the same rights and services as the general population. Or, the therapist might consider the creation or modification of a workplace's institutional, cultural, or physical environment to best enable an individual's participation. Using the PEO model, the focus is to improve the match between the client (person), workplace (environment), and work (occupation) in order to facilitate occupational performance (doing the work). In a redefined and evolving community practice domain, advocacy, knowledge of civil rights, and how to use them become a necessary aspect of occupational therapy practice. Partnerships are expanded beyond formal or traditional mental health services to include all stakeholders in the community resource base (see Figure 9-2).

Traditional Vocational Practice in Mental Health and Meaningful Work

Conventional approaches to assist individuals with mental illness into work roles have tended to focus exclusively upon interventions directed at increasing individuals' capabilities. Traditionally, vocational rehabilitation operated in a graduated step-wise fashion, beginning with vocational counseling, skills training with experiences in sheltered rehabilitation programs or work adjustment training programs, and advancing to a transitional placement in the community, sometimes supported with job clubs. The literature consistently reports that this step-wise "train" followed by "place" approach has not effectively resulted in sustained competitive employment for people with serious mental illness (Bond, Drake, Mueser, & Becker, 1997; Lehman, 1995; McGurrin, 1994). Lehman (1995) suggests that although these rehabilitation services have resulted in overall enhancement of vocational activities for participants, they do not have significant effects on competitive employment rates after participants leave the programs. Sheltered workshops, enclaves, and work crews have been criticized for training participants to deal with sheltered environments rather than competitive community work environments, creating dependency and inadequate remuneration (Trainor & Tremblay, 1992). In addition, self-directed strategies, whereby clients follow up on newspaper leads and participate in job clubs, have not proven to impact employment rates or tenure (Bond et al., 1997).

Increasingly critical of training programs in contrived environments, and "make work" projects, consumers of mental health services are demanding real and meaningful work as an integral aspect of a full continuum of the health care system (Rebeiro et al., 2001; Trainor et al., 1993). In order to be consistent with psychosocial rehabilitation philosophy, supporters advocate that work must be real (i.e., paid or integral to the organization's operation) and meaningful (i.e., a productive and valued activity) (Carver & Gaylor, 1990). However, it is clear from the literature that the value, meaning, and purpose of work will vary for different groups and individuals, particularly between clients and professionals and the systems that govern them (Fabian, 1989; Goldstein, Cohen, Lewis, & Struening, 1988; Powell, 1987). For example, the recovery process and meaningful work are often the focus of consumer concerns, whereas service providers and administrators tend to focus on reducing recidivism or savings in health care dollars.

The consumer literature explores how meaningful employment is believed to be essential to the recovery process (Leete, 1989), an internal change process in which disabled persons "experience themselves as recovering a new sense of self and of purpose within and beyond the limits of the disability" (Deegan, 1988, p. 11). During an 18-month ethnographic study, Strong (1998) explored the meaning of work for a group of people with persistent mental illness working at an affirmative business and found: a) each individual's experience is uniquely personal; b) meaning is influenced by the interplay of many variables (personal, occupational, workplace, external environment); and c) meaning varies with the individual's relationship with illness and his or her self-concept. All the themes that emerged concerning the meaning of work revolved around the recovery process, pointing to a relationship between meaningful work and the recovery process.

While there remains a need for vocational opportunities in sheltered settings (Bernardo, Kirkpatrick, Strong, Start, & Packer, 1996), there is consensus that there should be a full range of services available to reflect the preferences, needs, and capabilities of a diverse population (Cochrane, Goering, & Rogers, 1991; OMOH, 1993). A typology of mental health vocational programs outlines the considerable range of potential approaches to work and

illustrates the variations (Appendix A). For each client, the therapist must have the particular client identify what she or he considers to be meaningful work, and together plan the
best approach. Client-defined needs and client choices drive today's services.

ALTERNATIVE APPROACHES BASED ON
AN EXPANDED CONCEPTUALIZATION OF WORK

Given that ideas about what is considered work are changing, and traditional approaches have not been effective, alternative supported employment approaches are being considered. Referring to the Rehabilitation Act Amendments of 1986 (revised in 1992), Bond et
al. (1997) described the following characteristics of the new generation of supported
employment, "Clients work for pay, preferably the prevailing wage rate, as regular employees in integrated settings and in regular contact with non-handicapped workers, and receive
ongoing support" (p. 335). Supported employment services can be provided in a variety of
ways such as: job coach model, clubhouse model and transitional employment, assertive
community treatment model, or a "choose-get-keep" model (Appendix A). What is common to all supported employment programs is "a goal of permanent competitive employment, minimal screening for employability, avoidance of prevocational training, individualized placement instead of placement in enclaves or mobile work crews, time unlimited support, and consideration of client preferences" (Bond et al., 1997, p. 336).

Each of these aforementioned models has contributed to what is called individual placement and support (IPS) in today's practice. The key elements that define IPS are: a) direct
assistance with finding and keeping a job; b) integration of clinical and vocational services
versus the provision of brokered services; and c) attention to client preferences (Bond et al.,
1997). Research based on randomized controlled trials has shown IPS participants to be
more likely to obtain competitive jobs, average more hours working with more earnings, and
obtain a position in which they worked 20 hours or more a week without detriment to symptoms, self-esteem, or quality of life (Drake, McHugo, Becker, Anthony, & Clark, 1996).
Interestingly, research has found that IPS remains superior to the traditional "train" followed
by "place" approach for those individuals who are often considered poor candidates for vocational services because of poor work history, dual diagnosis and homelessness (Drake et al.,
1999). In these studies, the success of IPS has been attributed to the integrated approach,
intensive case management, and clients not having to switch to a new program that would
require new relationships and different systems of operation. In practice, therapists will use
this IPS approach to negotiate and modify existing work environments to provide opportunities for individuals with mental illness to successfully participate in occupations of choice
within chosen environments.

Increasingly therapists must be able to consider alternatives to full-time competitive
employment for their clients. Few or limited competitive employment opportunities exist,
and/or the effects and duration of illness, combined with individual variations preclude full-
time employment as a desired option for some consumers. Opportunities are expanded when
a more inclusive and broader definition of work is considered. This spectrum or continuum
of work extends beyond the new generation of supported employment (e.g., IPS) to include:
volunteering, self-employment, part-time or temporary commission work, home employment, affirmative businesses, and clubhouses. These work alternatives are differentiated
from supported employment, as a professional operating in a rehabilitation practice model

does not directly support them. The therapist has little control operating outside the formal health care system. The therapist relies on diplomacy, extended partnerships, role modeling, and resource linking to create supportive work environments. These actions in the creation of alternative supportive work environments are illustrated in the discussion of Northern Initiative for Social Action (NISA) later in this chapter.

Affirmative businesses are one example of innovative work alternatives being pursued by provider-consumer-community partnerships. Government-sponsored consumer co-operatives and affirmative businesses in Canada, the United States, and Great Britain have continued to operate since the 1980's, many with reduced reliance on government funding, and report positive outcomes during program evaluations (e.g., meaningful work, reduced use of formal health services, increased income) (Trainor & Tremblay, 1992; Hartl, 1992). These businesses provide opportunities for flexible employment with regular wages in a realistic work environment. They operate under an independent living model of practice rather than a service or rehabilitation model and range from informal, loosely organized groups to incorporated, formal operations (Trainor & Tremblay, 1992). They have varying degrees of support from professional staff. Originally conceived to provide self-help and peer support, these initiatives have expanded their role to include mutual support, economic development, knowledge production and skills training, advocacy, public education, contributions to professional education, and cultural activities (Consumer/Survivor Resource Centre, 1994). The Consumer/Survivor Information Resource Centre of Toronto has a Web site http://www.icomm.ca/csinfo/ with links to local, provincial, national, and international resources, including links to vocational services and sites.

Part-time, commission, and occasional work options should also be considered in the spectrum of opportunities for consumers. For example, NISA recently opened a gift shop in the lobby of the Northeast Mental Health Centre hospital. This economic venture resulted in the hiring of a full-time economics leader and seven to 10 part-time workers to run the shop. In addition, the gift shop's selling venue created the opportunity for artists and crafts persons to sell their wares on a commission basis, and at their own rate. By providing a variety of opportunities for paid work, therapists are more likely to address the diverse and unique needs of many individuals who are working toward goals of full time, competitive employment.

Volunteering is also a viable option that may lead to paid employment. Many of the perceived benefits from volunteering are similar to those experienced from competitive work (Rebeiro & Allen, 1998). In a single case study design, Rebeiro and Allen (1998) found that the occupation of volunteerism helped one individual with schizophrenia to construct a socially acceptable identity, to deal more effectively with the symptoms of schizophrenia, and to participate in a valued and meaningful work role in the community.

In summary, therapists can help create and develop a wide range of supportive environments to assist consumers to more fully and meaningfully participate in work and in their communities.

Creating Supportive Work Environments for People With Mental Illness

There are many important factors to consider when developing effective supportive work environments to facilitate the recovery process. Anthony (1993) emphasized that recovery-oriented services must focus on meaningful interventions that offer individuals success and satisfaction with life rather than focus on impairments, dysfunction, disability, and disad-

Table 9-1

Factors of Success

Person	Occupation	Environment
• Philosophy, approach to work and others	• Challenges and successes	• Sense of community, safety, teamwork
• Fit with skills, abilities, aspirations, interests	• A balance of variety and routine	• Flexibility, responsive to individuals' needs, openness to innovation
• Connection with activities, people, organization	• Valued contributions to the organization	• Effective communication and accepting attitudes
• Active agents in the recovery process	• Perceived control by workers, experience outcomes of the work	• Work experience, training and promotion opportunities

(Adapted from Ochocka et al, 1994; Strong, 1995)

vantages. Believing recovery can occur without professional intervention or formal services, the job of professionals is to facilitate consumers to do the work of recovery. The challenge is how to put into place structures and supports in work environments to facilitate this process. Anthony (1993) and Deegan (1992) identify the importance of having people who believe in them and who they can trust to be there in times of need. Therapists' actions and words can convey messages of hope, trust, and acceptance. More recently, Anthony (1994) advocated for "places to be and symptom free" as the parameters for successful outcomes in psychosocial rehabilitation. Davidson, et al. (2001) similarly highlighted an inclusive environment and community as essential ingredients in recovery.

The importance of workplace culture and key factors for success are identified in two studies: a) a 2-year qualitative research project, *Workplaces That Work* (Ochocka, Roth, Lord, & MacGillivary, 1994), examined successful community employment for people with disabilities from the perspectives of employees with disabilities, their employers, co-workers, and agency support persons; and b) Strong's (1995) ethnographic study of an affirmative business operated by individuals with persistent mental illness. Ochocka and colleagues focused on workplace integration and management style, while Strong analyzed differences in experiences and relationships with work and work environments. Remarkably, there is considerable agreement in findings concerning key factors for success. Although it is recognized that success can only be defined by the person involved, Table 9-1 summarizes the person, occupation, and environment factors related to optimal occupational performance and facilitation of the recovery process. Each of these factors can be considered and fostered by therapists when searching for or developing a niche where clients can successfully work. Both studies viewed successful employment as being largely determined by the quality of the fit between the worker and workplace (Ochocka et al, 1994; Strong, 1995).

Ochocka and colleagues' study recommended: workplaces with accepting, inclusive atmospheres in which management's style supported diversity and operated with clear, consistent communication; and assisting individuals with disabilities to become aware of particular workplaces' cultural values and standards and to strategize how to participate in the necessary patterns of interaction to integrate into the settings. Therapists can cue clients to particular workplace values, standards and patterns of interaction, and facilitate clients to connect with and communicate effectively with co-workers and supervisors. If needed, skills training can be conducted in parallel to a client working. Marrone, Balzell, and Gold (1995) offer many strategies to facilitate the client to access supports such as "natural" co-worker supports, personal network supports, and training supports in the workplace. An important strategy is to teach a client how to scan the work environment to identify supports.

Relationships and clear communication are important to successful employment. Perhaps equally important for this group in which relationships can be difficult are key institutional or organizational components of the work environment that form the basis for a low-stress, safe, accepting workplace culture (Strong, 1998). These institutional components are expanded and illustrated in this chapter's "Case Scenario: Aaron Working At Helping and Rehabilitating People."

The environment is an important contributor to the recovery process. Strong's (1998) study documented how engaging in work activities within a supportive environment can play a significant role in changing participants' self-concept, self-efficacy, and sense of well-being. For example, supportive environments can enable clients to participate in work, which in turn facilitates *connecting* (i.e., forming bonds with the work, people, place, organization's ideals to ascribe meaning), *being* (i.e., sense of themselves as individuals, experiencing making meaningful contributions), and *belonging* (i.e., sense of being part of a larger community or business). When there is a person-workplace-work match, work can become the medium through which a client's self-concept changes to one of *becoming* a capable person with a future, thereby supporting the recovery process. This suggests the quality of the fit between the person, occupation, and the workplace facilitates the recovery process (Strong, 1995).

The provision of an environment that is supportive of participants' being, belonging, and becoming needs, combined with the provision of the opportunity to participate in personally meaningful and socially valued occupation, assists individuals with serious mental illness to become more active participants in their community. This was illustrated in Rebeiro and colleagues' study (2001) of the benefits of participating in an occupation-based, consumer-governed initiative (known as NISA), in two studies that explored the development of consumer-run initiatives (Legault & Rebeiro, 2000; Wright & Rebeiro, in press), in Strong's (1995) and Renwick and Brown's (1996) studies on quality of life.

NISA illustrates how several aspects of the environment contribute to enabling greater participation in occupation (Legault & Rebeiro, 2000; Rebeiro & Cook, 1999; Rebeiro et al., 2001). For example, the element of choice and self-determination is important. Participants are told that they do not have to come to the program, and if they choose to come they can attend as often or as seldom as they want. Their level of participation is self-determined. It is also important that the environment is perceived to be both physically and emotionally safe for learning new skills, for forming relationships, and for the trial and error experimentation that often accompanies new learning (Rebeiro, 2000). A safe environment is crucial to the process of healing through addressing one's *being* and *belonging* needs, and essential to recovery by addressing one's *becoming* needs. Safety is attributed to the unconditional acceptance of all persons who attend the program and a focus on strengths versus illness. A final aspect of the environment that fosters a sense of belonging is the provision of both private and community space. Participants believe that private space is important when *being* needs require their attention, but that the opportunity to visit others and to meet in more public spaces is also important to their needs for social interaction. Such opportunities help them form a group identity and meet *belonging* needs. Participants speak of the importance of having their own space, and yet being involved with and having opportunity to interact with others.

In summary, the key features of supportive work environments that are particularly important for persons with mental illness are:

* Flexibility
* Responsiveness to individual needs
* Emotional and physical safety
* Choice and opportunity of occupation
* Shared decision-making and sense of teamwork
* Emphasis on the process of work for health and well-being
* A focus on personal strengths versus illness
* Accepting attitudes
* Effective, ongoing communication

The use of occupation as therapy and the enablement of occupational performance require occupational therapists to have an appreciation and understanding of the many factors that influence human occupation. To do this effectively, a broad appreciation of both micro- and macro-environments is necessary to enhance our knowledge and understanding of human occupation and the enablement of persons' occupational performance.

CASE SCENARIO: AARON WORKING AT HELPING AND REHABILITATING PEOPLE

The following example of an individual working in an affirmative business illustrates key elements for a supportive work environment, one that enables clients to feel safe, make connections, belong, contribute meaningfully, and experience challenges/successes.

The mission of Helping and Rehabilitating People (HARP) is to provide workers with psychiatric disabilities a range of opportunities for work, skills training, self-development, and employment through their active participation in the effective operation of a business. This business evolved from a group of service providers believing that outpatients involved in a tertiary care hospital-based ceramics rehabilitation program were capable of working in a community ceramics business. Given the lack of work opportunities that existed, and the successes of other consumer initiatives, they decided to create a consumer-driven business in Hamilton, Ontario. The developers intended that the venture would result in more consumer autonomy with decreased reliance on formal mental health services over time to make sure that they were not creating another institutional program in the community. Using government seed money in July 1992, an incorporated non-profit affirmative business was formed in partnership with consumers, community and business representatives operating through a board of directors. Additional individuals with experience as patients in the mental health system were hired with specific skill sets for the operation of a business (e.g., marketing, accounting, management). As HARP employees, approximately 35 consumers work in every aspect of the ceramics business and on working committees. Consumer supervisors, supplemented by government funding, receive minimum wage. Resource staff members employed by the sponsoring hospital facilitate individuals' participation by providing support, skills training, accommodation, and facilitating creation of a supportive work environment in partnership with consumers. Over the past 10 years, the ceramics affirmative business has expanded to become an umbrella organization or cooperative for several consumer initiatives known as Rainbow Ceramics and Gifts.

Aaron is a 28-year-old man with a schizoaffective disorder who began working at HARP in the finishing department cleaning products prior to shipping. The PEO model (Law et al., 1996) is used to explore his occupational performance experiences.

Person-Environment Interaction

Aaron reports making a commitment to HARP in contrast to past relationships to programs or organizations. Seeing himself as a marginalized consumer who has experienced discriminatory attitudes, he relates to and supports the business's ideals of creating work opportunities within a supportive community of consumers. The sense of community is supported by community business meetings, a committee structure for shared teamwork, and active solicitation of everyone's opinion for planning and evaluation. Resource staff members lower professional barriers and relate to members as individual people. For a few weeks at a time, Aaron has missed work complaining of depression. In HARP's safe, accepting environment he is able to talk about his illness experiences at coffee breaks and not worry about repercussions. He connects with other workers, making friendships with selected individuals. He appears to gain support and hope through observing others' examples, and gets practical information from people with a shared illness experience. Over time, he has become an outspoken leader in committees in which members appreciate his ability to voice their concerns.

Person-Occupation Interaction

Aaron reports that he works to stop himself from becoming depressed, to achieve something, and feel good about himself. His performance is inconsistent as he struggles with difficulties with memory, problem-solving, and organization of his work. His approach to dealing with problems as they arise in his work is a tendency to blame others. With offers to help, co-workers and consumer supervisors clearly convey the message that everyone makes mistakes as they learn, but every worker is responsible for making a competitive quality product for sale, and for maintaining standards of courtesy towards others. Resource staff members listen to concerns, express faith in his capabilities, and convey hope for the future. They also work with his consumer supervisor to provide repeated instruction, encouragement, and reorganization of the job into a series of steps. Over time, he has become better at discovering his mistakes for himself. After several months, he found the job of cleaning product boring, repetitive, and lacking challenge. He expressed that he wanted to be trained to do something else that he would find more interesting. When an opening arose, he was moved to the shipping-receiving department.

Occupation-Environment Interaction

The business is flexible toward scheduling work hours to meet workers' needs. Aaron can arrange to work during the afternoons, particularly when he is struggling with his illness. The work varies from week to week depending on customer orders. Errors are dealt with matter-of-factly. Expectations of competitive quality work are conveyed in quality control checks and reports about group production during business meetings. When there is pressure to get products out the door to customers there are others who can assist to meet deadlines.

Person-Environment-Occupation Relationship

In summary, HARP is a place where Aaron has made connections, feels he belongs, and is accepted in a community that emphasizes teamwork. Aaron fluctuates between a patient role and worker role. He has not yet become his own agent in the recovery process. After listening to other workers talk about taking control of their lives, and having observed several workers move successfully into their own apartments, he is considering doing the same. HARP provides some stability in his life and assists him to maintain his health. Many flexible institutional elements support his participation with clear communication of expectations. Support is given to help him manage his illness. For some individuals plagued by "voices," cognitive behavioral strategies are offered to help the person regain control. Individualized training is given with structure, support, and time to learn and experience. Accommodation is given for the back and forth progression of "good" and "bad" days.

CASE SCENARIO: DAVID WORKING AT NORTHERN INITIATIVE FOR SOCIAL ACTION

The following case example illustrates how occupational therapists can create alternative supportive work environments to enable occupational performance of persons with mental illness based on Legault and Rebeiro's study (2000). In particular, the case study illustrates how occupational therapists can assist a consumer-governed initiative to address people's occupational needs.

NISA is a consumer-governed, occupation-based mental health initiative operating from the site of the Northeast Mental Health Centre in Sudbury, Ontario, Canada. NISA has evolved from a small collective into a registered charitable status organization with 51 active participants and over 350 participants internationally. Initiated by a partnership between an occupational therapist and a group of mental health consumers, the program has developed through building partnerships with social service agencies and non-profit organizations that hold similar social visions of an inclusive community. NISA is guided by the belief that self-respect, an accepting community, and opportunities for meaningful occupation are keys to optimum wellness for all citizens, and not just for persons with disabilities. The organization seeks to offer an alternative yet interdependent link to traditional mental health services by providing a safe and supportive environment for mental health service consumers to develop social and occupation-related skills. NISA encourages its members to contribute to the overall wellness of the community while creating their own opportunities for a healthy and productive life. This reflects the belief that mental health consumers/survivors can function well in society and make valuable contributions when given the opportunity. Its mission is pursued through a variety of occupational initiatives, including research, writing and publishing, arts, crafts, computer recycling, gardening, and selling wares in a gift shop owned and operated by NISA.

The therapist's practice at NISA involves moving back and forth from micro- to macro-environments dealing with individual clients and group-based interventions. The therapist's role is client driven, based upon who attends the program and individuals' needs. Primary strategies to meet these needs involve developing extended community partnerships through advocacy, leadership training, or facilitating mentorship; identifying choices and supporting client decision-making; role modeling; resource linking; information sharing;

and taking risks to pursue innovations. Clearly, the requirements of the therapist in such a setting are nontraditional. A great deal of the therapist's work involves ensuring that the environment of NISA is supportive of all members and that participants learn the means to support each other through solution-focused peer support strategies. Pike (2001) described the role of occupational therapists in consumer-governed settings as one of coloring outside the lines. Pike observed therapists take risks, behave professionally in an unassuming way, employ non-traditional occupational therapy practices, be person centred, and engage in advocacy and actions to establish common ground among parties.

David is a 30-year-old man with a diagnosis of schizophrenia who began working at NISA in December of 1997. David worked in a variety of the program initiatives and is currently employed by the Dandelion Café as a cashier. The PEO model (Law et al., 1996) is used to explore his occupational performance experiences.

Person-Environment Interaction

David initially came to NISA to participate in a ParNorth research study exploring the experience of schizophrenia that was advertised in the local clubhouse. David was living independently in his own apartment but found that his day was long and he longed for meaningful work in his life. David has a European heritage and work has always been an important part of his culture. David stated that he had struggled with the effects of schizophrenia, abusing drugs and alcohol to cope with the symptoms of schizophrenia for the past 5 years. David wanted more for his life, but he had not been able to successfully connect with individuals or programs in the community. David found other programs did not offer him the opportunity for "real work" and his previous employment as a dishwasher did not provide him with the support and flexibility he needed to manage his illness.

David began his tenure at NISA in the Writer's Circle. He had had little experience with computers and initially had to learn the keyboard and basic computer functions. David mainly wrote about his experiences of schizophrenia, which were soon published in the *Open Minds Newsletter* of the Writer's Circle. David found that the support and encouragement provided by the occupational therapist and the editor of the newsletter were helpful to his writing and attendance at NISA. He soon was attending on a daily basis.

David initially found it difficult to get out of bed and to attend the program on a regular basis. Since he set his own schedule for attendance at NISA, he was able to gradually build his tolerance for work. His calls each morning to the occupational therapist discussing hallucinations that kept him from taking the bus or leaving his apartment dwindled as David participated more extensively at NISA, as he gained trust in the people at NISA and expanded his social support, and as he learned to persevere despite the auditory hallucinations that previously had prevented or limited his participation in the community. For David, the person-environment fit enabled him to gain a sense of control in his life and to determine what he wanted to do with respect to occupation. He clearly believes that without the support, community, friendships, acceptance, and opportunity provided at NISA, he would not be employed today or enjoying a satisfying social life in the community.

Person-Occupation Interaction

Prior to coming to NISA, David had a limited skill set and work-related experiences due to his dropping out of high school and working in low-skill, low-pay jobs such as dishwashing. However, David was motivated to participate in a variety of occupational opportunities as a volunteer at NISA because he was extremely motivated to become employed in the community. David experienced both auditory and visual hallucinations, and found that participating in occupations that required new learning and moderate concentration provided him with some respite from the hallucinations, thus helping him to be successful. Initially, David participated in a variety of occupational initiatives, but later came to focus most of his time in the Writer's Circle and in Warm Hearts, Warm Bodies programs. These programs offered David the new learning and concentration needed to manage his hallucinations, the flexibility to take breaks when needed, an environment in which to socialize with others, and immediate feedback about his performance; all contributed to his sense of self-confidence and competence. He has recently assumed a leadership role in orienting new members to the Warm Hearts, Warm Bodies program.

Occupation-Environment Interaction

Similar to Aaron's experience at HARP, David had found an environment that was flexible in meeting his own needs for occupation and that was conducive to belonging, meeting individual needs, supporting illness management, and which provided opportunities to experiment with a variety of occupational initiatives. For David, the environment at NISA provided him with the flexibility and support to create his own opportunities for participation. David stated that previously he had used artificial stimulation to offset the experiences of schizophrenia and to manage his illness, but now, NISA provided him with a high in life based upon his many successes. He now perceived himself as "a winner and as special." David has participated extensively at NISA and within many of the program initiatives, including making patchwork quilts in the Warm Hearts, Warm Bodies program, taking apart used computers, entering poetry contests and reading his poetry in public venues, publishing several of his poems in the *Open Minds Quarterly* and *Writer's Circle On-line*, completing his grade 12 equivalency, and more recently, obtaining a job as a cashier and assistant manager in the Dandelion Café. David actively participates in many social functions, now has several significant friendships from his connections at NISA, and enjoys an active social life outside of NISA.

Person-Environment-Occupation Relationship

In summary, David's experiences at NISA provided him with a good person-environment-occupation fit. A supportive work environment in which David was given unconditional acceptance as a person of worth, the opportunity to self-select occupations, and the flexibility to vary the frequency and duration of his participation all contributed to his recovery. These experiences have collectively helped him to better manage the effects of his illness and to gradually move into the world of competitive work. David's success is manifest in his recent position as an assistant manager in the café and in his appointment to the NISA Board of Directors.

Both the Aaron and David case scenarios illustrate how supportive work environments can improve the match between the client (person), workplace (environment), and work (occupation) in order to facilitate occupational performance (doing the work). Meaningful work is provided with the just right challenge for each individual by matching the individual's interests, skills, and abilities while providing support, training, and necessary accommodation for his or her illness. The occupation and environment correspond with the person's relationship with his or her illness and stage in the recovery process. The individual's needs for success, safety, and acceptance are met by the physical, cultural, and institutional aspects of the workplace environment. These scenarios demonstrate how by having a person-workplace-work match, work can become the medium through which a client's self-concept changes to one of becoming a capable person with a future, thereby supporting the recovery process.

CONCLUSION

Traditional approaches to assist individuals with mental illness into work roles—which focus exclusively upon interventions directed at the person with the illness in a graduated step-wise "train" followed by "place" approach—have not been supported by research or the consumer literature. Today's services are driven by client-defined needs, client choice, and focus on creating supportive work environments. A new generation of supported employment approaches has evolved, directly placing clients in community workplaces. Research supports the use of the IPS approach for individuals with prolonged mental illness. Whether working in a supported employment situation or creating work alternatives not in therapists' direct control, an examination of the literature identifies important considerations for effective supportive work environments to potentially facilitate the recovery process. Case scenarios illustrate important cultural and institutional elements of the workplace to consider and suggest strategies of how therapists can create alternative supportive work environments.

Summary of Environmental Intervention

The successful integration of mental health consumers into competitive work has been met with limited success. The literature attests to the fact that previous approaches to enhance the employment of persons with mental illness, being only 10 to 15%, have not helped to alleviate the poverty and social stigma that often accompanies mental illness (Church, 1997). A growing recognition by mental health professionals and the consumer movement is that innovative and creative solutions to the economic plight and lack of meaningful and sustained employment are required. This chapter has highlighted the need for therapists to work from an individualized client-driven approach in addressing work-related needs, and to consider the person-environment-occupation fit as a means to enable the occupational performance of persons with mental illness. Therapists need to be aware of key physical, cultural, and institutional aspects of the work environment that research has identified as facilitating or hindering clients' participation in work. Therapists must broadly conceptualize work in order to consider the full range and variety of occupational opportunities and possibilities, from volunteering to full-time competitive employment, in order to assist their clients to realize their work goals.

In this chapter, two programs were described to provide examples of innovations and the potential role for occupational therapists in creating supportive work environments. HARP and NISA maximize the person-environment-occupation fit and enable the occupational performance of a variety of individuals with long-standing mental health issues. It is hoped that these programs can stimulate the reader to consider program opportunities and possibilities beyond those currently funded or provided in existing mental health services. It is also hoped that occupational therapists, with their intent to be client-centred, may be encouraged to assume a leadership role in providing services that do not "fit within the box." Creative solutions are generated by collaborating with clients and also by thinking and acting outside of the current paradigms.

Application and Recommendations

Meaningful work, recovery, social stigma, and the opportunity for real work are important issues for mental health consumers. Environmental barriers have been described as more disabling than the illness itself (Rebeiro, 1999). For every client with a mental illness who has a work-related goal, therapists need to consider environmental interventions like the ones described in this chapter. Occupational therapists are encouraged to take a leadership role and to work collaboratively with persons with a mental illness in creating alternative and supportive work environments that help to realize consumers' goals for employment. Many of the strategies suggested to create supportive environments in this chapter are potentially transferable to many other socially marginalized groups with limited opportunities to participate in meaningful occupations.

Study Questions

1. What is the difference between the "train and place" and the "individual placement and support" models of vocational intervention with people with mental illness?

2. What do environmental interventions offer clients with mental illness who wish to work?

3. How does the occupational therapist's role change when the therapist moves from working in a traditional rehabilitation program to a supportive work program like the two described in the chapter?

REFERENCES

Anthony, W. A. (1993). Recovery from mental illness: The guiding vision of the mental health system in the 1990's. *Psychosocial Rehabilitation Journal, 16*(4), 11-23.

Anthony, W. A. (1994). Speaking out: Managed care outcomes—"Places to be and symptom free." *Psychiatr Rehabil J, 19*(2), 73.

Anthony, W. A., Cohen, M. R., & Farkas, M. (1990). *Psychiatric rehabilitation.* Boston, MA: Center for Psychiatric Rehabilitation.

Anthony, W. A., & Jensen, M. A. (1984). Predicting vocational capacity of the chronically mentally ill: Research and policy implications. *Am Psychol, 39*(5), 537-544.

Bernardo, A., Kirkpatrick, H., Strong, S., Start, S., & Packer, K. (1996). Long stay patients: Take another look. *Psychosocial Rehabilitation With and Within The Community, 1*(1), 10-14.

Bond, G. R., Drake, R. E., Mueser, K. T., & Becker, D. R. (1997). An update on supported employment for people with severe mental illness. *Psychiatric Services, 48*(3), 335-346.

Canadian Association of Occupational Therapists. (1997). *Enabling occupation: An occupational therapy perspective.* Ottawa, ON: CAOT Publications ACE.

Carver, J., & Gaylor, C. (1990). *Psychosocial Rehabilitation in Ontario, IAPSRS Ontario Chapter Working Paper.* Unpublished report.

Church, K. (1997). *Using the economy to develop the community: Psychiatric survivors in Ontario.* Ottawa, ON: Caledon Institute of Social Policy Publications.

Clutterbuck, P., & McKay, S. (1984). *Work and well-being, the changing realities of employment*. Toronto, ON: Canadian Mental Health Association.

Cochrane, J. J., Goering, P., & Rogers, J. M. (1991). Vocational programs and services in Canada. *Canadian Journal of Community Mental Health, 10*(1), 51-63.

Consumer/Survivor Information Resource Centre. (1994). *Consumer/survivor development initiative project descriptions*. Toronto. ON: Consumer/Survivor Development Initiative.

Corring, D. (1996). *Client-centered care means I am a valued human being*. Unpublished master's thesis. University of Western Ontario: London, Ontario, Canada.

Davidson, H. (1991). Performance and the social environment. In C. Christiansen & C. Baum (Eds.), *Occupational therapy: Overcoming human performance deficits* (pp. 144-171). Thorofare, NJ: SLACK Incorporated.

Davidson, L., Stayner, D. A., Nickou, C., Styron, T. H., Rowe, M., & Chinman, M. L. (2001). "Simply to be let in": Inclusion as a basis for recovery. *Psychiatr Rehabil J, 24*(4), 375-388.

Deegan, P. (1992). The independent living movement and people with psychiatric disabilities: Taking back control over our own lives. *Psychosocial Rehabilitation Journal, 15*(3), 3-19.

Deegan, P. (1988). Recovery: The lived experience. *Psychosocial Rehabilitation Journal, 11*(4), 11-19.

Diasio, K. (1971). The modern era —1960-1970. *Am J Occup Ther, 25*, 237-242.

Drake, R. E., McHugo, G. L., Becker, D. R., Anthony, W. A., & Clark, R. E. (1996). The New Hampshire study of supported employment for people with severe mental illness. *Journal of Consulting and Clinical Psychology, 64*(2), 391-399.

Drake, R. E., McHugo, G. L., Bebout, R. R., Becker, D. R., Harris, M., Bond, G. R., et al. (1999). A randomized clinical trial of supported employment for inner-city patients with severe mental disorders. *Arch Gen Psychiatry, 56*, 627-633.

Fabian, E. S. (1989). Work and the quality of life. *Psychosocial Rehabilitation Journal, 12*(4), 39-49.

Goldstein, J. M., Cohen, P., Lewis, S. A., & Struening, E. L. (1988). Community treatment environments: Patient vs. staff evaluations. *J Nerv Ment Dis, 176*(4), 227-233.

Hahn, H. (1984). Reconceptualizing disability: A political science perspective. *Rehabilitation Literature, 45* (11-12), 362-365.

Hartl, K. (1992). *A-Way Express: A way to empowerment through competitive employment*. Unpublished manuscript. Available from A-Way Express, 320 Danforth Ave., Toronto, ON, M4K 1N8.

Kirsh, B. (2000). Factors associated with employment for mental health consumers. *J Psychiatr Rehabil, 24*(1), 13-21.

Law, M., Cooper, B., Strong, S., Stewart, D., Rigby, P., & Letts, L. (1996). The person-environment-occupation model: A transactive approach to occupational performance. *Can J Occup Ther, 63*, 9-23.

Leete, E. (1989). How I perceive and manage my illness. *Schizophr Bull, 15*, 197-200.

Legault, E., & Rebeiro, K. L. (2000). Occupation as means to mental health: A single case study. *Am J Occup Ther, 55*, 90-96.

Lehman, A. F. (1995). Vocational rehabilitation in schizophrenia. *Schizophr Bull, 21*(4), 645-656.

Marrone, J., Balzell, A., & Gold, M. (1995). Employment supports for people with mental illness. *Psychiatric Services, 46*(7), 707-711.

McGurrin, M. C. (1994). An overview of the effectiveness of traditional vocational rehabilitation services in the treatment of long term mental illness. *Psychosocial Rehabilitation Journal, 17*(3), 37-54.

National Health and Welfare Canada. (1988). *Mental health for Canadians: Striking a balance*. Ottawa, ON: Minister of Supply and Services Canada.

Ochocka, J., Roth, D., Lord, J., & MacGillivary, H. (1994). *Workplaces that work*. Kitchener, ON: Centre for Research & Education.

Ontario Ministry of Health. (1993). *Putting people first: The reform of mental health services in Ontario*. Toronto, ON: Queens Printer for Ontario.

Ontario Ministry of Health. (1999). *Making it happen: Implementation plan for mental health reform*. Toronto, ON: Queen's Printer for Ontario.

Ontario Ministry of Health. (2001). *Making it work: Policy framework for employment supports for people with serious mental illness*. Toronto, ON: Queen's Printer for Ontario (ISBN 0-7794-0470-X, Cat. #7610-4232419).

Pike, S. (2001). *Coloring outside the lines: Defining the role of the occupational therapist in a consumer-run organization*. Unpublished manuscript, London, Ontario, Canada: The University of Western Ontario.

Powell, T. J. (1987). *Self-help organizations and professional practice*. Silver Springs, MD: National Association of Social Work.

Rebeiro, K. L. (1999). The labyrinth of community mental health: In search for meaningful occupation. *Psychiatric Rehabilitation Services, 23*(2), 143-152.

Rebeiro, K. L. (2000). Enabling occupation: The importance of an affirming environment. *Can J Occup Ther, 68,* 80-89.

Rebeiro, K. L., & Allen, J. (1998). Voluntarism as occupation. *Can J Occup Ther, 65, 279-285.*

Rebeiro, K. L., & Cook, J. V. (1999). Opportunity, not prescription: An exploratory study of the experience of occupational engagement. *Can J Occup Ther, 66,* 176-187.

Rebeiro, K. L., Day, D. G., Semeniuk, B., O'Brien, M. C., & Wilson, B. (2001). Northern Initiative for Social Action (NISA): An occupation-based mental health program. *Am J Occup Ther, 55,* 493-500.

Renwick, R., & Brown, I. (1996). The centre for health promotion's conceptual approach to quality of life. In R. Renwick, I. Brown, & M. Nager (Eds.), *Quality of life in health promotion and rehabilitation: Conceptual approaches, issues and applications* (pp. 75-86). Thousand Oaks, CA: Sage.

Strong, S. (1995). *An ethnographic study examining the experiences of persons with persistent mental illness working at an affirmative business.* Unpublished master's thesis. McMaster University, Hamilton, Ontario, Canada.

Strong, S. (1998). Meaningful work in supportive environments: Experiences with the recovery process. *Am J Occup Ther, 52,* 31-38.

Strong, S., Rigby, P., Stewart, D., Law, M., Cooper B., & Letts, L. (1999). The person-environment-occupation model: A practical intervention tool. *Can J Occup Ther, 66,* 122-133.

Trainor, J., Pomeroy, E., & Pape, B. (1993). *A new framework for support for people with serious mental health problems.* Toronto, ON: Canadian Mental Health Association.

Trainor, J., & Tremblay, J. (1992). Consumer/Survivor businesses in Ontario: Challenging the rehabilitation model. *Canadian Journal of Community Mental Health, 11*(2), 65-77.

Wright, C., & Rebeiro, K. (in press). Exploration of a single case in a consumer-governed mental health organization. *Occupational Therapy in Mental Health.*

Chapter 10

Enabling Young Children to Play by Creating Supportive Play Environments

Patty Rigby, MHSc, OT Reg. (Ont.); and
Lorrie Huggins, MA, EdS, BScOT

CHAPTER OBJECTIVES

* To encourage occupational therapists to value play, the primary occupation of childhood, and to identify strategies to enable children greater success with play.
* To describe a process for analysis of the child-environment-occupation relationship in play.
* To present occupational therapists with a variety of strategies to build environments that will support and encourage the play and playfulness of preschool children.
* To broaden the scope of occupational therapy practice with young children by demonstrating how occupational therapists can use environments to:
  a. Advocate for the value of play.
  b. Educate parents, educators, and other caregivers about creating environments that enable children of varying abilities to participate successfully in play.
  c. Establish play partnerships with children to enable the child's play and playfulness.

INTRODUCTION

With the emergence of a number of new theoretical models that are centred on occupational performance, as noted in Chapter 2, the focus for occupational therapy practice with children is shifting to one that places greater emphasis on the occupations of childhood (Case-Smith, 2001). Play is considered to be one of the primary occupations of childhood, particularly for preschool children (Bundy, 1997; Parham & Primeau, 1997). Play is what young children choose to do and what they are expected to do. Occupational therapists are now focusing more on enabling children to engage in the occupation of play (Stewart, et al., 1996).

Historically, occupational therapists have tended to use play as a modality for developing a child's performance components and skills (Stewart et al., 1996). Play was used to interest the child in the therapeutic activities chosen by the therapist (Couch, Deitz, & Kanny, 1998). Play is also considered by many to be the "work" of the child (Kielhofner & Barris, 1984; Stewart et al., 1996). Theorists and researchers have speculated that through play children can develop the skills necessary for success in adult life (Reilly, 1974; Stewart et al., 1996). In contrast to this, Bundy (1993) has advocated that occupational therapists should value play as a freely chosen occupation, rather than just the means for developing skills and abilities. Bundy (1997) also emphasizes that children engage in the "doing" of play, the process of playing, because that is what is rewarding and what motivates the child to initiate play and to continue playing.

This chapter focuses on play as an occupation, and on enabling preschool children, whether they have special needs or not, to play at home, in childcare centres and in the community by creating supportive environments. The strategies presented in this chapter are based on the recommendations arising from a variety of research studies on play and the clinical experiences of the authors. Many of the strategies recommended in this chapter apply to the play of all children, and are of particular importance for children who are at risk for compromised participation in play, due to developmental delays, physical impairments, or other challenges.

WHAT IS PLAY AND WHY VALUE IT?

"Play… is the way the child learns what no one can teach him."
(Lawrence K. Frank, 1890-1958. As cited in
Johnson, Christie, & Yawkey [1987].)

Among the issues of early childhood development, play is one of the most misunderstood. Sutton-Smith (1997), a well-known play theorist, points out that there has been a tendency for many professionals to see play as only fun, into which appropriate learning experiences need to be channeled. He goes on to describe how many professionals have treated play as something that precedes learning, while some see play as a sacred activity that should be free and not altered by adults. These views limit the importance and complexity of play and they fail to capture the changing contexts of play and the variations in play. According to Sutton-Smith, professionals usually recognize that play has value. They defend play as a means of socializing children but are hesitant to entrust cognitive development and other aspects of learning to play.

In recent years there has been a renewed interest in play and recognition of the value of play. Play experiences contribute to cognitive, social, emotional, and physical development. Learning through play facilitates the development of problem-solving ability, creativity, flexibility, achievement, motivation, morality, and social attitudes (Bundy, Nelson, Metzger, & Bingaman, 2001; Ferland, 1997; McCain & Mustard, 1999). Children's play is a fluid, elusive activity that resists definition. Nevertheless, a common understanding of play is essential for therapists' interactions with children's parents and other disciplines providing services to children. While no single definition exists for play, it is generally understood that play is pleasurable, self-motivating, engaging, a child's private reality, and an end unto itself, making the process of playing more important than an end product arising from a play experience (Bundy, 1997; Eden & Huggins, 2001; Parham & Primeau, 1997; Rubin, Fein, & Vandenberg, 1983).

PLAY AND SOCIETY

Until recently, many developmental specialists in health care and educational settings focused on promoting developmental skills acquisition (Shonkoff & Meisels, 1990). For example, occupational therapists have specialized in helping children to develop fine-motor, perceptual-motor, and cognitive skills. Occupational therapists and their colleagues have used developmental and components-focused assessments to identify the child's deficits and utilized interventions with the intent to "fix" and to normalize the child according to normal developmental performance parameters. Parents, too, have been anxious about developmental testing and how their child performs compared to the norm. Educators structured the curricula of early childhood programs to ensure that skills were taught. The curricula and programs developed for children were largely adult-centred and adult-directed. The intent was to provide structured therapy or teaching to achieve specific developmental goals (e.g., the therapist chooses the blocks, shows the child how to stack the blocks and directs the child to build a tower; the therapist's goal is to develop the child's fine motor control).

However, there is now consensus among developmental experts about the merits of promoting child-centred, play-based learning to encourage children to become self-directed, inquisitive problem-solvers who interact effectively with peers and adults (Bredekamp & Copple, 1997; Hanline, 1999; McCain & Mustard, 1999). These experts believe that stimulating experiences in play will positively impact upon early brain development and provide the critical foundations for success at school.

Happy, healthy, well-adjusted children are also considered to be equally important outcomes of play-based learning (Eden & Huggins, 2001; Ferland, 1997; McCain & Mustard, 1999). It is critical that children develop positive self-esteem, self-worth, and self-efficacy not only for a positive quality of life; research shows that these features are strongly associated with resiliency (the ability to recover from or effectively cope with adversity and/or stress) (Rutter, 2000). Supportive, stimulating play environments can enable young children to feel successful and competent with their play interactions, and this in turn builds their self-confidence to stretch and expand their skills and challenge themselves further.

Many early childhood programs are adopting play-based curricula to encourage experiential learning through play, and are teaching parents how to facilitate play in the home (Eden & Huggins, 2001; Rivkin, 1995; Scales, Almy, Nicolopoulou, & Ervin-Tripp, 1991). Play-based learning environments provide children with opportunities to explore, discover, and create. Professionals need to ensure that environments provide opportunities for problem-solving through play, as this provides the sensorimotor and cognitive stimulation that the young child absorbs and integrates into the core brain development. For example, block play provides opportunities to arrange, rearrange, and build things. This in turn offers the child the basic fundamentals of size and shape, an understanding of the physical world, and the cognitive weight of numbers necessary for mathematical thinking (McCain & Mustard, 1999).

Using Environments to Enable Play

The Role of the Environment in Enabling Play

The role of the environment in enabling play and in creating barriers to the play of young children has had little attention in the literature. However, there is agreement that the environment plays a role and that this role should be explored (Cameron, et al., 2001; Law & Dunn, 1993; Morrison & Metzger, 2001). The new International Classification of Functioning, Disability and Health (ICF) (WHO, 2001) emphasizes that aspects of environments can either facilitate or hinder a person's functioning and participation in daily activities, and consequently impact his or her health. Keating and Hertzman (1999) found that the brain developed in adverse ways leading to cognitive, social, and behavioral delays when children's early years were spent in relatively less stimulating environments that were not emotionally and physically supportive. Research on children's resiliency and development of coping abilities has also shown how environments positively or negatively influence children's mental health and the development of social relationships (Sandler, 2001; Werner, 1990). Zeitlan and Williamson (1990) found through a review of the literature that children with disabilities were less skilled at problem-solving, social interaction, and exploration of the environment.

Children with developmental, physical, and/or learning disabilities appear to be more vulnerable to environmental barriers during play, and consequently have been found to be less successful in play than their typically developing peers (Bronson & Bundy, 2001; Hanzlik, 1989; Rigby & Gaik, 2003). In a pilot study involving 16 children with cerebral palsy who were evaluated while playing in three different settings, half of the children were playful in one or two settings and not playful in the others (Rigby & Gaik). A thematic analysis of observations of the play in each setting demonstrated how some environmental factors supported the playfulness of these children, while other factors hindered their play. Consequently, in less supportive environments the children were not playful. When there was a good fit, or congruency between the child, the play setting, and the play opportunities, the children were playful. In a study that correlated playfulness with environmental supportiveness, many of the typically developing children and children with impairments were playful in spite of environments that were not rated as supportive (Bronson & Bundy). However, a number of children with impairments were not playful in the less supportive environments.

Findings from both studies raise questions about the child's capacity to be playful and the role of the environment in supporting or hindering the play experience. Children with developmental delays and impairments, in particular, need to have environments adapted and modified to ensure that they are congruent with their interests and abilities. It also appears that some children utilize environmental resources and opportunities, and handle environmental challenges better than others do. The development of this resourcefulness in children may be fostered in supportive, stimulating environments that offer opportunities to problem-solve and explore.

Accessing Resources and Removing Barriers to Play

The environment has many resources that can be accessed to promote and support the play experiences of young children. A number of the resources that we suggest in this chapter will appeal to your common sense, while others may surprise you, because their potential and value as a resource have been underestimated or not fully recognized. Environmental barriers are also identified here, and we suggest strategies for dealing with them.

Occupational therapists may find that environments are easier to change than a child's performance components in order to improve a child's play performance (Law & Dunn, 1993; Case-Smith, 2001). Furthermore, because play belongs to the player, it makes sense that to improve play, therapists should try to alter the play environment to create a better fit with the child's abilities and interests. Children generally do not play at the activities that are too hard or demand more than they have, nor those that do not interest them. For example, an occupational therapist can easily and quickly assist the child's caregiver who is part of the child's social environment to learn strategies that will facilitate a child's participation in play with blocks. This would likely take less time and energy for the therapist and the child, and meet with greater success than just focusing on reaching and strengthening exercises to improve the child's fine motor coordination for manipulating blocks, which for the child is likely not play.

In environments that provide opportunities and resources for play, children can be playful and challenge themselves to grow and develop their knowledge and skills. Through play, the child who has a disability and/or developmental delay can develop strategies for adapting and coping with his or her situation and build his or her competence and self-confidence.

Starting With Assessment: Examining Person-Environment-Occupation Fit in Play

It is important that occupational therapists who work with children adopt a family-centred approach, and view the child within the context of the family. In family-centred services, the family identifies the issues that their child is experiencing in occupational performance, identifies the priorities for occupational therapy intervention, and has the right to autonomous decision-making where it concerns their child and family (Rosenbaum, King, Law, King, & Evans, 1998). The occupational therapist develops a partnership to work collaboratively with the child's family. When the therapist is providing services to an institution such as a childcare centre, the therapist also involves those caregivers in the assessment and intervention planning process. The application framework from the person-environment-occupation (PEO) model, as shown in Figure 10-1, can guide the assessment process (Law et al., 1996) and facilitate the development of an intervention plan. This model has been described in greater detail in Chapter 2. The therapist can lead the caregiver team through the PEO analysis to examine what is contributing to, or is hindering a good PEO fit.

The outcomes of interest are focused here on the child's occupational performance in play. The parent(s) or other caregivers may identify performance issues with the child's play. For example, the child's parents may have noted that the child has a very small repertoire of interests in play and plays repetitively with the same toy. Or, in another example, the child's teacher may point out that the child does not engage in play with peers, but stays on the periphery watching.

Analysis of occupational performance issues using the PEO model builds upon familiar occupational therapy practices (Strong et al., 1999). Occupational therapists typically gather information from client records, interviews, observations, and standardized assessments to profile their client's strengths, competencies, and performance issues. In order to assess a child's performance in play, The test of playfulness (ToP) (Bundy, 1997) would be a good

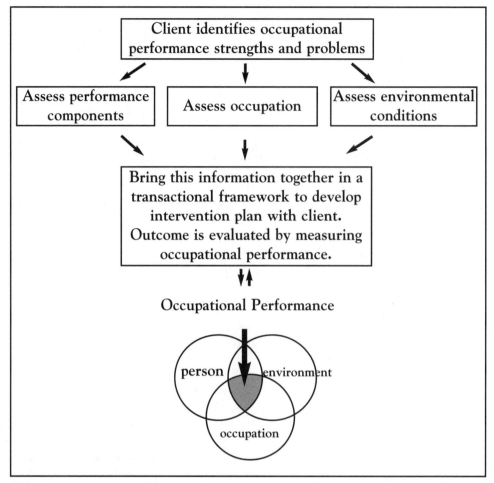

Figure 10-1. Application of the PEO model. (From Law, M., Cooper, B., Strong, S., Stewart, D., Rigby, P., & Letts, L. [1996]. The person-environment-occupation model: A transactive approach to occupational performance. *Can J Occup Ther, 63,* 9-23. Reprinted with permission from CAOT Publications ACE.)

measure to use. The ToP was developed to examine a child's performance in any play transaction (Bundy, 1997). Bundy asserts that a combination of three elements—intrinsic motivation, internal control, and suspension of reality—must be present in order for playfulness to occur. In addition, Bundy states that the concept of framing is critical to the play experience. Children will frame play by providing cues to each other about how they should act and react toward each other. These elements of playfulness can be observed in the child's behavior, and are influenced not only by the child's attitude, interests, and abilities, but also by environmental and occupation-related factors.

The ToP consists of 24 items that are measured in terms of the extent of time an item is observed during a play session, the intensity of the child's engagement in that aspect of play, and the skillfulness of the child with aspects of play (Bundy, 1997). For example, the evaluator rates the extent of time the child is actively engaged in the play; the intensity of the child's perseverance in order to overcome obstacles to continue playing with an activity; and the skillfulness with which the child actively changes the requirements or complexity of the activity in order to vary the challenge or degree of novelty. During assessment using the ToP, the occupational therapist should also assess the play environment and the various play options available in that setting.

Currently there are few standardized tools available to specifically evaluate the play environment or to examine the child-environment relationship during play. The test of environmental supportiveness (Harkness & Bundy, 2001) is a promising new tool, but is still under development and will be soon be available to use together with the ToP. The assessment of ludic behavior (Ferland, 1997) is a descriptive assessment of a child's play behavior and includes sections on the human environment and the sensory environment. It also addresses the child's interest in his or her environment. The home observation for measurement of the environment (HOME) (Caldwell & Bradley, 1984) is a useful tool to measure the relative supportiveness of the home environment in relation to child development, but includes only a few items that address play.

The observations of factors influencing playfulness form (Rigby, 2001) shown in Figure 10-2 uses the PEO model (Law et al., 1996) and is structured around the central elements of playfulness from the ToP (Bundy, 1997). It should be used together with the ToP. It can guide the therapist to make observations about both the play environment and play activities. This form also guides the occupational therapist through the analysis of PEO fit by pulling together the information and examining this transactional relationship. What is meant by "transactional" is that the relationship of the person(s), together with the environment and occupation are viewed as interwoven and difficult to tease apart. The analysis of fit across person, environment, and occupation is an important step, as it helps to illuminate what is helping and what is hindering the child's play experience. The information that is gained through this assessment process will provide the therapist and caregivers clear direction for intervention. At the end of the intervention program, the occupational therapist can use the ToP as an outcome measure of the child's occupational performance in play, in order to evaluate the effectiveness of the interventions.

Building an Environment That Enables Play and Playfulness

THE FOUNDATIONS THAT CREATE POSITIVE PLAY EXPERIENCES

It is important to set the stage for play. The first thing to attend to when setting the stage for children is to ensure that they feel safe and secure, as this will influence their willingness to explore a setting and engage in play. Knowledge about attachment theory is fundamental for understanding children's perceptions about their own safety and security, and their subsequent behaviors. *Attachment* is the deep emotional bond formed between a child and one or more adults, usually a parent and/or caregiver (Bowlby, 1988). It involves consistent nurturing and confidence that the child's cues will be read and that his or her needs will be met (Bowlby, 1988; Hardy & Prior, 2001). Attachment provides children with a sense of security that allows them to feel comfortable exploring their environment and taking risks during play, while using the attachment figure (who could be the child's parent, preschool teacher, or nanny) as their base of security. Children develop trust and confidence with their social world through experiences of secure attachment. Therapists as part of the child's environment must be aware of the child's need for security and attachment before the child feels comfortable venturing forth to play (Lieberman, 1993). The example below demonstrates how a child moves into play from her secure base.

Observations of Factors Influencing Playfulness

Name of Child: _____ Age: _____ Date: _____

Where is Play: _____ Who Supervised Play: _____ Time of Day: _____

Brief description of context of play (the setting, the players, the activities, and any occurrences preceding the play observation):

Test of Playfulness	Factors Influencing Person-Environment-Occupation Fit	Promote Playfulness	Hinder Playfulness
Intrinsic Motivation			
Internal Control			
Suspension of Reality			
Framing Play			

Summary:

Recommendations:

Figure 10-2. Observations of Factors Influencing Playfulness Form. Copyright © 2001 Patty Rigby, OT Reg. (Ont.).

Jill, a 16-month-old girl, has joined a new play group. She arrives with her mother. Jill holds her mother's hand and does not venture far from her side all morning. She hides her face in her mother's skirt when approached by teachers or other children yet she is a keen observer of the playgroup. Her mother shows interest in the toys and takes part in the singsong. The mother demonstrates, through her behaviors, that this is a safe and fun place to be. Slowly, Jill will take her mother's lead and explore the toys but will return to her mother frequently to be reassured and encouraged. This secure base is developmentally appropriate. Through continual exposure to this environment, Jill will explore and discover. She will begin to interact with others and experience pleasure in the new play opportunities.

Caregivers typically provide an environmental context for play by giving children permission to play and providing opportunities for play (Figure 10-3). They communicate that it is time to play, make resources available for play, and direct or take children to the place(s) where they can play. Children need to know what is allowed, what is acceptable, and the rules and expectations within the play setting. For example, children need to know if quiet play is expected, or whether they can run around and engage in rough and tumble play. Usually children generalize their knowledge of the cues to what is or is not permissible in various settings. Their developing ability to read environmental cues enables them to learn how to conform to society's rules and expectations. Climbing structures, slides, and tricycles in an open area cue them to active motor play, while an enclosed area with containers of blocks, cars, and small figures of animals and people might cue the children to engage in floor play involving construction and pretending. Caregivers should watch the unfolding play themes and produce additional materials that support the theme. At times, they need to give children permission to use materials that are not typically used in play. For example, when a caregiver observed that the children were initiating play as superheroes, he brought out towels and showed them how they could be used as capes, and helped them to construct masks from construction paper. This caregiver's actions enabled and supported the children's play theme.

Children with developmental delays and/or physical impairments have generally been found to assume passive, respondent roles during play, rather than autonomous roles in which they are self-directed and initiate play (Ferland, 1997; Hanzlik, 1989; Zeitlan & Williamson, 1990). Caregivers of children with developmental disabilities expect their child to be dependent versus autonomous (Ferland; Kellegrew, 2000) in most daily occupations, including play. Play settings that were directed and controlled by the caregiver/adult and were insensitive to the child's interests were not supportive of cognitive development (Mahoney & Powell, 1988), independence in play (Hanzlik, 1989), and playfulness (Okimoto, Bundy, & Hanzlik, 2000; Rigby & Gaik, 2003). In contrast, the environments that were more child-centred, responsive to child initiated cues, and sensitive to the child's interests were more supportive of the child's cognitive development, independence, and playfulness (Mahoney & Powell; Hanzlik; Rigby & Gaik).

Caregiver attitudes and expectations are vitally important in setting the stage to enable children with developmental delays and/or physical impairments to become more willing and able to self-initiate play and become more active participants in the play theme(s). Ferland (1997) stresses that the caregiver of the child with a disability must attend to and build the child's capacity for action. This means enabling the child to gradually move from passive participation to a more active role, as he is ready and willing. This involves capitalizing on the child's strengths and building in opportunities for the child to make choices and take the lead. If the child has a motor impairment, but is able to communicate effectively,

Figure 10-3. Caregivers can provide a context for exploration in play by modeling various ways to use play materials. (From Eden, S. T., & Huggins, L. [2001]. *YMCA playing to learn: A guide to quality care and education of young children*. YMCA of Greater Toronto. Reprinted with permission from YMCA of GTA.)

the caregiver can facilitate the child's participation in the activity by using talking and dramatic play, instead of movement. The therapist can help the caregiver learn how to modify the complexity of play activities, and grade the challenges, so that the child builds competency and success with the activity. Success fosters motivation to continue playing and to keep trying. Challenges can be increased as competencies grow. Ferland (1997) also reminds us that environmental barriers and challenges are a reality for children with disabilities, and therapists and caregivers can assist these children to approach these challenges with a sense of humor, and a view of problem-solving, rather than with a view of frustration and feelings of failure. Strategies to foster initiative, exploration, and self-directedness in play settings will be explored in the following sections.

THE DEVELOPMENTAL CONTEXT

Play can be enabled in environments in which the opportunities and play resources made available to the children are matched with their interests and abilities. Occupational therapists can use their knowledge about children's developmental abilities in play to assist parents and other caregivers to make age-appropriate toys, play equipment and other play materials available to the player. Through developmental assessment the therapist will learn about the child's skills and abilities. Developmental assessments of play, such as the revised Knox preschool play scale (Knox, 1997) or the transdisciplinary play-based assessment (Linder, 1990) involve observing the child during play sessions. These assessments will not only demonstrate the child's developmental level in various play dimensions; they also allow the therapist to identify areas of emerging interests and abilities.

Parents and other caregivers who are very familiar with the child can also be observant of cues that demonstrate a child's readiness for greater independence and for greater chal-

Figure 10-4. The therapist can facilitate a child to develop important skills by incorporating skill-building actions within the play theme. (From Eden, S. T., & Huggins, L. [2001]. *YMCA playing to learn: A guide to quality care and education of young children.* YMCA of Greater Toronto. Reprinted with permission from YMCA of GTA.)

lenges during play (Kellegrew, 2000). This knowledge is extremely helpful in guiding the selection of play materials to stimulate a child's interests, curiosity, and exploration. Knowledge of emerging abilities and of what to expect next of a child developmentally can assist the adult caregiver or playmate to build in challenges as part of a playful exchange during play, so the child has to stretch his cognitive, social, or motor skills to meet the challenge(s).

The therapist can also draw upon the concept of providing scaffolding for children's learning proposed by Vygotsky (1978), and described in some detail by Berk and Winsler (1995). Vygotsky believed that an adult or capable older child can facilitate a child's learning during play by building the child's problem-solving abilities and by incorporating skill-building actions shown Figure 10-4. For example, the therapist can model problem-solving strategies to the child during play to enable the child to meet challenges and experience successful play transactions. Within this approach the therapist can model a cognitive approach to the grading of activities and break the activity into smaller, achievable steps. The therapist can facilitate the child's use of language to talk him or herself through the step-by-step sequence of manipulating play materials. For example, when learning to use scissors, the prompt of keeping the "thumb up" to orient the scissors for cutting is a useful strategy. Vygotsky focused on the provision of environmental opportunities to match the child's potential to learn and experience success in play. The role of the therapist or other adults as a play partner who can create opportunities suited to the child's learning potential are explored later in this chapter in the section on play partners.

Social Aspects of the Play Setting and Social Resources to Support Play

The social environment in play includes other playmates (children and/or adults), caregivers, and any other person present. Strategies described in the section on setting the stage for play involve social resources. In particular, the therapist must be aware of the attachment needs and level of security of the child within the play setting. In this section the culture of the setting is included, as those present establish that. The organizational elements include the rules, expectations, schedules, and curricula of the setting. Not only are childcare centres structured around routines and schedules, this is also common in households. Routine is a good thing because it will foster children's independence and feelings of control, help them to learn what they are expected to do, and help them to anticipate what happens next. Flexibility in routines should also be established to provide opportunity for spontaneity.

Routines, rules, and schedules that support playfulness are responsive to the interests of the children. Settings that enable the children to develop their interests and fully engage in exploration use seamless, flexible scheduling and emergent themes. For example, in one preschool setting, a few girls wrapped silky scarves around their bodies and declared that they were "princesses." The therapist was also interested in facilitating the development of fine-motor skills for one of these girls, and offered to help them make crowns. Taking a piece of colored bristol board she cut a strip and showed the girls how it could be fastened as a band to fit around the head. The girls quickly became engaged in cutting a fancier edge on the band and adding gold stars. Soon, several other children joined in to make crowns, too. The theme of princesses and castles emerged and the children began using large blocks and a few cardboard boxes to create a castle and moat. They were encouraged to use markers and paints on the boxes to add windows and doors to the castle. The therapist assisted by cutting a door into a box. In this example, the interests of the children were built upon and expanded. Opportunities for socializing also grew, and the child of interest to the therapist was motivated to engage in fine-motor activities as part of this larger play theme.

As mentioned in the previous section, adult expectations and sensitivity to a child's interests and play cues can shape attitudes and the culture within a setting. By allowing the child to take the lead, the child learns to take initiative. Children can learn sharing and turn-taking during semi-structured play. This fosters cooperation in social interactions. The adult can move in and out of the play partner role, which is described in greater detail later in this chapter, to play alongside children and model to them how to include a child with a disability in their play activities.

Familiarity with playmates also fostered playfulness in the exploratory study of the play of children with cerebral palsy (Rigby & Gaik, 2003). When the children joined familiar adult or child playmates, they were able to build on common interests and experiences, establish familiar play themes very quickly, and become intensely engaged in the play. The social relationships appeared to be very motivating for the children. Therapists may find that intervening with pairs or small groups of children is more motivating for the children than intervening alone with the individual child.

The Physical Setting and Resources to Support Play

The occupational therapist can assist caregivers to identify and create physical spaces that are engaging for play yet ensure the safety and security of the children. Children need to know the physical boundaries for their play. For example, they need to know if they are confined to playing in only one room (e.g., the family room in a home), or are they allowed to move from room to room. While it is important that safe equipment and materials be made available for play, it is not always feasible to remove all potentially unsafe objects from a space used for play, such as large pieces of furniture that are tempting to climb. Supervision and rules help prevent children from undue risk-taking.

Young children love to crawl into and play in confined spaces like play tents, cardboard boxes, and cubbyholes. This is especially fun for dramatic play when the box becomes a house, a car, or a plane. Rigby and Gaik (2003) found that play spaces with natural or constructed boundaries facilitated the interactions of the children with physical disabilities with their peers. Within the confined space there is less movement and the demands for mobility are reduced. The reading and giving of play cues is also facilitated when children play closely together. Plus, distractions can be diminished by the physical boundaries. For example, three young girls, including a peer with spastic diplegia, engaged in doll play for over 30 minutes in a small space between a sofa and a wall. They kept their own faces and their dolls within a 2-foot diameter of each other. Structures such as fences or walls can also support the physically active play of some children. For example, when playing a ball game within a contained space, the ball would bounce back into play from the wall and wouldn't bounce so far away as compared to playing in an open field.

Having a variety of play opportunities and options, and play materials that are open-ended and flexible to use, are also important features for a play setting, particularly for children with impairments or developmental delays. Children should have choices in what they use for play, and how they use the toys/materials. For example, Karen, a child with a mild motor impairment was too afraid to ride a toboggan during the outdoor play at her preschool. After she carefully made her way down the slope, she then found it impossible to keep up physically with her peers when they pulled their toboggans back up the hill. This was the only play option that day, and she was excluded because of her physical impairment. Had there been more options than tobogganing available, or had the caregivers graded this activity and allowed Karen to get used to riding a toboggan on a flat surface rather than down the hill, this activity might have been more inclusive for her. Typically the activities that are more open-ended in how they can be played are more inclusive of children with varying abilities. These include water and sand play, play with building blocks, and dress-up play. These types of activities can be graded to facilitate a child's participation. These activities also allow children with varying abilities (which includes children of different ages) to play together.

Child-sized furniture, play structures, and play materials are conducive to safe, engaging play. For example, young children begin to use representational play and will imitate adult routines with toy related mimicry around the age of 2 years (Knox, 1997). These children enjoy using child-sized play houses, kitchen sets, and dolls. Small chairs and tables allow children to sit comfortably and safely to enable them to focus on play with smaller materials, such as puzzles or crafts. A wide variety of special floor sitters and adapted chairs are now available for children with motor impairments. Information about adaptive furniture can be accessed from http://www.abledata.com or http://www.ablenetinc.com.

Play materials and toys must be accessible to children. As much as possible, play materials should be stored at heights and in containers that are easy to access by children. On the other hand, caregivers may wish to keep some materials out of reach, if their use requires close supervision. The therapist can also advise caregivers about what features of toys would make them more accessible for children with developmental delays, or other types of impairments. For the therapist, this involves knowledge of the child's impairment and analysis of the child's abilities in order to determine the physical properties of toys that would best match that child's abilities. For example, a cobbler's bench that has balls that fit into the holes, rather than the version that uses dowels in the holes, is easier for a child with poor eye-hand coordination to use successfully. A variety of adapted, accessible toys and play materials are commercially available, usually from specialty companies. A listing of sources can be found at the ERIC Clearinghouse on Disabilities and Gifted Education at http://ericec.org/fact/toys.html. The National Lekotek Center provides guidance to care-givers for the selection of toys for children with disabilities through their Web page: http://www.lekotek.org. The therapist's knowledge of the child's developmental level in play, and needs for adapted, accessible play materials can also be shared with family and friends to guide the selection of gifts of toys and other play materials for birthdays and other holi-days.

Electronic and computer technology has also provided greater play opportunities for chil-dren with motor, cognitive, and learning impairments (Brodin & Jonson, 2000). Through switch access, such as a large, single-switch button, children with limited hand function can still turn on and off a motorized toy, a tape recording of music, or other cause-effect toys. Computer games and activities are mainstream and very motivating to children. Many can be easily accessed by children with cognitive and learning impairments, and can be modi-fied for children with various types of sensory or motor impairments. Many of these activi-ties can be played with a peer or on one's own. Resource information on computer and elec-tronic technology can be found through the ERIC Clearing House and through Abledata at http://www.abledata.com.

Rigby and Gaik (2003) found that children with physical disabilities showed interest in and became engaged in the novelty of new toys and play materials within their familiar play environments. Caregivers may find benefit in rotating their child's play materials/toys, so that every few weeks or so, some toys are reintroduced from storage to provide some novel-ty again. Interestingly, in our clinical experience children gravitate to materials that they are familiar with in new environments, such as a therapy clinic. Possibly, children seek a sense of security and reassurance from familiar toys/materials when in a new, unfamiliar environ-ment. To foster a child's comfort within an unfamiliar environment, therapists can ask care-givers to advise them about their child's play preferences, and/or suggest that the child bring along a favorite toy.

The environment can also be modified or enriched to meet the sensory needs of children. Occupational therapists are knowledgeable about sensory processing problems that some children experience (Dunn, 2001). Therapists can help to explain a child's sensory-seeking or sensory-avoidance behaviors and help the child's caregiver make appropriate sensory experiences available in the play environment. For the child who is avoiding some sensory experiences, the environment can be modified to minimize exposure to those experiences. The therapist can also show the caregiver how to grade a child's exposure to the avoided sen-sory input so that it is acceptable and tolerated by the child. The therapist can also advise on how to enrich the environment with sensory opportunities for children who seek senso-ry input and need it to maintain an optimal level of nervous system arousal (Dunn, 2001). For example, the environment could be set up to include many options for tactile explo-ration for the child who seeks tactile input.

Occupational therapists have traditionally played a role in consulting on the design of play spaces for children with physical impairments. Accessible playgrounds have become more commonplace. Accessible playgrounds typically include ramped climbing structures, rails on ramps, elevated sand and water play areas, and cushioned, safe ground surfaces. In the United States, the new Americans With Disabilities Act (ADA) *Accessibility Guidelines for Buildings and Facilities: Play Areas* can be accessed from the Web site at http://www.access-board.gov/play/finalrule.htm. These guidelines will ensure that newly constructed and altered play areas meet the requirements of the ADA.

EXTENDING OCCUPATIONAL THERAPY ROLES WITH CHILDREN, FAMILIES, AND OTHER CAREGIVERS

Role of Advocate in Promoting Play Opportunities

Through advocacy, the occupational therapist can encourage others to enhance environments to support the play of children. A major obstacle that could be the focus for advocacy is the attitude shared by many individuals, including parents, childcare programs, schools, and the developers of children's programs and services about play. As mentioned earlier in this chapter, play has often been viewed as unimportant and frivolous, and has been undervalued. Parents, particularly parents of children with disabilities, may feel that structured skill building routines offered through educational and therapeutic programs are more important than ensuring that the child has time to play. Through advocacy, the occupational therapist can facilitate changes in attitudes about play. He or she can promote greater understanding of the value and importance of play, using arguments made earlier in this chapter. Parents may also need reassurance and permission from therapists that it is okay to play together with their child and experience the mutual enjoyment of a playful transaction, rather than feeling obliged to work on a therapy routine.

The occupational therapist could readily assume the advocacy role during routine services to families, childcare centres, schools, and to other community programs. The occupational therapist could also establish him or herself in the broader community as a strong voice for the promotion of children's play. Through advocacy, the occupational therapist can lobby at the individual and local levels for changes to enable children's programs to adopt a child-centred, play-based focus and to promote the establishment of community resources to support the education of parents, teachers, and other child caregivers about enabling children's play. Advocacy can also be directed at policy-makers and groups who develop community and/or children's services for the development of and expansion of accessible, inclusive environments and programs for all children, regardless of their special needs. This would include such ideas as barrier-free community playgrounds and play groups and programs with schedules that allow working parents the opportunity to participate. When advocating at the macro-environmental level, the therapist would likely join groups of parents who share the common goal of making government and children's services more responsive to the needs of children and families.

Role of Educator

Occupational therapists can support and educate parents and teachers about how to create supportive environments within the home, classroom, and community settings to enable play (Figure 10-5). Parents do not always understand the link between play and learning, and the therapist can provide information about this significant duality. This can be achieved through informal and formal meetings where information about development and play can be shared with the parents and other involved caregivers such as grandparents and babysitters. The therapist could provide information at parenting groups, school meetings, parent/teacher meetings, and/or home visits. This needs to be done sensitively and well so that parents don't get the idea that the therapist is referring to "educational toys" and the learning of academic or pre-academic concepts. Rather, the therapist is reinforcing the

Figure 10-5. Occupational therapists can support and educate parents about how to create supportive environments within the home. (From Eden, S. T., & Huggins, L. [2001]. *YMCA playing to learn: A guide to quality care and education of young children.* YMCA of Greater Toronto. Reprinted with permission from YMCA of GTA.)

importance of children playing, with peers or adult playmates, to explore and learn through a mutually enjoyable play exchange.

Developing a family-centred relationship with families requires acknowledgement of the family's unique nature and culture, and sensitivity to the many other priorities, needs, and responsibilities of the family. A family's parenting style, values, culture, and religious background are factors that will shape the relationship and the interventions. It should be understood and fully acknowledged that the family is the significant and primary caregiver in a child's life. Educational material to help families understand and value play and learn how to facilitate positive, supportive play experiences should be pragmatic and focused, so as not to be overwhelming or create unnecessary burden. The occupational therapist can also assist a family that is selecting a preschool program for their child by discussing with parents the benefits and the limitations of the various curricula in the available programs. A program that offers a child-centred, play-based curriculum would be preferred, based on arguments made in this chapter.

The therapist may have suggestions that would enhance the playscape and these should be shared with the parent and/or educator and incorporated in a way that would not interrupt the play. It is essential to observe and discuss the observations about the child's play experiences prior to intervening. Through observation the adults can determine ways to make the program more effective for individual children, to support the less able child, and to encourage overly dominant children to become more cooperative and ultimately more collaborative and social. These observations could also become part of a comprehensive developmental review of each child.

Occupational therapists working in child-care centres and schools would collaborate together with teachers, as a team, with each member respecting each other's contributions.

The philosophy of client-centredness would also apply when the therapist is consulting to the program, and the educational team becomes the client. The goal, creating enabling environments for play, could apply to the whole program and could also be tailored to meet the needs of individual children. The therapist could share how to use the PEO model to frame observations of children at play to analyze and identify factors that facilitate or hinder play (an approach that therapists commonly refer to as an occupational or activity analysis) (Strong et al., 1999). This approach would illuminate what could be done to make the setting more conducive to successful play experiences. The therapist could go on to model and/or guide educators to adopt the role of play partner, as described below, to facilitate children's play.

Role of Play Partner in Creating Environments That Support Play

The play partner is part of the child's social environment and is an enabler of occupational performance in play. The play partner assumes the role of player with an individual child or group of children. Parents, teachers, occupational therapists, and other caregivers can assume the role of play partner. This is an important role for fostering a child's participation in play, and it also enhances the child-adult relationship, as the child feels supported and respected by the adult play partner. The adult will gain greater cooperation from a child and have an enjoyable interaction with the child by partnering in play.

The play partner can use resources available in the environment to enhance the participation of children in play, and can guide children to use their own resources to problem-solve in order to overcome challenges in play. The play partner can also facilitate social interactions among peers. The effective play partner knows when to be "in-role" as a player, when and how to extract oneself from the play after providing facilitation, and when not to interfere with children's play.

The occupational therapist who uses the play partner role to provide intervention should adopt reflective practices to hone his or her expertise. This involves a systematic approach to observation and analysis of what is happening or what has the potential to happen in the play setting (Bridge & Twible, 1997). The occupational therapist must develop strong observational skills to examine the various complex factors that influence child-environment-occupation fit and integrate this information with what she already knows about the child and his behavior. This includes being attentive to the child's cues and interests and to follow his lead. Use of a mental image of the PEO model (Law et al., 1996) can help the therapist broaden and organize her analysis of what is impacting upon play or the child's potential to play. The therapist can assume the role of player to facilitate better child-environment fit and optimize the child's play experience.

Using the approach first described by Vygotsky (1978), the therapist also uses her knowledge of the child's learning potential in her role as play partner, to facilitate the child to stretch his abilities and to expand his repertoire of interests and skills. The therapist should be knowledgeable about child development in play, and must be observant of the child's readiness to be challenged. During play together with the child, the therapist can introduce new and different ways to use and play with the familiar materials, to enable the child to build upon his skills and broaden his mastery of the environment. As mentioned earlier in this chapter, the preschool play scale (Knox, 1997) is a helpful observational tool of the development of play skills and guides the therapist about the types of play materials, themes, and activities in which the child may develop interests.

Through interactive play, the play partner models to the child how to interact with and use the environment successfully. An illustration of a therapist in-role as a play partner is

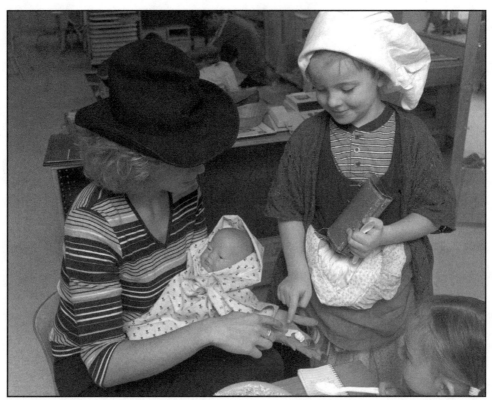

Figure 10-6. The therapist in-role as a play partner creates a supportive play environment for a child with a disability. (From Eden, S. T., & Huggins, L. [2001]. YMCA *playing to learn: A guide to quality care and education of young children.* YMCA of Greater Toronto. Reprinted with permission from YMCA of GTA.)

shown in Figure 10-6. One way is to model the reading of contextual cues so that both the play partner and child successfully enter and join the play of a group of children who are already engaged in interactive play. This is particularly important for the child who is interpreting play cues in such a way that either he doesn't know how to respond effectively or doesn't feel welcome to join in. Another way is to show the child how the environment can be modified to support him in joining in the play, as shown in the following example. We discussed earlier that children with physical disabilities are especially vulnerable to environmental obstacles and benefit by learning how the environment can be adapted to enable their participation in interactive play. In the following example the play partner modeled for Sam how the resources in the environment could be used to adapt the game to facilitate his involvement. Sam could then draw upon his strengths in imaginative play and have more control over the physical demands of the play activity.

Sam, a child with spastic diplegia, tried to keep up with a group of 3-year-old peers in a game of chase. He couldn't keep up physically as the children ran about the playroom. The play partner joined Sam and said loudly enough for the others to hear, "Yikes, that monster is going to get us! Come on Sam, I know a place where the monster won't find us" and pulled a bed sheet over herself and Sam, and said, "we must hide" as they both crouched down. When Sam was safely hidden under the sheet, the play partner emerged and called out, "I wonder where Sam went? He's disappeared Mr. Monster!" A few of his peers in the game of chase came to investigate and the play partner encouraged them to hide under the sheet, too. When the monster came to look, Sam called out, "You can't see us!" The play partner then looked all around and said, "They're gone, I can't see them!" One child emerged from the sheet and made a quick dash around and returned to crawl under the sheet again. The game then changed, and the sheet became the safe place from the monster. Sam helped it evolve to a game of water and land, with the monster becoming a shark and the kids used a floor mat as the island that was safe from the shark.

CONCLUSION AND RECOMMENDATIONS

In this chapter we have emphasized the value and importance of the occupation of play, and play-based learning opportunities for young children. Occupational therapists can enable children in their play performance by optimizing the PEO fit, which in this case is the child-environment-play fit. The occupational therapist can evaluate a child's occupational performance in play by using the ToP (Bundy, 1997). This assessment, used together with semi-structured observations of the play environment and play opportunities, as described in this chapter, will provide the therapist with an understanding of what is influencing the child's playfulness and will aid in the selection of appropriate intervention strategies.

Research has provided evidence that the environment can support, prevent, or inhibit children's play and playfulness. This is especially true for children with developmental delays, and physical, cognitive, or learning impairments. In this chapter we have provided many suggestions and described a breadth of environmental resources that the therapist and caregivers can utilize to enable children to achieve greater success in play. We have also identified environmental barriers and strategies to reduce them.

In addition, we described three important roles for occupational therapists to adopt in which they can help create more optimal play environments for children. The role of advocate can be viewed in a health-promoting framework and used at various levels of environment from working with individual parents and caregivers through the level of community development to advocate for better play opportunities for children and families. The role of educator can be used to prepare caregivers to recognize and utilize environmental resources more fully to support children's play. Finally, being a play partner provides a key role in enabling a child's playfulness. This element of the child's social environment can be used by the therapist and can be taught to children's caregivers. The play partner provides scaffolding to support and build a child's success in play, through implementing many of the strategies provided within this chapter.

The environment is rich with opportunities that support play, and the environment is amenable to change, to reduce barriers that impede play. The therapist will find that using environments to enable play is both successful and highly rewarding.

Study Questions

1. What is the value of a child-centred, play-based learning environment?

2. Identify four strategies to build environments that enable children's play and playfulness.

3. You have been invited to speak to a group of new mothers who have formed a parenting support group within their community. Their children are all preschoolers. They would like you to advise them about developmentally appropriate toys to foster their children's fine-motor and gross-motor development.

   How would you address their interest in fine-motor and gross-motor development while refocusing the meeting to emphasize the value of play? What role(s) would you use, and why? What would be your main message for the mothers?

4. Amir, a 4-year-old with a severe motor impairment, has just started an integrated preschool program. During your first occupational therapy visit to the program the teacher explains that she has set up an individualized educational plan for Amir that involves one-to-one programming directed by either a teaching assistant or one of the program volunteers. Amir will be taught school readiness skills and the teacher has requested input from you, the occupational therapist serving this program. During a 15-minute observation of the program, you noticed that Amir sat with a volunteer who helped him complete a puzzle with hand-over-hand assistance, while the other children appeared to freely choose where and what to play.

   What are potential environmental barriers that could hinder Amir's participation in play, and the environmental opportunities to facilitate his participation in play?

REFERENCES

Berk, L. E., & Winsler, A. (1995). Scaffolding children's learning: Vygotsky and early childhood education. In *NAEYC Research into Practice Series, Vol 7*. Washington, DC: National Association for the Education of Young Children.

Bowlby, J. (1988). *A secure base: Clinical applications of attachment theory*. London: Routledge.

Bredekamp, S., & Copple, C. (Eds.). (1997). *Developmentally appropriate practice in early childhood programs*. Washington, DC: National Association for the Education of Young Children.

Bridge, C. E., & Twible, R. L. (1997). Clinical reasoning: Informed decision making for practice. In C. Christiansen & C. Baum (Eds.), *Occupational therapy: Enabling function and well-being* (2nd ed., pp. 158-181). Thorofare NJ: SLACK Incorporated.

Brodin, J., & Jonson, U. (2000). Computer play centres for children with disabilities. *Int J Rehabil Res, 23*, 125-128.

Bronson, M. J., & Bundy, A. (2001). A correlation study of a Test of Playfulness and a Test of Environmental Supportiveness for play. *Occupational Therapy Journal of Research, 21*, 241-259.

Bundy, A. C. (1993). Assessment of play and leisure: Delineation of the problem. *Am J Occup Ther, 47*, 217-222.

Bundy, A. C. (1997). Play and playfulness: What to look for. In L. D. Parham & L. S. Fazio, (Eds.), *Play in occupational therapy for children* (pp. 52-66). St. Louis, MO: Mosby.

Bundy, A., Nelson, L., Metzger, M., & Bingaman, K. (2001). Validity and reliability of a Test of Playfulness. *Occupational Therapy Journal of Research, 21*, 277-293.

Caldwell, B. M., & Bradley, R. H. (1984). *Home Observation for Measurement of the Environment administration manual* (Rev. ed.). Little Rock, AR: University of Arkansas.

Cameron, D., Leslie, M., Teplicky, R., Pollock, N., Stewart, D., Toal, C., & Gaik, S. (2001). The clinical utility of the Test of Playfulness. *Can J Occup Ther, 68*, 104-111.

Case-Smith, J. (2001). Development of childhood occupations. In J. Case-Smith (Ed.), *Occupational therapy for children* (4th ed., pp. 71-94). St. Louis, MO: Mosby.

Couch, K. J., Deitz, J. C., & Kanny, E. M. (1998). The role of play in pediatric occupational therapy. *Am J Occup Ther, 52*, 111-117.

Dunn, W. (2001). The sensations of everyday life: Empirical, theoretical, and pragmatic considerations. *Am J Occup Ther, 55*, 608-620.

Eden, S. T., & Huggins, L. (2001). *YMCA playing to learn: A guide to quality care and education of young children.* Toronto, ON: YMCA of Greater Toronto.

Ferland, F. (1997). *Play, children with physical disabilities and occupational therapy: The Ludic Model.* Ottawa, ON: University of Ottawa.

Hanline, M. F. (1999). Developing a preschool play-based curriculum. *International Journal of Disability, Development and Education, 46*, 289-305.

Hanzlik, J. R. (1989). The effect of intervention on the free-play experience for mothers and their infants with developmental delay and cerebral palsy. *Phys Occup Ther Pediatr, 9*, 33-51.

Hardy, C., & Prior, K. (2001). Attachment theory. In L. Lougher (Ed.), *Occupational therapy for child and adolescent mental health* (pp. 48-66). London: Churchill Livingstone.

Harkness, L., & Bundy, A. (2001). The Test of Playfulness and children with physical disabilities. *The Occupational Therapy Journal of Research, 21*, 73-89.

Johnson, J.E., Christie, J. F. & Yawkey,T. D. (1987). *Play and early childhood development.* New York: Harper Collins.

Keating, D. R., & Hertzman, C. (Eds.). (1999). *Developmental health and the wealth of nations: Social, biological, and educational dynamics.* New York: Guilford.

Kellegrew, D. H. (2000). Constructing daily routines: A qualitative examination of mothers with young children with disabilities. *Am J Occup Ther, 54*, 252-259.

Kielhofner, G., & Barris, R. (1984). Collecting data on play: A critique of available methods. *The Occupational Therapy Journal of Research, 4*, 150-180.

Knox, S. (1997). Development and current use of the Knox Preschool Play Scale. In L. D. Parham & L. S. Fazio (Eds.), *Play in occupational therapy for children* (pp.35-51). St. Louis, MO: Mosby.

Law, M., Cooper, B., Strong, S., Stewart, D., Rigby, P., & Letts, L. (1996). The person-environment-occupation model: A transactive approach to occupational performance. *Can J Occup Ther, 63*, 186-192.

Law, M., & Dunn, W. (1993). Perspectives on understanding and changing the environments of children with disabilities. *Phys Occup Ther Pediatr, 13*, 1-17.

Lieberman, A. F. (1993). *The emotional life of the toddler.* New York: The Free Press.

Linder, T. (1990). *Transdisciplinary play-based assessment.* Baltimore, MD: Paul H. Brookes.

Mahoney, G., & Powell, A. (1988). Modifying parent-child interaction: Enhancing the development of handicapped children. *Journal of Special Education, 22*, 82-96.

McCain, M. N., & Mustard, F. (1999). *Reversing the real brain drain: Early years study final report.* Toronto, ON: Publications Ontario ISBN: 0-7778-8953-6. Retrieved February 10, 2002, from http://www.gov.on.ca/CSS/page/services/oey/oey.html.

Morrison, C. D., & Metzger, P. (2001). Play. In J. Case-Smith (Ed.), *Occupational therapy for children* (4th ed., pp. 528-544). St. Louis, MO: Mosby.

Okimoto, A. M., Bundy, A., & Hanzlik, J. (2000). Playfulness of children with and without disability: Measurement and intervention. *Am J Occup Ther, 54*, 73-82.

Parham, L. D., & Primeau, L. (1997). Play and occupational therapy. In L. D. Parham & L. S. Fazio (Eds.), *Play in occupational therapy for children* (pp. 2-21). St. Louis, MO: Mosby.

Reilly, M. (1974). An explanation of play. In M. Reilly (Ed.), *Play as exploratory learning* (pp. 117-149). Beverly Hills, CA: Sage.

Rigby, P. (2001). *The Observations of Factors Influencing Playfulness form.* Toronto: Author.

Rigby, P., & Gaik S. (2003). *Playfulness of children with physical disabilities: A pilot study.* Manuscript submitted for publication.

Rivkin, M. S. (1995). *The great outdoors: Restoring children's right to play outside.* Washington, DC: NAEYC.

Rosenbaum, P., King, S., Law, M., King, G., & Evans, J. (1998). Family-centred service: A conceptual framework and research review. *Phys Occup Ther Pediatr, 18*, 1-20.

Rubin, K., Fein, G. G., & Vandenberg, B. (1983). Play. In P. H. Mussen (Ed.), *Handbook of child psychology: Socialization, personality and social development* (4th ed., pp. 693-774). New York: Wiley.

Rutter, M. (2000). Resilience reconsidered: Conceptual considerations, empirical findings, and policy implications. In J. P. Shonkoff & S. J. Meisels (Eds.), *Handbook of early childhood intervention* (2nd ed., pp. 651-682). New York: Cambridge University.

Sandler, I. (2001). Quality and ecology of adversity as common mechanisms of risk and resilience. *American Journal of Community Psychology, 29*(1), 19-53.

Scales, B., Almy, M., Nicolopoulou, A., & Ervin-Tripp, S. (1991). *Play and the social context of development in early care and education.* New York: Teachers College.

Shonkoff, J. P., & Meisels, S. J. (1990). Early childhood intervention: The evolution of a concept. In S. J. Meisels & J .P. Shonkoff (Eds.), *Handbook of early childhood intervention* (pp. 3-32). New York: Cambridge University.

Stewart, D., Pollock, N., Law, M., Ferland, F., Toal, C., Sahagian, S., Harvey, S., & Rigby, P. (1996). CAOT practice paper: Occupational therapy and children's play. *Can J Occup Ther, 63*(2), insert.

Strong, S., Rigby, P., Law, M., Cooper, B., Letts, L., & Stewart, D. (1999). Application of the person-environment-occupation model: A practical tool. *Can J Occup Ther, 66,* 122-133.

Sutton-Smith, B. (1997). *The ambiguity of play.* Cambridge, MA: Harvard University.

Vygotsky, L. S. (1978). Mind in society: The development of higher psychological processes. In M. Cole, V. John-Streiner, S. Scribner, & E. Souberman (Eds.), *Mind in Society.* Cambridge MA: Harvard University Press.

Werner, E. (1990). Protective factors and individual resilience. In S. J. Meisels, & J. P. Shonkoff (Eds.), *Handbook of early childhood intervention* (pp. 97-116). New York: Cambridge University.

World Health Organization (WHO). (2001). *International Classification of Function, Disability and Health.* Geneva: Author.

Zeitlan, S., & Williamson, G. G. (1990). Coping characteristics of disabled and nondisabled children. *Am J Orthopsychiatry, 60,* 404-411.

Enabling Student Participation Through Occupational Therapy Services in the Schools

Mary Muhlenhaupt, BS OT, OTR/L, FAOTA

CHAPTER OBJECTIVES

* To identify the multiple influences that may impact student performance within the school context.
* To describe and illustrate a collaborative process through which occupational therapists work with other school-based team members to plan individualized programs that enable students' participation in school routines and activities.
* To illustrate strategies that enable student participation in the context of school routines and activities.

INTRODUCTION

The school system in developed countries assumes an important responsibility in educating all students to become contributing members within a society. Varying interests, motivations, and capabilities within student populations require attention when planning experiences that promote learning and accomplishment. Individualized programs that are needed to meet the unique educational needs of students with disabilities are designed through a variety of approaches employed by multidisciplinary school-based teams. Occupational therapy is one of a number of disciplines that contributes to this aspect of educational programming. For many youth with disabilities, a major portion of the activity they pursue centres around school experiences. The education system is positioned to play a significant role in promoting functional outcomes that prepare them for satisfying adult lives.

In many countries, occupational therapy services in the schools are guided by education or civil rights legislation, social, or health policies. Regional and local laws or practices uphold and interpret these provisions. Some jurisdictions maintain certification requirements for therapists working in their school districts. An understanding of pertinent regulatory and policy underpinnings is necessary for school-based therapists. National and local occupational therapy associations are a good starting point for information and referral to learn more about credentialing that is unique to school system practice within a region. These influences combine with features of the education system and with characteristics of the profession itself to define the provision of occupational therapy in a region's schools.

When enabling students with disabilities, a broad range of potential and actual roles, functions, and practices exist for therapists. Yet, across diverse regions the goal of school-based therapy services is similar. Therapists in the schools support the education system's mission to prepare youth for adult roles and responsibilities, post-school work, and community living. School-based therapy services include evaluation, program planning, and intervention for students whose performance and level of participation interferes with accomplishment of desired educational and social outcomes. Therapists provide a variety of different interventions that are designed to enable students' occupational performance on the school campus and in extracurricular activities. Practice guidelines and reports in the literature (Kalpogianni, Frampton, & Rado, 2001; Mulligan, 2001; Scott, McWilliam, & Mayhew, 1999) indicate that school-based occupational therapy services generally focus intervention on the students and their immediate surroundings as a means to this end. While these therapy services are provided to support educational outcomes, historically, the approaches therapists use in schools have been grounded in a medical model paradigm (McEwen & Shelden, 1995).

Goal setting within the medical model approach is discipline-specific and is led by each service provider. Typically, the therapist's targeted outcomes are skill acquisition and remediation of performance components. As a result, the intended occupational performance is not clarified for parents or other school personnel, nor is it obvious within the intervention plan. For example, occupational therapy services to increase a child's participation in self-dressing prior to the daily recess period may include goals to develop upper body and arm strength, to express directed hand movements and to refine precise pinch. These goals reflect component skills the student uses to put on a coat and manage the fasteners. Medical model service delivery rises from an impairment-based approach that seeks to cure, care for, or reduce the impact of pathologies. Occupational therapy interventions that are based on a medical model focus on reducing deficits or differences within the student as a means to develop increased capacities (Muhlenhaupt, Hinojosa, & Kramer, 1999). Using the example just presented, the therapist may engage the student in propping and pulling activities for upper body and arm muscle strengthening. Finger painting, placing tasks, and peg board games may be selected to increase visually directed hand movements and finger dexterity. These interventions are terminated when the targeted deficiencies and incapacities are reduced or eliminated or, when this resolution is not possible through the continued provision of specific services. This approach is compatible with traditional views of special education (Slate & Jones, 2000) that direct intervention toward remediation, reteaching, and changing students' characteristics in order to increase their function.

SHIFTING PARADIGMS IN OCCUPATIONAL THERAPY AND IN THE EDUCATION OF CHILDREN WITH IDENTIFIED NEEDS

Evolving Views of Health, Disability, and Occupation

Changing views about concepts such as ability and function and newer models of health direct attention to how an individual's functional performance is interpreted and supported in his or her daily life. The Independent Living Movement (ILM) and social models of disability advocate person-environment fit in order to increase successful participation in activities that are important to the individual and that promote an improved quality of life. (See Chapter 1 for more details concerning these models of health.)

The recently published International Classification of Functioning, Disability and Health (ICF) (World Health Organization, 2001) is useful when considering how to promote functional outcomes for students in the schools. The ICF has been described in more detail in Chapter 1. The ICF highlights the impact that a variety of changing environmental factors have on a student's ability to participate and derive satisfaction from self-management, work, and play in the classroom and cafeteria, on the playground, in school sports, and during extracurricular activities. The analysis of what influences a student's participation should expand beyond those environmental variables that come quickly to mind such as the nature of visual and auditory stimuli in the classroom, size of the classroom desk and chair arrangement, and other aspects of the student's immediate physical surroundings. Some other examples of environmental factors that influence a student's participation across school settings include school rules and expectations, the availability of a multi-level curriculum in an academic subject area, teaching philosophies and styles, peer supports, and organized leisure opportunities.

Models of practice that are advocated for and applied in a variety of practice areas assume an interdependent link between individual, task, and environment in shaping behavior (Christiansen & Baum, 1997; Dunn, Brown, & McGuigan, 1994; Law et al., 1996; Mathiowetz & Haugen, 1994). Occupational therapists working with children and adults recognize a need to alter practice approaches and move away from a focus that views the characteristics of the individual as the source of disability and as a priority for intervention (Coster, 1998; Hemmingsson & Borell, 2000; Muhlenhaupt et al., 1999; Rebeiro, 2001). The concept of occupation, both as the goal of occupational therapy intervention, and as the treatment modality, represents the profession's unique focus (Gray, 1998).

Changing Views About Education

Contemporary views of disability and function emphasize the dynamic interrelationship between a person, what he or she wants to accomplish, and the context in order to understand and influence the individual's level of participation in a variety of activities. From the latter part of the 1990s, a number of shifting perspectives in education also reflect this assumption. Beliefs that underlie segregated classrooms, integration, and educational practices known as "mainstreaming" are questioned and challenged as these newer perspectives gain attention.

One expanded view based on a sociocultural framework (Rueda, Gallego, & Moll, 2000), embraces the assumption that learning and development are dynamic processes that are social in nature. Rueda and his colleagues argue against competence as an attribute that resides solely within an individual. They reason that competence results from the interaction of the individual with others during activity. Thus, the analysis of learning and development should incorporate the interaction of these three dimensions. Activity-based instruction and flexible intervention approaches that fit within this framework are designed to accommodate the unique and changing needs of the student across a variety of situations and experiences.

Inclusion is an often-heard term in contemporary education. The belief that all children belong together is inherent in the concept of inclusion and in the practices of inclusive schooling (Sailor, Gee, & Karasoff, 2000). Programs that are influenced by inclusive philosophies value the membership and participation of students with disabilities in their communities and realize beneficial outcomes from heterogeneous groupings. An ethics paradigm of inclusion (Paul & Ward, 1996) looks beyond the student as the source of disability and the target of intervention. Schools and institutions are encouraged to accommodate the diverse needs of all learners, rather than "change the child" to meet traditional education standards and expectations. The movement toward inclusive education provides opportunities for therapy personnel, special and general education school teachers, and administrators to partner with each other as they create learning experiences in new settings. A variety of alternate instructional materials and approaches are encouraged to achieve a diverse range of outcomes among students.

There has been doubt expressed whether some educational and therapeutic approaches for students with special needs lead to functional outcomes (O'Shea & O'Shea, 1997; Schwartz, Sandall, Garfinkle, & Bauer, 1998). Questions about special education's efficacy related to post-school work and independent living outcomes encourage a variety of reform efforts (Kavale & Forness, 2000). In the United States, federal law that supports special education services (Federal Register, 1997) has changed in an effort to promote improved services. The Individuals With Disabilities Education Act Amendments of 1997 (commonly known as IDEA '97) places a new emphasis on assuring outcomes and accountability for services provided under this legislation (Huefner, 2000). In addition, provisions that encourage parent participation in goal setting, program design, and implementation for their children are strengthened through these amendments. Language that enables all students to access the general education curriculum is clarified, along with the provision that special education is a service rather than a place for those students requiring learning and instructional accommodations. Regulations to implement IDEA '97 in the United States took effect in 2000, providing another impetus for change to traditional educational approaches for students with special needs.

CURRENT PERSPECTIVES IN SCHOOL-BASED OCCUPATIONAL THERAPY PRACTICE

Occupational therapists rethink traditional practices and create innovative approaches based on updated knowledge. Rourke (1996) has suggested this particular strategy as a way for therapists to meet the unique needs of students with disabilities and support the profession's growth in the education system. Practice in the schools is changing as therapists apply new models and designs to the services that they provide. Inclusion philosophies and a need to increase students' function with their same-aged peers in a variety of typical school situations underlie many efforts that transform occupational therapy services in the schools (Kellegrew & Allen, 1996; Strong et al., 1999; Szabo, 2000).

In its current guide to therapy in education programs, the American Occupational Therapy Association (1999) promotes broader views of the "client" and the intervention "context." Individuals and groups of students, parents, teachers, other school personnel, and the education system itself are potential clients for school-based therapy services. The naturally occurring activities and routines within the daily school schedule are valued by school-based occupational therapists. These situations provide a wealth of opportunities to facilitate student performance. Intervention in these contexts is preferred over contrived situations or the implementation of "therapy activities" that are not typical within the school setting. Enabling students' participation through intervention in a variety of activities and settings in which all children learn and develop together represents best occupational therapy practice (Dunn, 2000a) in school-based services.

Mancini, Coster, Trombly, and Heeren (2000) have identified some factors that predict participation levels of students with disabilities in inclusive education settings. They confirm the value of functional performance assessment over measures of impairment. Based on their research, they argue that a student's physical disabilities may only partly limit participation, and that a combination of physical and cognitive-behavioral variables will better predict full participation. Occupational therapists understand how these domains are interrelated and impact students' participation across school settings. They use their expertise to help educators plan effective programs and assure that students with diverse needs are prepared for important adult roles and responsibilities.

The profession has recently been promoting the adoption of evidence-based practice (Cusick, 2001; Holm, 2000), and this is significant for therapists in the education system. Information that substantiates the impact of interventions on a student's school function informs practice and increases valid decision-making by team members. The existing database of research to confirm the extent to which occupational therapy strategies enable student participation in the schools is limited. Methods to assure that interventions support outcomes that are valued by consumers (Law & Baum, 2001) are needed in school-based occupational therapy practice. Therapists should use systematic approaches (Hawkins & Matthews, 1999) to measure the variety of services that are provided in order to demonstrate the impact of occupational therapy for individual students and for program or community development initiatives.

Multiple Interdependent Influences on a Student's School Function

According to the person-environment-occupation (PEO) model (Law et al., 1996) occupational performance is the outcome of the transaction between a person engaged in an occupation within an environmental setting. The PEO relationship is unique for each student and defines his or her occupational performance in school and in extracurricular pursuits. Occupational performance and participation across the campus is optimized when there are harmonious relationships, or a good "fit" between person, environment, and occupation dimensions. Knowledge about person, environment, and occupation variables and their interrelationships is essential in order to enable a student's function across school settings.

PERSON

In school settings the student represents the *person* dimension. Student characteristics include his or her changing physiological, neuromusculoskeletal, cognitive, and psychosocial attributes, learning styles, skills, and work habits. A student's qualities, traits, and skills are unique to the individual. These variables interact with each other and contribute to the student's attention and engagement in activity, as well as to a sense of achievement and the degree of satisfaction derived from school experiences. The student's adaptation over time is considered when analyzing school function.

ENVIRONMENT

The school *environment* is multi-faceted, including cultural, physical, temporal, social, and institutional dimensions. Classrooms and schools provide a vast array of sights, sounds, materials, relationships, expectations, and opportunities that contribute to the individual experiences of each student in attendance. Contextual features give behavior its meaning. A student sitting and coloring a picture in the corner of the classroom evokes a different response from the teacher when this behavior occurs during the outdoor recess period than when the same behavior is observed along with several peers at a table in the classroom during periods of "quiet choice" activities.

Environments help to define standards for desirable student behavior. Carrying on loud conversations with peers and frequent movements around the area are expected performances when lunch or afternoon snack is served in the elementary school cafeteria. These same behaviors are not required, and their expression may be problematic, when meals occur in the classroom. School rules and customs, specific learning materials, and media are some of the environmental features that are adopted to facilitate a targeted range of skills and student behaviors.

OCCUPATION

A third influence on a student's occupational performance is the *occupation* itself, along with features of its associated tasks and activities. Law and her colleagues have defined activities, tasks, and occupations in relation to the PEO model of occupational performance (1996). Similar concepts are included with related terminology in both the American and Canadian Occupational Therapy Associations' practice frameworks (American Occupational Therapy Association, 2002; Canadian Association of Occupational Therapists, 1997). Specific occupations that are important for students vary between school cultures and change across the lifespan from early childhood through adolescence. However, typical areas of student occupations include the following:

* Establishing and maintaining friendships
* Participating in academic and special subject areas
* Pursuing cocurricular and extracurricular activities
* Managing classroom and school work roles

PROMOTING SCHOOL FUNCTION THROUGH ENABLING PERSON-ENVIRONMENT-OCCUPATION FIT

The changing nature of person, environment, and occupation elements and their interrelationships with each other are key considerations for occupational therapists when they enable students' participation in the schools. The dynamic influence of these elements and processes on occupational performance means there are potentially many ways to influence and support student performance across school settings. For example, games that students select and their play activity during a rainy day recess period in the classroom are very different from those that characterize recess on a sunny spring day in the playground. The presence or absence of peers and the nature of peer exchanges influence opportunities and student behaviors. Features and conditions within an environment impact a student's level of participation. When the school playground accommodates a swing that holds a wheelchair, the student who uses a wheelchair is able to participate with peers in recess activity. When a teacher encourages peer modeling during clean-up time in the classroom, participation by the student who has difficulty initiating and persisting through the task is supported.

Temporal aspects of environments also impact student participation. For example, during the winter months in cold climates, the types of clothing that a student wears are bulkier and require that garments be fastened for warmth. Classroom routines may accommodate for this through schedule alterations that increase the time available for dressing before an outdoor activity. As an alternative, a student's clothing may be modified with fasteners that are easier to secure. Over the course of the school term, teachers naturally reduce their cues and directions in order to increase students' independent work habits. Picture cards, written schedules, or lists may be added to a student's desktop or on the bulletin board, either on an interim basis or for an extended period, to support different levels of self-direction. Each of these examples illustrates how student participation is increased through environmental strategies. The following scenario illustrates intervention directed toward the environment in order to influence a student's school function.

> *Danielle, a 6th-grade student, tires frequently during the school day and only partially completes assignments. Often, she sits at her desk and watches others as they carry out science lab experiments. Mr. Parker, her teacher, has a limited knowledge base related to Danielle's unique needs. He is not aware that he could change the materials that Danielle uses or he could make adjustments to her schedule to increase her participation within his class. The occupational therapist discusses Danielle's functional needs with Mr. Parker to increase his understanding of how the stress and exertion levels associated with specific classroom tasks and activities can have a negative influence on Danielle's participation at school. The therapist makes recommendations to modify specific physical demands in the classroom and in the school itself, and provides guidelines to enable Mr. Parker to adjust the sequence of activities within Danielle's day. Tasks requiring exertion are alternated with less strenuous experiences. Danielle has opportunities to recoup her physical reserve so that she doesn't tire within specific lessons. Her stamina is increased for the afternoon's learning opportunities and experiences. These intervention strategies promote her occupational performance by enabling her to participate in school activities and assignments that further increase her competence and satisfaction.*

Environments in a broader sense also impact student participation in the schools. District-wide curriculum standards and teacher preferences help to define student roles and create expectations that students master particular subject matter. For example, the deliberations of a curriculum committee that is approving revisions to the district's physical education standards ultimately have implications for the gymnasium activities that are included in the instructor's weekly lesson plans.

Occupational Therapy and the Environment in Schools

The occupational therapist has the potential to bring a unique perspective to the school team about the various ways environments can support an individual student's participation in school activities. This knowledge can be of great importance for student evaluation, program planning, and program implementation processes within the schools.

Program Planning Considerations

When school-based occupational therapists plan programs, they typically focus on a particular student, what the student wants to accomplish, and where, when, and with whom relevant roles and behaviors occur. A range of variables impacts a student's participation in school experiences, and the student's expression and utilization of both academic and functional life skills. Some of these influential features receive more attention than others do when teams think about how to help students learn and develop. Often person characteristics and elements of the individual's immediate material environment are prominent. For example, the teacher and occupational therapist may focus on the obvious factors that influence Jack's reading abilities, such as his visual discrimination and memory, his phonetic skills and comprehension, the curriculum used and the teacher's instructional activity. They may overlook the positive influence of classmates who read aloud to Jack as a means to support his achievement in reading.

It is the broader perspective of the occupational therapist that enables the school team to plan interventions that are considerate of personal, environmental, and occupational influences on a student's participation in school activities. In some situations, the occupational therapist may help the school team to identify the various environmental barriers that limit a student's participation. In other situations, environmental resources will be identified that can support and facilitate a student's participation at school. When the school team integrates this broader perspective into student evaluation and planning processes, they will consider a wider variety of options and strategies that can help the student to accomplish relevant outcomes. What works for one student may also benefit a group of students within the classroom, the school building, or across the district.

The Educational Planning Team

Within the education system, the educational planning team is responsible to design a program that leads to valued outcomes for the student. Legislation mandates team involvement in some countries, while others implement team approaches without regulatory oversight. This group is known by different names between school districts and programs, such as the Child Study Team, Instructional Support Team, Individualized Education Program Team, or others. The team's composition varies according to student needs. Parent(s) or caretakers(s) and, if appropriate, the student are primary participants. The classroom teacher and school personnel who are familiar with the student, including physical, speech, and occupational therapists, psychologists, personal assistants, or others, play important roles within the team. Additional members may include the school principal or other supervisory or administrative personnel from the district. Pediatricians, specialized medical or health care personnel, or child protection and advocacy representatives from health care and social service settings may be involved as extended members on a short-term or continuing basis.

Table 11-1

Examples of Tasks and Focused Questions for Educational Planning Team

Planning Steps for Team	Driving Questions
1. Set goals/outcomes for student	1. What does the student want and need to do?
2. Identify student's current level of performance	2. What does the student's current participation across a variety of school activities look like?
3. Specify special services, supports, accommodations that this student requires to achieve goals	3. What intervention is needed to move this student from the status revealed in step #2 to achievements the team has targeted in step #1?
4. Evaluate impact of services and programs implemented	4. How and when will we monitor this student's progress and use the information we collect?

They may respond to specific requests, attend review and planning meetings, or send information and reports for the team's consideration.

The collective work of the educational planning team members is critically important in relation to what the student ultimately accomplishes. The group gathers and reviews information about the student's strengths and needs from a variety of sources. They work together to synthesize this data into a plan that leads to relevant and meaningful outcomes for the student. Teams make decisions regarding each student's educational program, including content and implementation. They are responsible to monitor the impact of interventions, making adjustments as needed when anticipated results are not evident (Table 11-1).

Several different processes and approaches are used in education programs to coordinate the multiple viewpoints of teachers, therapists, parents, and administrators (Lamorey & Ryan, 1998; Ogletree, Bull, Ruby, & Lunnen, 2001; Rainforth & England, 1997). In any of these, no one person's perspective is viewed as the authority. Isolated and autonomous decision-making has been identified as a negative factor in planning and providing programs (Giangreco, 1991) and thus, can limit the efficiency through which student outcomes are achieved. In contrast, collaborative practices between the many different persons concerned with an individual student's program are featured in effective education programs for children (Demchak & Morgan, 1998; Friend, 2000). A process of consensus decision-making enables the team to bring different perspectives together when planning for students with special educational needs. Clear communication and good interpersonal skills assist team members as they work together to integrate diverse viewpoints. Using their expertise in group interaction and process skills, occupational therapists can be instrumental in facilitating collaborative practices within the educational planning team.

Goal Setting: A First Step in Program Planning

One of the most important foundations for effective educational planning is the development of a set of unified goals for the student (Pretti-Frontczak & Bricker, 2000). These team-generated goals are the driving force that underlies all decisions about a student's individualized school program. They need to be determined before any classroom placement or service decisions are concluded (Giangreco, 1995). The team members need to know what

Peter *wants and needs to do* before they determine what type of program is suitable and create an intervention plan that's designed *to get him there.* Outcomes that are defined within
individualized goals help to clarify projected program areas and frame intervention planning
for the student (Dunn, 2000b). The expectations that are reflected in the team-generated
goals are a significant environmental variable that influences Peter's occupational performance outcomes.

The goal-setting process requires the team to project how much a student's participation
and function will improve over a school year. Asking questions is a practical strategy that
prompts team members to think about outcomes and target specific, relevant behaviors as
they establish goals. What do we want Peter's level of participation in math and social studies to look like at the end of the school year? What kinds of performance will reflect the
improved communication and social skills that Peter needs to develop? Questions that probe
"*where?*" and "*with whom?*" focus attention on a variety of contextual features that impact a
student's experiences. This also helps the team to appreciate the influence that PEO processes have on student performance. This statement represents one of the performance areas
that the team expects will improve by year's end:

> Peter will initiate, sustain, and respond to social exchanges with peers without dis-
> abilities in the classroom and on the playground.

As a next step, the team identifies and documents a current picture of Peter's performance in each outcome area and generates interim objectives. The description of his "present
performance levels" defines the baseline against which Peter's progress is monitored
throughout the programming period.

FAMILY AND STUDENT PERSPECTIVES

Families hold a long-term vision for their child and their view is an important part of the
goal setting process. Family or caretaker priorities are incorporated into educational program
planning activities when they are identified and available for each team member to use during their data-gathering activities. Here are some examples of one family's goals for their
son's school experience:

* We want Matthew to have friends and relationships with other children in the community.
* We want Matthew to have a sense of self-esteem, worth, and contribution in the general education program.

When this information is reviewed before, or at the time evaluators initiate procedures,
they are able to direct their approaches to address the priorities and concerns the family has
for their child.

An interview with the family completed by any member of the team with the necessary
expertise is a useful way to learn about parent perspectives and goals for the future. During
this part of the goal-setting process the team may rely on the therapist's knowledge and
interview skills. The school-based occupational therapist has valuable insights concerning
self-care, work, play, and leisure across the lifespan. These domains are important in relation
to the education system's mission of preparing youth for productive living in society (Brown
& Snell, 2000). Several tools and frameworks are available to help both family members and
school personnel as they collect this important information, or teams may devise their own
interview format (Table 11-2). Strategies that enable parents to articulate priorities that
reflect their own cultural and personal values are exemplified in the tool, Making Action
Plans (MAP) (Pearpoint, Forest, & O'Brien, 1996). Information about MAP is available at
www.inclusion.com. Strategies are also shown in choosing options and accommodations for

Table 11-2

Key Points for Initial Data-Gathering Interviews

Questions for Student's Teacher	Questions for Student's Family
1. What is the focus of your classroom curriculum?	1. What areas of your child's performance in school are of the most concern to you?
2. How is this student managing in that curriculum?	2. What is your child's behavior like in similar activities at home or in the community?
3. What has been tried to date to increase this student's participation?	3. What was the focus of any previous occupational therapy your child received and how do you feel the service was beneficial?
4. What was the impact/effect of these strategies?	
5. How do you think occupational therapy services may help this student succeed in school?	4. How do you think occupational therapy services may help your child to succeed in school?

children (COACH) (Giangreco, Cloninger, & Iverson, 1998). Information about COACH is available at www.brookespublisher.com. These tools are valuable additions to the educational planning team's customary data-gathering and evaluation procedures.

Students should be included throughout the data gathering and goal-setting process as much as they are able. Teams can incorporate a variety of methods to assure student participation. Some possibilities include interviews with the student, activity surveys, audio or videotape recordings of student interests, submission of written and drawn material generated by the student, and inclusion of the student at all planning meetings that concern his or her educational program.

Occupational Therapy Interventions in the Schools

The PEO model (Law et al., 1996) can guide occupational therapists to promote the participation and school function of students with disabilities. Using this framework, school-based therapists focus on how the PEO transaction influences a student's occupational performance in relation to the targeted outcomes identified by the school team. When applying this model to practice the therapist assesses dimensions of the person, environment, and occupation transaction to analyze what is contributing to the challenges or issues in occupational performance experienced by the student. This analysis leads naturally to the identification of potential interventions to facilitate greater PEO fit and consequently improves the student's occupational performance.

ASSESSMENTS TO INCREASE SCHOOL FUNCTION

The therapist begins with information about the student's overall goals, as explained in the previous section. Teacher and family perspectives about the student's current perform- ance across relevant activities are important components when developing effective thera- py programs in the schools. Therapists accomplish this data gathering through face-to-face interviews, telephone or email communication, or written questionnaires. Observation in the classroom and across a series of transitions between school subjects and locations pro- vides the therapist with first-hand information about many of the expected roles for students in the class. Data about the performance of the particular student is noted. Standardized measures may also be used to collect information about a student. For example, the school function assessment (Coster, Deeney, Haltiwanger, & Haley, 1998) is a helpful tool for the team to use to analyze the student's participation in non-academic activities and tasks across the school environment.

Data collected about individualized goals and about person, environment, and occupa- tion dimensions is used to make decisions during the assessment. This is demonstrated in the following example about a student, Sabrina.

> Mrs. Quigley, the art teacher, reports that although Sabrina seems to be working in class, she rarely completes an assignment with the detail and quality that other students produce. Sabrina doesn't ask Mrs. Quigley for any assistance with her assignments. From several 10-minute observation periods across a few days, the occupational thera- pist has already observed that Sabrina handles the classroom writing tools and manages to manipulate materials without difficulty. She sets up, eats, and cleans up after lunch without any help from the cafeteria monitor. There's no evidence that Sabrina's visual- motor skills are a factor in relation to her limited participation in art class, so further evaluation in this area is not justified.
>
> Mrs. Quigley's art curriculum is reviewed in more detail in order to understand the types and sequences of instructional units that have been provided. Activities planned for the remainder of the year are also surveyed. During the process the therapist learns that in each grade level every month, one student's artwork is displayed in the "Artist of the Month" showcase. The showcase is in a visible location near the school's main entrance and always attracts interest from visitors and students as they enter and leave the building. Students enjoy having their work featured and receive a printed certificate to recognize this award. Sabrina's work has not been represented during the past year.
>
> According to her mother, Sabrina enjoys crochet, embroidery, and weaving pothold- ers as hobbies at home. Sabrina reports that she's not a very good artist in school and that the assignments "aren't much fun." She perceives that two classmates who sit at her worktable are much more artistic than she is. They often draw attention to the qual- ity of their own work in relation to the projects completed by other students in the group. The therapist makes observations in the art studio when Sabrina and her classmates are there. Sabrina anticipates that nearby students will evaluate her work and often glances at them. She immediately covers her project when they direct attention toward her activ- ity. She turns her paper over and begins again, or gets another clean sheet when she hears these two students praise their own work. During the class, Mrs. Quigley moves around the studio and offers to help students at their workstations but Sabrina refuses any assistance.

INTERVENTION IN SCHOOL ROUTINES AND ACTIVITIES

Sabrina wants to and needs to increase her participation in art class. With this desired outcome in mind, the therapist uses the CAPE tool to analyze the person-environment, person-occupation and occupation-environment relationships influencing Sabrina's occupational performance, as illustrated in Figure 11-1. The CAPE is useful for documenting the PEO analysis in the schools (Muhlenhaupt, 2000), and adapts the clinical application of the PEO model (Strong et al., 1999). The therapist identifies whether the current relationships are supportive, neutral, or limiting influences on Sabrina's occupational performance. The therapist may ask how do Sabrina's interests, her psychosocial, cognitive, and motor abilities, and modes of learning fit with the required art tasks? This question probes areas within person-occupation relationships. Does Mrs. Quigley's teaching style and the activities of other students enable Sabrina's participation? This generates information about the occupation-environment transaction. Is positive reinforcement and reward available for Sabrina's efforts? This question taps into person-environment processes.

The therapist is ready to plan interventions that will increase Sabrina's occupational performance in relation to the targeted outcome. Options that increase PEO fit need to be identified. As a first step, intervention strategies that facilitate the person-occupation-environment relationships that support Sabrina's participation are selected. Second, strategies that reduce the impact of, or eliminate the relationships that interfere with her participation are considered. Intervention may be directed to any, or all, of the dimensions in this model. When multiple, compatible interventions are implemented, opportunities for enablement are increased and reinforced throughout the learning context. The therapist should consider the availability of resources within the program and which intervention options represent the most efficient and effective means to support the student's participation across several environments when selecting from multiple intervention options. In some situations therapists seek a balance of interventions across person, environment, and occupation dimensions. Student preference among the proposed options should also be considered whenever possible.

Consider some of the supporting and limiting factors that were identified in Sabrina's PEO analysis (see Figure 11-1). Sabrina has the cognitive and sensorimotor abilities to participate in the required art tasks, a finding that reflects a good person-occupation match. However, Sabrina's self-perception of her artistic ability in relation to the art class assignments is not a compatible person-occupation fit. Peer behaviors toward Sabrina, her response, and a limited amount of positive feedback from the environment reflect a negative person-environment relationship. The small variety of art tasks used within the teacher's curriculum, despite the availability of physical space and multiple art media supplies is a limiting occupation-environment interaction. The following is a list of potential intervention options and strategies to improve Sabrina's occupational performance in art class. Most use readily available environmental resources (e.g., physical, social, and institutional aspects of the environment) that would be quite easy to implement with cooperation from the teacher and parents.

* Move Sabrina's seat assignment to another table.
* Encourage the teacher to schedule a class lesson that includes examples of how diverse abilities and contributions have enriched the arts in general, with reference to how the range of skills by students is valued in the school.
* Include opportunities for choice into curriculum, with Sabrina and classmates selecting their own projects related to the current art instruction and theme.

Sabrina wants and needs to: *participate in activities within the art studio and complete assignments*

Dimensions for consideration in school-based services	Probe areas (consider already existing and potentially existing)	What supports or enables student to accomplish the team-generated outcome?	What limits or prevents this student from accomplishing the team-generated outcome?
C CHILD/STUDENT (characteristics, abilities, experiences, learned skills)	• Feelings of self-esteem • Physical, cognitive, sensorimotor, psychosocial attributes, and communication skills • Work habits • Personality traits and styles • Self-determination	• Visual-fine motor skills—sufficient for range of manipulation activities for art tasks • Wants a "better grade in art"	• Feeling that peers are better "artists" • Self-esteem related to school's current art program
A ACTIVITIES, TASKS, OCCUPATIONS/ CURRICULUM (includes extracurricular)	• Meaningful and relevant student roles • Specific curricular materials, methods • Daily/weekly scheduling (temporal aspects and content) • Homework • Testing and grading • Extracurricular pursuits	• From activities inventory: Needlework is her hobby, enjoys crafts at home • Art studio is well-supplied with variety of media	• Curriculum provides limited opportunity for choice and self-selection in media assignments
P PEERS/PERSONS (family and school social factors and context)	• Relationships, interactions with peers and adults • Use of peer buddies • Teacher/other adult behaviors • Teaching style, attitudes • Specific social opportunities and experiences • Play opportunities • Use of personal assistance	• Positive reinforcement for Sabrina's work efforts and finished projects • Multiple other grouping options available in studio	• Peers praise their own work and compare it to Sabrina's efforts • Teacher has been unsuccessful in attempts to help Sabrina complete art assignments • Teacher's grading system emphasizes effort • Family doesn't see her art projects
E ENVIRONMENT (physical, sensory, institutional, cultural, and temporal environments)	• Aspects of the built environment (student's immediate environment, whole school, campus, and community) • Use of assistive technology • School rules, norms, customs • Family/at-home resources	• Student art assignments are displayed in school's showcase throughout the year • Studio includes space, areas for variety of different media in simultaneous use	• Sabrina's work has not been included in showcase this year

© 2000, Mary Muhlenhaupt, OTR/L, FAOTA

Figure 11-1. CAPE. A tool to facilitate team problem-solving.

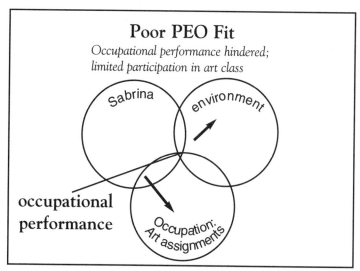

Figure 11-2. Sabrina demonstrates poor PEO fit during art class. She has difficulty participating during art class.

* Suggest that the art teacher identify alternate selection/scheduling options for monthly showcase to assure that each student's work is featured over the school term.
* Include a project completed by students outside of school that reflects their hobbies or extracurricular interests in the monthly showcase.
* Sabrina's parents could provide specific positive reinforcement for Sabrina's art projects when they are brought home.
* Implement interventions to increase Sabrina's self-esteem.

Figure 11-2 illustrates Sabrina's poor PEO fit during art class before intervention, and Figure 11-3 demonstrates how using the environment in intervention can increase PEO fit and improve Sabrina's occupational performance by facilitating her participation in art class.

INTERVENTION DECISION-MAKING

Depending upon the planning and review approach that is used within specific school programs, occupational therapists follow different procedures to communicate their findings and recommendations to other members of the student's educational planning team. For example, in the United States a multidisciplinary review and team decision is mandated to approve individualized goals, services, supports, and accommodations within a student's school program. In other regions, following the therapist's evaluation, the range of intervention strategies may be reviewed and approved in a meeting with the teacher and parent. The opportunity to consider potential intervention options with other members of the team enriches the analysis and decision-making process. Related intervention ideas from other disciplines are coordinated into a unified plan when this collaboration between disciplinary perspectives occurs. The team is able to attend to areas of overlap and duplication, and consider possible omissions when intervention plans from different disciplines are reviewed together before service decisions are finalized for implementation.

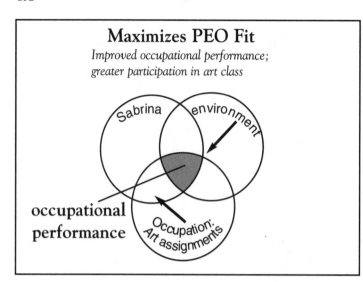

Figure 11-3. Environmental resources and opportunities can be used to increase PEO fit and improve Sabrina's occupational performance in art class.

New Frontiers in School-Based Occupational Therapy Practice

The Person-Environment-Occupation Model in a Broader Application in the Education System

School-based occupational therapists generally join the education planning team when a specific student's need for occupational therapy is initially questioned. Typically, the primary concern is to determine *if* and *how* an occupational therapy service supports the student to accomplish desired outcomes. A therapist's ongoing participation as a team member is sustained as long as the team agrees that occupational therapy is required for the particular student. When the service is no longer needed, the therapist leaves the planning team. Unless other students in the school building continue to receive occupational therapy services, the therapist's involvement in the school ends.

Occupational therapists enable student participation through contributions that extend beyond service delivery within a student's individualized program, or "therapy services for a particular student." The application of the PEO model that is illustrated in this chapter can be integrated within the collaborative processes undertaken by teams as they plan programs for students with a range of educational needs. When planning with a team the occupational therapist can take the lead to facilitate the team through the problem-solving process. Individualized assessment results from multiple disciplines are gathered and the findings are categorized according to person, environment, and occupation dimensions. The occupational therapist identifies areas of PEO fit and mismatch. Then the therapist clarifies for the team how these processes impact the specific student's function across school settings and in relevant tasks.

Together, the team considers potential interventions, using the strategy to generate solutions that is described in this chapter (see section, "Intervention in School Routines and Activities"). Planning team members select any or all of the proposed options and design a cohesive intervention plan that coordinates multiple perspectives. Their actions are influenced by the team's collective view of each strategy's efficiency and success in relation to the desired outcome, and by their standard practices for designing classroom programs and related support services. Through the use of problem-solving strategies within a collaborative process, the school-based team identifies, creates, and supports a context that matches the student's unique strengths and needs in order to facilitate increased school participation.

A variety of additional factors influence students' participation in the schools beyond the local environments that constitute their everyday experiences within the classroom or across the campus. The institution's curriculum resources, staff professional development programs, team processes, planning approaches, and student performance standards are some examples that represent layers of influence in the broad perspective of the environment. A therapist's planning and programming expertise makes a valuable contribution within this larger context of the education system. Occupational therapy representation can be incorporated within a variety of school district or building-level curriculum committees and commissions that investigate, design and adopt standards and programming in academic, social-behavioral, and extracurricular areas. District-wide staff training and professional development programs are another opportunity for occupational therapists to contribute their knowledge. These alternatives offer ways for therapists to enhance factors within the environment and occupation dimensions that influence occupational performance of individuals and groups of students.

Through these approaches, all team members work together, contributing their own expertise to identify, create, and sustain a context that matches each student's unique and changing strengths and needs. As a result, schools use multiple resources to design and provide programs that help students achieve necessary levels of occupational performance to support their accomplishment of desired outcomes. Occupational therapists depend upon their self-advocacy skills to inform school district administration and other school personnel about more innovative uses of occupational therapy services within their programs. Collaborative practices between occupational therapists and administrators, families, and school personnel from a variety of disciplines are increasingly important as these new approaches are implemented on a broader scale within programs, and change is integrated within systems.

CONCLUSION

School-based therapists offer a variety of assistance to the education system in order to support the occupational performance of students with special needs. In view of the many changing variables and forces that influence behavior, PEO fit is an important criterion to enable satisfactory occupational performance for students with special educational needs. The PEO model (Law et al., 1996) supports planning and implementation approaches that target relevant variables to impact successful student participation across the campus and in valued extracurricular activities.

Occupational therapists are in a key position to contribute a valuable perspective to the education system that is adapting to philosophical shifts related to disability and education, increased knowledge, and regulatory and policy changes. The contemporary practice of occupational therapy in the education system concerns the needs of an expanded group of consumers that includes general education administrators and teachers and other school support personnel, as well as the institution itself. These perspectives suggest a variety of new roles and responsibilities for therapists in the schools, beyond those associated with the established student-therapist dyad in traditional practice.

Study Questions

1. Describe an occupational therapy perspective of student participation in the schools by incorporating contemporary views of health, disability, occupation, and function.

2. Eating meals with peers in the cafeteria and completion of weekly homework are two occupations that are relevant for many students. List and differentiate several activities and tasks that compose each of these student occupations. Identify at least three environmental variables that may influence a student's performance with both of these occupations.

3. Identify a sequence of program planning steps that an occupational therapist uses with other team members to design education programs that influence student participation across settings and in valued activities.

4. Identify two strategies that an occupational therapist could consider to enable the development of friendships on the school playground when one student in the elementary school class uses a power wheelchair for mobility.

5. Discuss ways to increase an educational planning team's awareness that an occupational therapy perspective:

   a. Values teacher, student, and parent views in goal-setting.

   b. Is concerned with how multiple environmental variables influence student participation.

   c. Incorporates a problem-solving approach that is directed toward enabling all students to be successful within the school curriculum.

REFERENCES

American Occupational Therapy Association. (1999). *Occupational therapy services for children and youth under the Individuals With Disabilities Education Act* (2nd ed.). Bethesda, MD: Author.

American Occupational Therapy Association. (2002). Occupational therapy practice framework: Domain and process. *Am J Occup Ther, 56*, 609-639.

Brown, F., & Snell, M. (2000). Meaningful assessment. In M. Snell & F. Brown (Eds.), *Instruction of students with severe disabilities* (5th ed., pp. 67-114). Upper Saddle River, NJ: Prentice-Hall.

Canadian Association of Occupational Therapists. (1997). *Enabling occupation: An occupational therapy perspective.* Ottawa, ON: CAOT Publications ACE.

Christiansen, C., & Baum, C. (1997). Person-environment occupational performance: A conceptual model for practice. In C. Christiansen & C. Baum (Eds.), *Occupational therapy: Enabling function and well-being* (2nd ed., pp. 46-71). Thorofare, NJ: SLACK Incorporated.

Coster, W. (1998). Occupation-centered assessment of children. *Am J Occup Ther, 52*, 337-344.

Coster, W., Deeney, T., Haltiwanger, J., & Haley, S. (1998). *School function assessment.* San Antonio, TX: The Psychological Corp./Therapy Skill Builders.

Cusick, A. (2001). OZ OT EBP 21C: Australian occupational therapy. Evidence-based practice and the 21st century. *Australian Journal of Occupational Therapy, 48*, 102-117.

Demchak, M., & Morgan, C. (1998). Effective collaboration between professionals and paraprofessionals. *Rural Special Education Quarterly, 17*, 10-15.

Dunn, W., Brown, C., & McGuigan, A. (1994). The ecology of human performance. *Am J Occup Ther, 48*, 595-607.

Dunn, W. (2000a). Best practice philosophy for community services for children and families. In W. Dunn (Ed.), *Best practice occupational therapy: In community service with children and families* (pp. 1-9). Thorofare, NJ: SLACK Incorporated.

Dunn, W. (2000b). Designing best practice services for children and families. In W. Dunn (Ed.), *Best practice occupational therapy: In community service with children and families* (pp. 109-134). Thorofare, NJ: SLACK Incorporated.

Federal Register. (1997). *Individuals With Disabilities Education Act Amendments of 1997.* 20 U.S.C., 1400-et seq.

Friend, M. (2000). Myths and misunderstandings about professional collaboration. *Remedial and Special Education, 21,* 130-132.

Giangreco, M. (1991). Common professional practices that interfere with the integrated delivery of related services. *Remedial and Special Education, 12,* 16-24.

Giangreco, M. (1995). Related services decision-making: A foundational component of effective education for students with disabilities. *Phys Occup Ther Pediatr, 15,* 47-67.

Giangreco, M., Cloninger, C., & Iverson, V. (1998). *Choosing outcomes and accommodations for children: A guide to educational planning for students with disabilities* (2nd ed.). Baltimore: Paul H. Brookes.

Gray, J. (1998). Putting occupation into practice: Occupation as ends, occupation as means. *Am J Occup Ther, 52,* 354-364.

Hawkins, R., & Matthews, J. (1999). Frequent monitoring of clinical outcomes: Research and accountability for clinical practice. *Education and Treatment of Children, 22,* 117-135.

Hemmingsson, H., & Borell, L. (2000). Accommodation needs and student-environment fit in upper secondary schools for students with severe physical disabilities. *Can J Occup Ther, 67,* 162-172.

Holm, M. (2000). Our mandate for the new millennium: Evidence-based practice. *Am J Occup Ther, 54,* 575-585.

Huefner, D. (2000). The risks and opportunities of the IEP requirements under IDEA '97. *The Journal of Special Education, 33,* 195-204.

Kalpogianni, E., Frampton, I., & Rado, T. (2001). Joint working between occupational therapy and clinical psychology in a school setting: A neurorehabilitation case study of a child with acquired developmental disability. *British Journal of Occupational Therapy, 64,* 29-33.

Kavale, K., & Forness, S. (2000). History, rhetoric, and reality. *Remedial and Special Education, 21,* 279-296.

Kellegrew, D., & Allen, D. (1996). Occupational therapy in full-inclusion classrooms: A case study from the Moorpark model. *Am J Occup Ther, 50,* 718-724.

Lamorey, S., & Ryan, S. (1998). From contention to implementation: A comparison of team practices across service delivery models. *Infant-Toddler Intervention, 8,* 309-331.

Law, M. & Baum, C. (2001). Measurement in occupational therapy. In M. Law, C. Baum & W. Dunn (Eds.), *Measuring occupational performance: Supporting best practice in occupational therapy* (pp. 3-19). Thorofare, NJ: SLACK Incorporated.

Law, M., Cooper, B., Strong, S., Stewart, D., Rigby, P., & Letts, L. (1996). The person-environment-occupation model: A transactive approach to occupational performance. *Can J Occup Ther, 63,* 9-23.

Mancini, M., Coster, W., Trombly, C., & Heeren, T. (2000). Predicting elementary school participation in children with disabilities. *Arch Phys Med Rehabil, 81,* 339-347.

Mathiowetz, V., & Haugen, J. (1994). Motor behavior research: Implications for therapeutic approaches to central nervous system dysfunction. *Am J Occup Ther, 48,* 733-745.

McEwen, I., & Shelden, M. (1995). Pediatric therapy in the 1990's: The demise of the educational versus medical dichotomy. *Phys Occup Ther Pediatr, 15,* 33-45.

Muhlenhaupt, M. (2000). *CAPE: A tool to facilitate team problem-solving.* Unpublished manuscript.

Muhlenhaupt, M., Hinojosa, J., & Kramer, P. (1999). Perspective of context as related to frame of reference. In J. Hinojosa & P. Kramer (Eds.), *Frames of reference for pediatric occupational therapy* (2nd ed., pp. 41-65). Baltimore: Lippincott Williams & Wilkins.

Mulligan, S. (2001). Classroom strategies used by teachers of students with attention deficit hyperactivity disorder. *Phys Occup Ther Pediatr, 20,* 25-43.

Ogletree, B., Bull, J., Ruby, D., & Lunnen, K. (2001). Team-based service delivery for students with disabilities: Practice options and guidelines for success. *Intervention in School and Clinic, 36,* 138-145.

O'Shea, D., & O'Shea, L. (1997). Collaboration and school reform: A twenty-first century perspective. *J Learn Disabil, 30,* 449-462.

Paul, P. & Ward, M. (1996). Inclusion paradigms in conflict. *Theory Into Practice, 35,* 4-11.

Pearpoint, J., Forest, M., & O' Brien, J. (1996). MAPs, Circles of Friends, and PATH: Powerful tools to help build caring communities. In S. Stainback & W. Stainback (Eds.), *Inclusion: A guide for educators* (pp. 67-86). Baltimore, MD: Paul H. Brookes.

Pretti-Frontczak, K., & Bricker, D. (2000). Enhancing the quality of individualized education plan (IEP) goals and objectives. *Journal of Early Intervention, 23*, 92-105.

Rainforth, B., & England, J. (1997). Collaboration for inclusion. *Education and Treatment of Children, 20*, 85-104.

Rebeiro, K. (2001). Enabling occupation: The importance of an affirming environment. *Can J Occup Ther, 68*, 80-89.

Rourke, J. (1996). Roles for school-based occupational therapists: Past, present and future. *Am J Occup Ther, 50*, 698-700.

Rueda, R., Gallego, M., & Moll, L. (2000). The least restrictive environment: A place or a context? *Remedial and Special Education, 21*, 70-78.

Sailor, W., Gee, K., & Karasoff, P. (2000). Inclusion and school restructuring. In M. Snell & F. Brown (Eds.), *Instruction of students with severe disabilities* (5th ed., pp. 1-29). Upper Saddle River, NJ: Prentice-Hall.

Schwartz, I., Sandall, S., Garfinkle, A., & Bauer, J. (1998). Outcomes for children with autism: Three case studies. *Topics in Early Childhood Special Education, 18*, 132-143.

Scott, S., McWilliam, R., & Mayhew, L. (1999). Integrating therapies into the classroom. *Young Exceptional Children, 2*, 15-24.

Slate, J., & Jones, C. (2000). Assessment issues in special education. In M. Winzer & K. Mazurek (Eds.), *Special education in the 21st century: Issues of inclusion and reform* (pp. 68-83). Washington DC: Gallaudet University.

Strong, S., Rigby, P., Stewart, D., Law, M., Letts, L., & Cooper, B. (1999). Application of the person-environment-occupation model: A practical tool. *Can J Occup Ther, 66*, 122-133.

Szabo, J. (2000). Maddie's story: Inclusion through physical and occupational therapy. *Teaching Exceptional Children, 33*, 12-18.

World Health Organization. (WHO)(2001). *International classification of functioning, disability and health*. Geneva: Author.

Peer Mentorship as an Environmental Support for Adolescents and Young Adults With Disabilities

Debra Stewart, MSc, OT Reg. (Ont.)

CHAPTER OBJECTIVES

* To describe peer mentorship as an environmental support.
* To outline the key elements of a peer mentorship program for youth with physical disabilities.
* To explore the role of the occupational therapist in using the social environment to enable occupational performance of youth with physical disabilities during adolescence and the transition to adulthood.

INTRODUCTION

This chapter describes an innovative program of peer mentorship that was developed and implemented by occupational therapists who worked with a group of adolescents and young adults with physical disabilities in rehabilitation services in Ontario, Canada. The peer mentorship program emerged from a research study about the experiences and needs of adolescents as they prepared for the transition to adulthood.

BACKGROUND INFORMATION

Occupational Therapy and the Transition to Adulthood

Occupational therapists working in pediatrics have traditionally worked with young people with disabilities from birth until their discharge to adult services. The age of discharge from pediatric services varies from 18 to 21 years. Most occupational therapists work within the rehabilitation and medical care system, which separates pediatric services from adult services. This results in a forced transition from one service system to another for young people with disabilities when they reach the age of discharge from pediatric services. A key role of the occupational therapist working in the pediatric service system is to prepare an adolescent and his or her family for this transition (Spencer, 2001). The amount of literature on this topic has increased in the past decade, but occupational therapists and other health care professionals are still struggling to plan and implement services that meet the needs of this population (Stewart, Law, Rosenbaum, & Willms, 2001).

PEER MENTORSHIP

Peer support and mentorship are environmental strategies that can enable occupational performance for people of all ages and situations. *Mentorship* is described as a relationship, usually between two people. Literature on mentorship describes it as a partnership between a "mentor" or a "mentoring volunteer" and a "mentee" or a "mentoring partner" (deRosenroll, Saunders, & Carr, 1993). Modern mentoring is characterized as an equal or horizontal relationship (deRosenroll et al., 1993) and the essence of the relationship is one of sharing, guiding, and supporting (Darling, 1985; Dodgson, 1986). Traditionally, mentorship programs involve a volunteer forming a relationship with a younger, "at-risk" member of the community. However, mentoring has expanded to become an effective method for orientation, career planning, problem-solving, coaching, and peer support (Carr & deRosenroll, 1999). Mentorship programs now exist in businesses, schools and universities, and recreation and community agencies.

The amount of literature about mentorship for different populations and communities has grown in the past decade. Although there is relatively little written about mentorship programs for young people with disabilities, evidence is mounting about the benefits of such programs. Programs described in the literature cover all age ranges and types of disabilities, including preschool children (Balenzano, Agte, McLaughlin, & Howard, 1993; Goldstein, Kaczmarek, Pennington, & Schafer et al., 1992), school-aged children (Cole, Vandercook, & Rynders, 1988), adolescents (Haring, Breen, Pitts-Conway, Lee, & Gaylord-Ross, 1987; Powers, Sowers, & Stevens, 1995; Roberts & Cotton, 1994), young adults (Anderson & Reardon, 1981; Cullen & Barlow, 1998; Orzek, 1984; Trach & Mayhall, 1997), parents of children with disabilities (Searcy, 1995), and older persons (Shipman, 1991). The evaluations of most of these programs demonstrate benefits for the "mentees" (the people receiving mentorship), such as increased social interaction and play, improved communication skills, improved academic performance, additional workplace supports, positive changes in behaviors, improved self-efficacy, and reduced stress. There is less evidence about the benefits of a mentorship relationship for the mentors themselves. One study found that mentors viewed themselves as a vital source of support following training (Cullen & Barlow, 1998).

Some of the benefits that have been identified for mentors in other studies include a sense of satisfaction and sense of community cooperation (Hals-Eisbrenner & Shipman, 1997), development of new interpersonal skills, and a greater understanding of disabilities (Saxton, 1983) and increased self-efficacy (Rhodes, 1994).

OCCUPATIONAL THERAPY AND MENTORSHIP

The influence of family-centred services in the 1990's resulted in more attention being paid to what clients and families identified as important to them (Dunst, Trivette & Deal, 1988; Humphrey & Case-Smith, 2001; Rosenbaum, King, Law, King, & Evans, 1998). Health care professionals learned, by listening to families and young people, about the activities that were important in their lives and how service providers could best help them to achieve their goals. The environment was identified by children, adolescents, and their families as an influence on their participation in daily occupations (Law et al., 1999). They recognized that as they got older, their environments and occupations changed. Young people with disabilities identified environmental barriers, including physical, social, cultural, and institutional barriers that limit their participation in vocational, recreational, and social activities, especially as they prepare to make the transition into the adult world (Stewart, Law, Rosenbaum, & Willms, 2001).

As client-centred and family-centred services became a reality, the service approach of occupational therapy in pediatrics shifted from direct, hands-on services to include compensatory and environmental approaches (Law, Missiuna, Pollock, & Stewart, 2001). Occupational therapists are working in schools, homes, and communities to improve physical accessibility and to educate teachers, coaches, caregivers, and other people about the abilities and goals of young people with disabilities. The profession of occupational therapy is also paying more attention to a young person's occupational performance and participation in daily activities, rather than performance components alone (Case-Smith, 2001). This is resulting in greater emphasis being placed on vocational readiness, life skills, community management, and the all-important adolescent need for peer relationships and socialization (Healy & Rigby, 1999). Research by occupational therapists and other rehabilitation professionals is supporting this focus, as researchers are learning from adolescents and young adults that the environment is an important influence on daily occupation, social participation, and ultimately, their quality of life (King & Cathers, 1996; Stewart et al., 2001).

Occupational therapy research about mentorship with young people with disabilities has not been published to date. Traditionally, occupational therapists have developed mentorship relationships within clinical and academic settings to facilitate the growth and development of clinicians, academic faculty, and students (Nolinske, 1995; Schemm & Bross, 1995). A recent publication on mentorship and supervision within the profession of occupational therapy describes mentoring as a process that is characterized by reciprocity, sharing, and a relationship that is beneficial to both members of a partnership (Baptiste, 2001). It is a flexible concept "that can be adapted to the needs of the parties involved" (p. 12). This concept of mentoring appears to fit well with current occupational therapy practice, which strives to assist young people with disabilities to meet their self-identified needs for peer support and socialization as they make the transition from adolescence to adulthood. It also fits well with current health promotion approaches, which focus efforts on the environments around the person such as developing networks and collaborative planning (Letts, Fraser, Finlayson, & Walls, 1993).

A Peer Mentorship Program
for Youth With Physical Disabilities

The need for peer mentorship and support was identified by a group of young people with physical disabilities who were part of a research project at a children's rehabilitation centre (Stewart & Sahagian Whalen, 1995) in Ontario, Canada. The research project explored the experiences, perceptions, and needs of adolescent clients as they made the transition from adolescence into adulthood. For several years, therapists had been hearing stories from clients who were discharged from pediatric services about the numerous barriers they were experiencing in the adult world, including the lack of employment opportunities (Pollock & Stewart, 1990).

In order to address this growing concern using a client-centred approach, two occupational therapists requested and received funding to study the experiences and needs for service and support of young people with physical disabilities, from their perspective. One summer, we invited a group of young adults ages 19 to 29 years old who were clients and former clients, to participate in some focus group sessions. The participants were informed that the purpose of these sessions was to hear their stories and learn from them, in order to plan services that would address the needs of adolescents with physical disabilities as they prepared for and experienced the transition through adolescence and into adulthood. This group of seven young adults met four times over the summer. During these focus group sessions, they told us that services needed to include all occupations of preparing for and living through the transition from adolescence into adulthood, including moving away from the family home, socialization and peer relationships, recreation, education, and work. The participants focused on their ideas for services to support youth with physical disabilities through adolescence and the transition into the adult world. One of their ideas was peer mentorship. They recognized the potential power of young people supporting each other during periods of transition. They all wished that they had known a mentor to help them through the difficult times, and they felt that they had much to offer young people with their experiential wisdom. They were all willing to continue as volunteers in an advisory capacity to develop a peer mentorship program. Their ideas for peer mentorship fit well with current occupational therapy service models and beliefs, and we requested facility support for two therapists to participate in the development of a program.

The development and implementation of a peer mentorship program was thus planned by this small group of young adults with physical disabilities, and facilitated by the occupational therapists. The group received special start-up funding that enabled them to hire a coordinator. The first year focused on learning more about peer mentorship and training the members of the advisory group to be the first mentors. A 2-week training program was run in the summer months, and the occupational therapists were the trainers. The key elements of the training program were:

* Defining peer mentorship for our program
* Communication skills: interviewing, active listening, feedback, etc.
* Dealing with the expression of feelings
* Problem-solving
* Conflict management
* Self-evaluation
* Developing guidelines for mentors

At the end of the training session, the advisory group developed a "marketing plan" to inform the younger clients at the children's rehabilitation centre and their families about the new mentorship program. One member of the advisory group became the first coordinator of the program on a part-time basis, which was her first paid job. The occupational therapists acted as advisors to the coordinator, and assisted her in developing referral forms and procedures for the program. As referrals came in, the coordinator matched mentors and mentees together. Monthly peer mentorship meetings were held to support each other as the program became a reality and to continue the training. The peer mentors themselves needed a great deal of support from each other and the occupational therapists in the beginning, as their roles were changing from being the recipients of services to being part of the social environment (providing services) for younger clients. The roles of the occupational therapists also changed from providing "therapy services" to acting as facilitators and trainers.

At the end of the first year of operation, an evaluation of the mentorship program was completed by a student occupational therapist, as part of a research project for her final year of study. Qualitative methods were used to interview some of the mentors, mentees, and the parents of mentees, to identify benefits and disadvantages of the program, and suggestions for improvement (Carr & Stewart, 1998). Three main themes emerged from a qualitative analysis of the interview transcripts:

1. *Been there, done that*—The participants placed great value on the personal experience and on the knowledge gained about oneself through the mentorship experience. Parents and "mentees" found that the mentor could offer insight into an issue as a result of having experience with it, for example, using the community transit system for people with disabilities.

2. *Not being alone*—All participants referred to the feeling of "being alone" at times, which changed through the mentorship experience. They expressed appreciation for being with others, and feeling supported by people who were like them and were experiencing similar challenges and experiences. Participating in positive social relationships was clearly an important issue for these young people.

3. *Exploring and risk-taking*—All participants, mentors, mentees, and their parents who were involved in the program experienced an increased sense of exploring and risk taking. Inner exploration was described in terms of learning more about one's abilities, and outward exploration included increased opportunities to get out in the community. Exploration led to an increased desire to take risks—physically, emotionally, and socially—and to try new things.

The results of this evaluation were presented to the advisory group, and they used the information to make changes and plan the next year's mentorship program. Furthermore, the occupational therapists sought feedback from the mentors regarding their role in the program. The majority of the mentors felt that the primary functions of the therapists were listening and guiding. One mentor used the phrase "a safety net" to describe the therapists' role. As time went on, and the mentors developed their skills, the occupational therapists were able to take a less active role in the formal training and supervision of the program, and did most of our work "behind the scenes." This gave the message to everyone that the mentors were in charge of this program.

Case Study: Daniel

Daniel was 13 years old when his mother requested a mentor through the Peer Mentorship Program. Daniel was in grade seven, and doing well academically. He had a diagnosis of cerebral palsy, and used a power wheelchair efficiently for mobility. He used to be quite active in Boy Scouts, but quit this year. His mother was concerned that he was spending a lot of time sitting on the couch and watching television. She wanted him to get out into the community more. Daniel was matched up with Tim, who was 20 years old and living on his own in an accessible apartment. Tim had a diagnosis of cerebral palsy and also used a power wheelchair for mobility. He was attending community college in a business administration course. When Tim arrived at Daniel's house the first time via specialized wheelchair transportation, both Daniel and his mother were impressed with how well Tim got around by himself. Tim told Daniel about his current activities and plans for the future. He helped him apply to the specialized wheelchair transportation system, and they began to go out to the local shopping mall and for lunch. Later on, Daniel came by himself to Tim's apartment to see what it was like. Tim spent a lot of time supporting Daniel as he explored different types of recreational and social activities in the community that he was interested in. During the evaluation of the mentorship program, Daniel's mother was interviewed and she talked about how Tim helped Daniel and his parents "see the possibilities" for the future in a positive way. She stated: "…[at the beginning] he would just sit on the sofa and not say anything. It was bothering me because he wasn't ever like that before…[and after] he says 'Mummy, can I do this now?'… I said 'yeah. There's no reason why not. If you want to you can do it.' I cannot explain. He was totally different."

Conclusion

The process of developing and implementing a peer mentorship program to support youth with disabilities through adolescence and into adulthood is an example of how occupational therapists can use the social environment to enable occupational performance. We focused our program on the powerful support that peers can provide each other. This program was innovative, as few programs have included youth with disabilities as mentors. Furthermore, most occupational therapists who work with young people with disabilities still tend to focus their efforts on the client (e.g., helping the young person to develop specific skills. Occupational therapy services have often included group sessions, but the focus is usually on the social (or other) skill development of the clients. The peer mentorship program shifted our focus to using environmental supports to promote our clients' participation in daily occupations.

By adopting a client-centred approach, two occupational therapists were inspired and guided by a group of young adults with physical disabilities to develop and implement this peer mentorship program. The success of the program was due to the participatory nature of the initial study and the ongoing planning of the program. This approach proved to be a great challenge to the occupational therapists, as we had to often sit quietly in group planning meetings, and let the mentors problem-solve and make decisions for themselves. This meant that we had to let them make mistakes and learn from their mistakes, and this was appreciated by the mentors. The importance of letting young people with disabilities learn from their mistakes was a message that the young people in the initial research study gave to us.

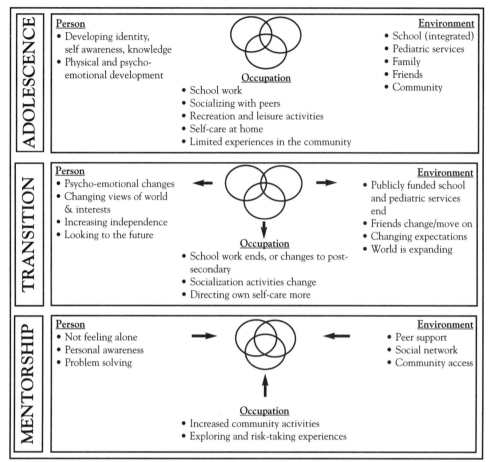

Figure 12-1. Application of the PEO model to a mentorship program.

The peer mentorship program was evidence based, as we referred to the literature about mentorship and the transition to adulthood during the study and development phases. The advisory group found this information very helpful in planning their program. We presented key points from different sources and the advisory group analyzed the information within the context of their program for youth with disabilities. Furthermore, our evaluation at the end of the first year of the program provided evidence to the advisory group and to the funders of the program that it was working. The qualitative approach to evaluation was useful in identifying strengths and weaknesses of the program, and to better understand the experience of peer mentorship.

This approach fits well with current occupational practice and models, which recognize the mutual influence of person and environment on occupational performance. The person-environment-occupation (PEO) model (Law et al., 1996) can be used by occupational therapists who are working with adolescents and young adults with disabilities to see how a change in one element can influence a young person's occupational performance overall. Figure 12-1 illustrates how peer mentorship can increase the fit between person and environment over time.

A mentoring relationship, as an environmental strategy, provides young people with the social support they need to feel good about themselves, to explore, and take some risks with new occupations in their community. For some adolescents, such as Daniel in the case study, a lunch at McDonald's with their peer mentor was the first time they went somewhere in the community without their parents. These experiences helped the young people and their parents begin to see future possibilities for participation in community activities.

A peer mentorship approach can be generalized to other settings and other populations, as a valuable way to provide social support. For example, peer mentorship programs can be used in schools, work settings, and community settings to help a young person develop confidence and abilities for future occupations. Peer mentorship fits well with current views about health promotion and prevention, which acknowledge the critical role of social support and other environmental influences on the quality of life and well-being of all people (Berkman, 1995; Patrick, 1997; Renwick, Brown, & Raphael, 1996).

Occupational therapists can make an important contribution in the area of peer mentorship and health promotion, as they can identify person, environment, and occupation elements that are important for a person's growth and development. For example, therapists can assist a client to identify those occupations that he or she will need to, want to or be expected to do in the future. As the client develops the ability to set goals for him or herself, the occupational therapist and client together can explore different environmental supports that are available and can link peers with each other for ongoing support and mentorship. Occupational therapists can also apply their knowledge of, and experience with, the teaching-learning process, human development and the value of activities and participation to the training of peer mentors. They can broaden the scope of traditional groups for young people to include social support and mentorship elements, which can have an impact on their quality of life and well-being. As one mentor stated during the evaluation at the end of the first year of the program, "You have helped us be the best we can be." This fits well with our current beliefs about occupational therapy assisting clients in "doing, being, and becoming" (Wilcock, 1998, p. 248).

The development of a peer mentorship program for young people with disabilities required the role of the occupational therapist to shift from providing therapy to facilitating support. We acted as facilitators to enhance the mentorship relationship, and we provided ongoing support, when requested, to both mentors and mentees. This program demonstrates how occupational therapists can use the power of the environment to facilitate a strong PEO fit as young people with disabilities make the journey through adolescence and into adulthood.

ACKNOWLEDGMENTS

The author would like to thank Sandra Sahagian Whalen for her invaluable contribution to the peer mentorship program; and Jennifer Carr, who agreed to evaluate the mentorship program for her research project in her final year of study. Also, the dedication and efforts of the members of the advisory group who made the initial peer mentorship program a success is acknowledged.

Study Questions

1. What is the value of peer mentorship as a social support for young people with disabilities?

2. How does the role of the occupational therapist change in planning and developing peer support programs for young people?

3. How could an occupational therapist use a peer mentorship program with other groups of people with different types of disabilities?

REFERENCES

Anderson, J. A., & Reardon, R. C. (1981). Self-directed career planning for persons with disabilities. *Journal of Employment Counseling, 18,* 73-81.

Balenzano, S., Agte, L. J., McLaughlin, T. F., & Howard, V. F. (1993). Training tutoring skills with preschool children with disabilities in a classroom setting. *Child and Family Behavior Therapy, 15,* 1-36.

Baptiste, S. (2001). *Mentoring and supervision: Creating relationships for fostering professional development.* Ottawa, ON: CAOT Publications ACE.

Berkman, L. F. (1995). The role of social relations in health promotion. *Psychosomatic Medicine, 57,* 245-254.

Carr, J., & Stewart, D. (1998). *Evaluation of a mentorship program for youth with physical disabilities: An exploratory study using qualitative methods.* Toronto, ON: Unpublished manuscript.

Carr, R., & deRosenroll, D. (Eds.). (1999, Winter). *Compass: A magazine for peer assistance, mentorship and coaching.* Victoria, BC: Peer Resources.

Case-Smith, J. (2001). An overview of occupational therapy for children. In J. Case-Smith (Ed.), *Occupational therapy for children* (4th ed., pp. 2-20). St. Louis, MO: Mosby.

Cole, D. A., Vandercook, T., & Rynders, J. (1988). Comparison of two peer interaction programs: Children with and without severe disabilities. *American Educational Research Journal, 25,* 415-440.

Cullen, L. A., & Barlow, J. H. (1998). Mentoring in the context of a training program for young unemployed adults with physical disability. *Int J Rehabil Res, 21,* 389-391.

Darling, L. A. W. (1985). "Mentors" and "Mentoring". *Journal of Nursing Administration, 15,* 42 – 43.

deRosenroll, D., Saunders, G., & Carr, R. A. (1993). *The Canadian stay-in-school mentor strategy.* Victoria, BC: Peer Systems Consulting Group.

Dodgson, J. (1986). Do women in education need mentors? *Education Canada, 26,* 28-33.

Dunst, C., Trivette, C., & Deal, A. (1988). *Enabling and empowering families: Principles and guidelines for practice.* Cambridge MA: Brookline Books.

Goldstein, H., Kaczmarek, L, Pennington, R., & Shafer, K. (1992). Peer-mediated intervention: Attending to, commenting on, and acknowledging the behavior of preschoolers with autism. *J Appl Behav Anal, 25,* 289-305.

Hals-Eisbrenner, C., & Shipman, M. I. (1997). *The Ontario mentoring experience.* Toronto, ON: United Generations Ontario.

Haring, T. S., Breen, C., Pitts-Conway, V., Lee, M., & Gaylord-Ross, R. (1987). Adolescent and peer tutoring and special friend experiences. *Journal of the Association for Persons with Severe Handicaps, 12,* 280-286.

Healy, H., & Rigby, P. (1999). Promoting independence for teens and young adults with physical disabilities. *Can J Occup Ther, 66,* 240-249.

Humphrey, R., & Case-Smith, J. (2001). Working with families. In J. Case-Smith (Ed.), *Occupational therapy for children* (4th ed., pp. 95-135). St. Louis, MO: Mosby.

King, G., & Cathers, T. (1996). What adolescents with disabilities want in life: Implications for service delivery. *Keeping Current, 96-2.* Hamilton, ON: McMaster University, Neurodevelopmental Clinical Research Unit.

Law, M., Cooper, B., Strong, S., Stewart, D., Rigby, P., & Letts, L. (1996). The person-environment-occupation model: A transactive approach to occupational performance. *Can J Occup Ther, 63,* 186-192.

Law, M., Missiuna, C., Pollock, N., & Stewart, D. (2001). Foundations for occupational therapy practice with children. In J. Case-Smith (Ed.), *Occupational therapy for children* (4th ed., pp. 39-70). St. Louis, MO: Mosby.

Law, M., Haight, M., Milroy, B., Willms, D., Stewart, D., & Rosenbaum, P. (1999). Environmental factors affecting the occupations of children with physical disabilities. *Journal of Occupational Science, 6,* 102-110.

Letts, L., Fraser, B., Finlayson, M., & Walls, J. (1993). *For the health of it! Occupational therapy within a health promotion framework.* Ottawa, ON: CAOT Publications ACE.

Nolinske, T. (1995). Multiple mentoring relationships facilitate learning during fieldwork. *Am J Occup Ther, 49,* 39-43.

Orzek, A. M. (1984). Special needs of the learning disabled college student: Implications for interventions through peer support groups. *Personnel and Guidance Journal, 62,* 404-407.

Patrick, D. L. (1997). Rethinking prevention for people with disabilities part 1: A conceptual model for promoting health. *American Journal of Health Promotion, 11,* 257-260.

Pollock, N., & Stewart, D. (1990). A survey of activity patterns and vocational readiness of young adults with physical disabilities. *Canadian Journal of Rehabilitation, 4,* 17-26.

Powers, L. E., Sowers, J., & Stevens, T. (1995). An exploratory randomized study on the impact of mentoring on the self-efficacy and community-based knowledge of adolescents with severe physical challenges. *Journal of Rehabilitation, 61,* 33-41.

Renwick, R., Brown, I., & Raphael, D. (1996). *Quality of life: Linking a conceptual approach to service provision.* Toronto, ON: Quality of Life Research Unit, University of Toronto.

Rhodes, J. E. (1994). Older and wiser: Mentoring relationships in childhood and adolescence. *Journal of Primary Prevention, 14,* 187-195.

Roberts, A., & Cotton, L. (1994). Note on assessing a mentor program. *Psychol Rep, 75,* 1369-1370.

Rosenbaum, P., King, S., Law, M., King, G., & Evans, J. (1998). Family-centered service: A conceptual framework and research review. *Phys Occup Ther Pediatr, 18,* 1-20.

Saxton, M. (1983). Peer counselling. In N. Crewe & I. K. Zola (Eds.), *Independent living for physical disabled people* (pp. 171-186). San Francisco, CA: Jossey-Bass.

Schemm, R. L., & Bross, T. (1995). Mentorship experiences in a group of occupational therapy leaders. *Am J Occup Ther, 49,* 32-37.

Searcy, S. (1995). Mentoring new leadership roles for parents of children with disabilities. *Remedial and Special Education, 16,* 307-14.

Shipman, M. (1991). *The elder mentors project. A descriptive/evaluative report.* Toronto, ON: Ryerson Polytechnical Institute.

Spencer, K. (2001). Transition services: From school to adult life. In J. Case-Smith (Ed.), *Occupational therapy for children* (4th ed., pp. 878-894). St. Louis, MO: Mosby.

Stewart, D., Law, M., Rosenbaum, P., & Willms, D. (2001). A qualitative study of the transition to adulthood for youth with disabilities. *Phys Occup Ther Pediatr, 21,* 3-22.

Stewart, D., & Sahagian Whalen, S. (1995). *A pilot study using qualitative research methods to plan client-centered vocational readiness services for adolescents and young adults with physical disabilities.* Unpublished manuscript.

Trach, J. S., & Mayhall, C. D. (1997). Analysis of the types of natural supports utilized during job placement and development. *Journal of Rehabilitation, 63,* 43-48.

Wilcock, A. (1998). Reflections of doing, being and becoming. *Can J Occup Ther, 65,* 248-256.

An Environmental Approach to Evaluation and Treatment in an Upper Extremity Injury Clinic

Carol Anderson, MA, OTR, CHT; and Jean Spencer, PhD, OTR, FAOTA

CHAPTER OBJECTIVES

* To illustrate how the process of rehabilitation following upper-extremity injury can be developed based on social participation of the client in local environmental contexts.

* To describe through application of the community adaptive planning assessment the process of client-centred evaluation, treatment planning, and goal-setting as well as outcomes evaluation with clients following upper-extremity injury.

* To provide a systematic means for documentation of those areas that are normally elusive to traditional component-based upper-extremity evaluations.

* To encourage readers to consider a client-centred approach incorporating evaluation of environments when working with individuals with upper-extremity injuries.

PROBLEM AND NEED

Occupational therapists working within the specialty area of upper-extremity injury, better known as "hand therapists," often become indistinguishable from their physical therapy counterparts due to a predominance of exercise-based and protocol-based treatment methods (Cooper & Evarts, 1998). "Hand therapy" in occupational and physical therapy has been guided since its inception by the biomedical model. Traditional occupational therapy evaluation and treatment with clients following upper-extremity injuries have been component-based, involving assessment and measurement of wounds, scar, edema, range of motion (ROM), dexterity, and strength. Client-centred goals and relevant occupationally based treatment plans have not been documented due to the perceived importance of the more obvious impairment of physical performance components. In addition, the design of traditional occupational therapy upper-extremity evaluations has not allowed for more holistic documentation of contextual and psychosocial factors (Dunn, Brown, & McGuigan, 1994; Foto, 1997; Toth-Fejel, Toth-Fejel, & Hedricks, 1998). Thus, objective measurements of physical performance components are documented over functional measurements, resulting in confusion for the clients, the therapists, and other third parties involved in the client's care in terms of the purpose and meaning of occupational therapy services.

Few occupational therapy assessments include the environment and/or social role expectations of clients (Pollock, 1993). In particular, assessments that elicit narrative data to identify the valued occupations of individuals are rare (Simmons, Crepeau, & White, 2000). This is obvious in occupational therapy evaluation of individuals following upper-extremity injuries, and it is true despite the fact that in the last 20 years, occupational therapy itself has undergone a resurgence. Occupational therapists have taken a renewed interest in client-centred therapy, as well as a renewed interest in psychosocial and contextual factors as important contributors to overall health.

Concomitantly, the World Health Organization (WHO) has revised its terminology in regards to disability. The old system of the International Classification of Impairments, Disabilities and Handicaps (ICIDH) is now structured around social perspectives rather than pathology. The new system of International Classification of Functioning, Disability and Health (ICF) (WHO, 2001) specifies several levels at which functioning and disability can occur: the body level, the person level, and the societal level (see Chapter 1). From the perspective of occupational therapy practice, upper extremity injuries affect bodily structure and function. However, the influence of the injury can occur at activity and participation levels, both of which are also influenced by environmental factors. Therefore, all of these components need to be addressed in practice.

In light of these changes in terminology by the WHO, the American Occupational Therapy Foundation (AOTF, 2000) has recently proposed specific overriding research questions for occupational therapy. These questions specifically address occupational therapy Interventions in relation to body structure/body function, activity, participation, and interaction with the environment. Moreover, questions of how to effectively measure and document outcomes in these areas have been proposed (AOTF, 2000). Thus, a client-centred approach to occupational therapy evaluation and treatment has recently taken on a greater significance in the overall health care system.

THEORY GUIDING ENVIRONMENTAL INTERVENTIONS

In thinking about the approaches that occupational therapists can use to incorporate considerations of the environment into their work with clients who have sustained injuries to their upper extremities, it is useful to consider guiding principles or theories that contribute to practice. Two models have guided the work that is described in this chapter: the model of human occupation (Kielhofner, 2002) and the occupational adaptation model (Spencer, Davidson, & White, 1996). When designing an effective occupational therapy treatment plan and related intervention, two factors are paramount for consideration: motivation and adaptation. Both of these concepts can be understood within the context of the environment.

Within the occupational therapy literature, motivation for engagement in occupations is best explained in the model of human occupation (Kielhofner, 2002). In this model, humans are viewed as dynamic systems interacting in both positive and negative ways with their environment. Specifically, the model suggests that humans are composed of several subsystems—volition, habituation, and performance, which work together in conjunction with the environment to produce occupational behavior. Motivation is interpreted within the volition subsystem. It is described as a process of willing choice and behavior based on human need to master and explore the environment. This view is in direct contrast to classic behavioral and psychoanalytic literature, which has presented human motivation as causal behavior resulting from innate urges or drives.

Motivation within the volition subsystem is further described in terms of personal causation, values, and interests. Specifically, *personal causation* is defined as "a collection of dispositions and self-knowledge concerning one's capacities for and efficacy in occupations" (Kielhofner, 1997, p. 190). In other words, as humans we realize our present abilities, those things we can and cannot do, based on our past experience. Control over these behaviors directly stems from our past successes. *Values* are "a coherent set of convictions that assign significance or standards to occupations, creating a strong disposition to perform accordingly" (Kielhofner, 1997, p. 190); that is, the emotions in terms of right or wrong ascribed to a particular occupation by an individual or group strongly influence future action. Finally, *interests* are defined as "dispositions to find pleasure and satisfaction in occupations, and the self-knowledge of our enjoyment of occupations" (Kielhofner, 1997, p. 191). Occupations that we enjoy and derive pleasure from in terms of the challenge, human interaction, and environmental setting are anticipated in the future as pleasurable, instilling in us a willingness to participate in those occupations again. Overall, these three factors together—personal causation, values, and interests—guide our motivational process for choosing and engaging in specific occupations.

Rehabilitation with clients who have acquired injuries involves the concept of adaptation as well as motivation. Spencer et al. (1996) illustrate adaptation in terms of an *adaptive repertoire* that is the use of past experience to help shape the future. They point out that the concept of adaptation is not only one of change, but it is also one of maintaining continuity in the lives of our clients.

The concept of adaptive repertoire is described within the context of the environment, the person, and the processes of change. First, occupations are viewed as inextricably tied to the environmental contexts in which they are performed. The term *local world* describes this relationship of occupations embedded within a particular context or location. Occupations may be understood through the environmental contexts in which they occur. Second, adaptive repertoires are behaviors that are learned over time through engagement in occupations. An important component of this learning process is the acquisition of a sense of meaning of occupations. This acquisition of the meaning of particular occupations impacts future understanding and decision-making regarding the engagement in occupations. Finally, adaptive repertoires are directly influenced by major life change. It is at times of major life change that people are faced with ambiguity and must re-evaluate their circumstances thus drawing on an acquired adaptive repertoire. This illustration of the use of an adaptive repertoire suggests that past experience enables one to adapt in ways that are within "one's ability of competent performance and appropriate within one's local world" (Spencer et al., 1996, p. 531).

A Tool for Collaborative
Planning of Environmental Intervention

In order to engage clients with upper extremity injuries in a client-centred occupational therapy approach that includes consideration of the environment, the community adaptive planning assessment (CAPA) (Spencer & Davidson, 1998) was implemented for purposes of evaluation and treatment plan design. Specifically, the CAPA is a client-centred evaluation tool that is grounded in the client's local world. It was originally developed to address the need for a future-planning tool within the field of occupational therapy. The need for future planning is not new and certainly not confined to the practice of occupational therapy. It has also been identified by the Commission on Accreditation of Rehabilitation Facilities (1992) and the Joint Commission on Accreditation of Healthcare Organizations (1992), as a requirement for discharge from skilled nursing and rehabilitation facilities. The underlying premise of the CAPA is that:

> "If occupational therapy is to sustain its grounding in helping clients return to mean-ingful occupation in daily community life as the fundamental goal of our profession, then methods are needed both to structure substantive future planning with clients and to document the nature of this planning in clinically practical ways that can also be used in studies of intervention outcomes" (Spencer & Davidson, 1998, p. 20).

According to Spencer and Davidson, review of the relevant occupational therapy, rehabilitation, and gerontology literature indicates that the following features are necessary for a future planning tool:
1. Grounding in the community life of the client.
2. Longitudinal orientation that connects future planning to past life experience.
3. A focus on occupational performance.
4. Consideration of performance contexts and their impact on client occupations.
5. Examination of the meaning of occupations to the person. The CAPA is an attempt to organize and address these features through gathering both qualitative and quantitative data.

Data obtained through use of the CAPA are organized on occupation cards, according to the specifics of each occupation; this includes activities within each occupation, time involved, participants involved, physical and cultural aspects of the environment, and value given the occupation by the participant. Each occupation card also includes "previous status," "expected changes," and "outcomes" for each occupation. Previous status documents the individual's engagement in the particular occupation prior to onset of disability or life change. "Expected changes" documents the individual's perceptions in terms of the losses and gains of the particular occupation. It also illustrates the individual's expectations of how the occupation may be performed in the future. Moreover, the "expected changes" section represents and documents the collaborative problem-solving process between client and therapist, which is the underlying purpose of the CAPA. Finally, the "outcomes" column documents the extent to which future plans and goals may or may not have been met based on follow-up evaluation.

In addition to the narrative data, the CAPA also gathers quantitative data. Specifically, client perception of occupation(s) is recorded on a 1 to 10 scale of satisfaction with time spent, current degree of participation, ability to negotiate or perform the occupation within the context of the personal environment, and importance of continued engagement in the occupation. The ratings are documented along with the narrative information on the occupation cards. Data from the CAPA are used to plan and implement occupational therapy interventions, which are designed to enable clients to meet their goals to perform desired occupations within their natural environments.

Client-centred evaluation tools which address occupational performance that have already been field-tested and used in occupational therapy include the occupational performance history interview (OPHI) (Kielhofner & Henry, 1988) and the Canadian occupational performance measure (COPM) (Law et al., 1998). The OPHI is a client-centred evaluation tool organized around specific interview questions addressing five content areas: 1) organization of daily living routines; 2) life roles; 3) interests, values, and goals; 4) perceptions of ability and responsibility; and 5) environmental influences. Scores are obtained for each content area. Separate scores are obtained for past and present. An overall summary of the content areas pertaining to the individual (content areas 1 to 4) provides a score that represents the individual's overall adaptation. In addition, the individual's life history pattern is recorded on a five-item ordinal scale in terms of adaptation or maladaptation. The goal of the OPHI is to provide an accurate and clinically useful picture of an individual's past and present occupational behavior (Kielhofner & Henry, 1988).

The COPM is another type of client-centred evaluation tool. It is based on the Canadian model of occupational performance (CAOT, 1997), which recognizes the individual as a fundamental part of the therapeutic process. The model itself describes individual occupational performance as a balance of three areas: self-care, productivity, and leisure. It also describes influencing factors including individual mental, physical, sociocultural, and spiritual characteristics. In order to capture these influencing factors on the three individual areas of performance, the COPM incorporates roles and role expectations within the client's own environment. It takes into account individual perception of the importance of a particular skill or activity. In addition, the COPM considers the client's satisfaction with present performance. The COPM as an evaluation tool is designed to establish occupational performance goals based on the client's perceptions of need. It is also designed to objectively document change in the defined problem areas via individual rating of activity importance, performance, and satisfaction on a 1 to 10 scale (Law et al., 1998).

The community adaptive planning assessment (CAPA) (Spencer & Davidson, 1998) is similar to the OPHI and COPM in that it is a client-centred evaluation tool. It too involves a semi-structured interview process in order to gain understanding of the individual's occupational behavior and performance. In comparing the CAPA to the OPHI and COPM, the CAPA appears to share more commonalities with the OPHI. Both the OPHI and the CAPA examine and document individual past history and then utilize this information when planning future goals. However, the OPHI examines past history and the environment in a broader context, looking mainly at overall adaptive or maladaptive behaviors. It does not examine specific occupations in relation to these variables. Moreover, the OPHI does not include environmental influences when assessing an overall score for adaptation.

The CAPA and the COPM both differ from the OPHI in that these tools provide a format for assessment of client perception and satisfaction with occupational roles and performance. Unlike the CAPA, the COPM does not examine occupations that the individual may find rewarding or support systems that may benefit the individual.

Overall, the advantage of the CAPA in comparison to the OPHI and COPM is that it provides an objective format for documenting individual past history and present occupational performance for establishment of future goals. In essence, the CAPA is a future-planning tool. It provides a format that goes beyond present evaluation of disability and goal structure based on current performance. Theoretical roots of the CAPA are grounded in the occupational adaptation frame of reference (Schkade & Schultz, 1992). This orientation recognizes that people seek to maintain continuity of the past and present in their daily lives. Following onset of disability, this continuity is disrupted. Moreover, the nature of individual interaction with the environment is changed, making it more difficult to visualize and set goals for the future.

One Client and His Story

The following is a case scenario that illustrates using the CAPA to plan and implement intervention using an approach that considers and uses the environment. The participant, Merrill, sustained an acute upper-extremity injury. Merrill's case example was taken from a larger set of case studies of six individuals with upper-extremity involvement, three with acute and three with chronic involvement. All six individuals were participants in an exploratory study involving use of the CAPA for the development of collaborative goals, intervention, and future planning based on individual environments and primary life occupations.

CASE SCENARIO: MERRILL

Merrill is a 60-year-old man who is right upper-extremity dominant. Originally from Texas, Merrill has lived his whole life in the same small community in which he was raised. For the last 18 years Merrill has been working full-time in the maintenance department of a hospital. Five weeks prior to his referral, while cutting an access hole for a large air-conditioning unit, Merrill sustained partial amputation of his left index, long, and ring fingers. Merrill's index and ring fingers were amputated just distal to the proximal interphalangeal joints (PIP), and his long finger was amputated proximal to the PIP joint. Both Merrill's thumb and small finger on the left hand remained intact.

When Merrill was first evaluated, he was a little over a month post injury. He had transferred into the clinic after receiving treatment for his injured hand by another occupational therapist at another clinic. His reason for transferring clinics was that he felt that he had not been receiving good treatment in the other clinic. When the other treating therapist was consulted regarding Merrill's case, she stated that Merrill was a difficult patient who did not want to work. In her words, he was basically a pain management problem. She reported that a referral had been made for Merrill to see a pain specialist. The conversation closed with her wishing "good luck" in getting Merrill to do anything.

Merrill did have a pain problem. In order to understand the dynamics of this problem and plan appropriate treatment, it was important to understand Merrill's associations with his environment (i.e., how he defined his pain within the context of his environment). Through administration of the CAPA, several things were learned about Merrill. First, he associated immediate physical pain with his injury and lack of motion (body structure/function). He associated emotional and psychological pain to his treatment or lack of treatment by others in recovery from his injury (social environment). He also associated emotional and psychological pain with his inability to interact with the environment in both work and everyday activities. Specifically, he was distressed due to his inability to play guitar (activity/participation).

Through qualitative interview (semi-structured format), Merrill identified playing guitar as one of his major occupations pre-injury. Merrill reported that he had been playing guitar since he was 9 or 10 years old, and it had significant positive associations for him with family. Throughout his youth he had played with his brother in a musical duo. More recently Merrill reported that he had been performing with a local band. Prior to his accident, he also had recorded and planned to copyright a song in which he played guitar and keyboards. In addition to his performance history, Merrill also reported that he enjoyed teaching others to play guitar, both professionals and nonprofessionals.

It became clear following the qualitative interview process of the CAPA that playing guitar served many purposes for Merrill within his environment. It was an activity that provided him with personal and social connections as well as economic benefit due to its part-time career status.

Because the CAPA format is client-centred and empowers the client to plan and foresee his or her abilities in the future, Merrill was encouraged to develop his own long-term goals. At the initial assessment, Merrill rated his ability to negotiate the environment through guitar playing at a 3 out of 10, compared to a rating of 10 pre-injury. Merrill's long-term goals included:

* Independence playing guitar left-handed (using his intact right hand for manipulating chords and left hand for strumming)
* Independence using his left hand to play an alternative instrument
* Modified independence using prosthetic fingers to play a musical instrument

Following review of Merrill's goals, the second, independence using his left hand to play an alternative instrument, appeared most attainable for Merrill early on in his rehabilitation. Merrill chose the slide guitar as an alternative instrument.

In order to implement an occupational therapy treatment plan conducive to Merrill's goal of learning to play an alternative instrument, the occupational therapy hand clinic was modified to fit Merrill's needs for intervention. First, Merrill was asked to bring his slide guitar to clinic. In addition, Merrill initiated bringing information into the clinic on the slide guitar and how to play it. He would read and highlight important points while seated at the table with his left hand warming up in heat prior to his actual practice sessions. Merrill was allowed use of a private room, one that would normally be used for a shoulder patient, for his actual practice sessions (Figure 13-1). The private room was helpful to Merrill because he could hear how he was playing without distraction of the larger clinic sounds. These modifications to the normal routine of the hand clinic had more than the obvious effect of teaching an adaptive technique on a musical instrument. First, it turned the decontextualized environment of the "hand clinic" into a more familiar environment for Merrill. Second, Merrill was no longer the client or patient, he was the teacher. He was showing and teaching the therapist how to play slide guitar drawing from his past experiences as a musician and a teacher of guitar music.

Initially, Merrill was unable to correctly hold the slide bar (the tool held in the left hand, which is pushed up and down the neck of the slide guitar to create the various chords that can be played on the instrument). Therefore, adaptive techniques were incorporated for purposes of manipulating the instrument.

Merrill's first re-evaluation with the CAPA was completed at week four. Merrill's perception of his ability to manipulate the guitar within his current environment had increased from 3 to 5. He had adjusted somewhat to the fact that even with prosthetics, he probably would not be able to manipulate his left hand on the neck of a guitar to play the chords necessary for accomplished music. His ability to re-engage in music via playing the slide guitar, however, motivated Merrill to consider playing other instruments including keyboards and piano. Merrill felt that he could still be affiliated with a band if he was proficient in playing

Figure 13-1. Merrill playing slide guitar in a private room of the clinic. (Printed with permission from Texas Woman's University.)

Figure 13-2. Merrill playing piano at friend's house. (Printed with permission from Texas Woman's University.)

one of these instruments. In terms of the original goals set, the goal of using the left hand to play an alternative instrument had been partially met.

Merrill was discharged from occupational therapy services a month later (two months after referral). Prior to his official discharge, a home visit was completed to assure his fit with his environment. A date was arranged with Merrill and a couple of his friends, with whom he had previously played both professionally and socially. These friends were also influential in Merrill's rehabilitative process through their social support via verbal encouragement and provision of a slide guitar for practice. The visit was arranged at their home. During the visit, Merrill demonstrated his ability to play the slide guitar correctly. At this point, Merrill was able to hold the slide bar in a technically correct position as well as maneuver his left hand correctly while playing the instrument. In addition, Merrill demonstrated his ability to play the piano as well as sing during the visit (Figure 13-2). This visit affirmed Merrill's connection with an important social environment.

Through use of the CAPA, the identification of meaningful occupation pre-injury and the grounding of that particular occupation into the rehabilitation process for purposes of recovery and maintenance of identity were accomplished. The original goal of gaining independence using the left hand to play an alternative instrument had been met. At the time of discharge, Merrill was able to play the slide guitar proficiently and correctly. Merrill's perception of his ability to negotiate slide guitar had increased at this time to an 8. Again, ratings for importance of the activity remained a 10. Finally, because the recovery process had been grounded in meaningful activity and individual environment, Merrill's continued participation in playing slide guitar allowed for a natural compliance with a continued home exercise program.

INCORPORATING ENVIRONMENT INTO
UPPER EXTREMITY OCCUPATIONAL THERAPY

Although more occupational therapists are becoming aware of the importance of incorporating environmental context into client-centred therapy, it continues to be difficult and somewhat foreign to therapists who specialize in rehabilitating clients with upper-extremity injuries. Often, clients may be encouraged to engage in activities that appear to be beneficial or important; however, individual meaning may be missing. Ultimately, the activity recommended may or may not be beneficial to individual recovery. This scenario demonstrates that it is beneficial for therapists and clients to devote some time during assessment to identify areas of occupational performance important to clients. This information can then be used throughout the intervention process, and can provide a rationale for using resources (such as private rooms) for purposes different than those originally intended.

Implementation of the CAPA, as illustrated through the case example of Merrill, demonstrates the implementation of evidence-based practice. According to Upshur (2001) the goal of evidence-based practice is to integrate research and practice. Holm (2002) describes evidence-based practice as an awareness of the multiple levels of existing evidence underlying a given intervention approach and careful evaluation of that evidence as it applies to a specific patient encounter. Through Merrill's rehabilitation we can see this process unfold.

The organizational scheme of the CAPA was useful in applying the concepts of motivation and adaptation as described in the model of human occupation (Kielhofner, 2002) and the occupational adaptation frame of reference (the concept of adaptive repertoire, [Spencer et al., 1996]) to the occupational therapy intervention. The quantitative portion of the CAPA documented Merrill's perception of his abilities to interact with his environment reflecting the premises of the model of human occupation—that is the composite of human subsystems working together in conjunction with the environment to produce occupational behavior. Specifically, in Merrill's case, the ability to document his perception of negotiability with the environment in chosen occupations, satisfaction with those occupations, in addition to previous status and expected changes, provided an objective understanding of current ability and his expectations for the future.

The qualitative portion of the CAPA integrated the concept of adaptive repertoire by providing insight into Merrill's past experience in order to help shape his future. Specifically, the narrative format provided an understanding of his past experience including the values he had attributed to occupations over time. Because of Merrill's past successes with music and because of the value he attributed to the activity, his ability to play an instrument again became a very motivating goal. In addition, Merrill's self-identification as a musician connected his past identity with a potential identity for the future.

Documentation of outcomes occurred at various levels of treatment. First, initial perceptions were recorded. Merrill's initial perceptions of his ability to negotiate playing the guitar in his immediate environment were low, revealing poor perception of his physical ability. Additional qualitative information revealed a low degree of personal satisfaction secondary to his pain and emotional state. Final recordings prior to Merrill's discharge from occupational therapy services demonstrated an increase in Merrill's perception of his ability to negotiate his environment and successfully play the slide guitar. Qualitative data appeared to relate to Merrill's perceived increase in ability by revealing a higher degree of personal satisfaction. This increase in satisfaction can be attributed to intervention focused on the person, the environment, and the occupation in which he engaged.

Central to evidence-based practice is client participation (Law, 2002). The client-centred format of the CAPA provided the necessary member checks to assure trustworthiness of the final documented outcomes or evidence of practice. The CAPA has proven effective in previous trials including acute care, rehabilitation, home health, and community support programs (Spencer, Hersch, Eschenfelder, Fournet, & Murray-Gerzik, 1999). Finally, clinical effectiveness of the CAPA has been examined through the use of clinical follow-up studies. Analyses of the outcomes of the CAPA-based intervention indicated that when used by experienced therapists, the CAPA format allowed for a future planning process that led to the establishment of community oriented goals that participants were able to implement roughly three-fourths of the time (Spencer & Davidson, 1998).

The challenge presented to occupational therapists working in the area of upper-extremity rehabilitation is understanding the various environmentally based evaluation and treatment planning tools, as well as incorporating them into practice. Often, evaluation and treatment of a client's physical deficits overrides what may be the more subjective goals of the client. Unfortunately, this can actually be counterproductive because the reality is that our hands are truly nonfunctional without an environment in which to interact. The CAPA is one environmentally-based evaluation and treatment planning tool that can be used with clients following upper extremity injury to understand the client's environment and bring to the forefront that which is most important to the client for functional return.

In conclusion, treatment of physical performance components with clients with hand or upper-extremity injuries is medically necessary and should not be ignored or dismissed. The key to therapeutic success, however, is understanding clients' environments first, pre-injury, then bringing to fruition the clients' goals for return to their environments, and ultimately providing clients with opportunities to re-engage in their environments.

Occupational therapists working as hand therapists or within the realm of upper-extremity rehabilitation have the unique opportunity to implement a client-centred evaluation tool that incorporates consideration of the environment. The affinity of our hands with our environment should be captured in rehabilitative efforts. The CAPA provides a blueprint, or road map, for documenting some of the complexities of our clients and establishing a basis for intervention.

Study Questions

1. Besides physical components (such as edema, ROM, strength, sensation), what should occupational therapists assess when working with clients who have sustained upper extremity injuries? Why is it important to assess these areas?

2. Name three ways in which results of the CAPA can be used to plan occupational therapy intervention.

3. You are an occupational therapist, certified as a hand therapist. You have just applied for a position in a newly forming clinic that will focus on treating clients who have sustained traumatic upper extremity injuries. The plastic surgeon who is leading the development of the clinic has asked you to describe the unique contribution occupational therapy can provide in such a clinic. How will you respond?

4. Julie is a 70-year-old right, upper extremity-dominant female. She sustained a distal radius fracture to her left upper extremity after a fall at home. Medical treatment of the fracture included a closed reduction and positioning of her left wrist and forearm in a cast. Julie's cast was removed at 6 weeks. At that time she demonstrated symptoms of chronic regional pain syndrome (CRPS) including excessive redness, swelling, pain and hypersensitivity through her hand, wrist, and forearm. At initial evaluation, Julie did not want to move her hand. She was observed to carry it as if it were not her own. Initial CAPA findings identified Julie's primary occupations as (1) volunteer, and (2) homemaker. Table 13-1 provides a completed occupation card for Julie based on her role as a volunteer. Taking into account Julie's local world, establish occupational therapy goals and plan intervention that would be meaningful to her as well as address the physical needs of her injury.

Table 13-1
Completed Occupation Card for Julie

	Previous Status	Expected Changes	Outcomes
PRIMARY OCCUPATION: VOLUNTEER Time–9/10 **PERSONS**	Part-time: Working 3 days per week doing volunteer archives work. Also, volunteer work at local high school and with church, homemaking, taking care of convalescent mother.	Losses: Inability to use left hand for functional activities including archives work, homemaking (particularly baking, and self-care); husband "doing everything."	Goals:
Participation– 9/10 **ENVIRONMENT**	Leisure—enjoys books, volunteering, collecting things, baking, socializing with family and friends.	Gains: Ability to return to archives work three times per week; Ability to care for self and home	
Negotiability– 2/10	Identity—"I'm a teacher," "Once a teacher, always a teacher," connecting current self-perception with known		
VALUE Importance–9/10	past identity.		

REFERENCES

American Occupational Therapy Foundation (AOTF). (2000). *Practice parameters and research priorities for occupational therapy developed through a grant from the RGK Foundation.* Retrieved October 14, 2000 from: http://www.aotf.org/html/practice_parameters.html.

Canadian Association of Occupational Therapists (CAOT). (1997). *Enabling occupation: An occupational therapy perspective.* Ottawa, ON: CAOT Publications ACE.

Commission on Accreditation of Rehabilitation Facilities. (1992). *The standards manual for organizations serving people with disabilities.* Tuscon, AZ: Author.

Cooper, C., & Evarts, J. (1998). Beyond the routine: Placing therapeutic occupation at the center of upper-extremity rehabilitation. *OT Practice, 3,* 18-22.

Dunn, W., Brown, C., & McGuigan, A. (1994). The ecology of human performance: A framework for considering the effect of context. *Am J Occup Ther, 48,* 595-607.

Foto, M. (1997, March). *Trends, tools, technology.* A presentation to the Los Angeles Occupational Therapy Leadership Forum. Los Angeles, CA: University of Southern California, University Hospital.

Holm, M. (2002). Our mandate for the new millenium: Evidence-based practice. *Am J Occup Ther, 54,* 575-584.

Joint Commission on Accreditation of Healthcare Organizations. (1992). *The 1993 accreditation manual for hospitals: Volume I, standards.* Chicago: Author.

Kielhofner, G. (1997). *Conceptual foundations of occupational therapy* (2nd ed., pp. 187-192). Philadelphia: F. A. Davis.

Kielhofner, G. (2002). *A model of human occupation: Theory and application* (3rd ed., pp. 13-27; 44-62). Baltimore: Lippincott Williams & Wilkins.

Kielhofner, G., & Henry, A. D. (1998). Development and investigation of the occupational performance history interview. *Am J Occup Ther, 42,* 489-498.

Law, M. (2002). Introduction to evidence-based practice. In M. Law (Ed.), *Evidence-based rehabilitation* (pp. 3-12). Thorofare, NJ: SLACK Incorporated.

Law, M., Baptiste, S., Carswell, A., McColl, M. A., Polatajko, H., & Pollock, N. (1998). *Canadian occupational performance measure* (3rd. ed.). Ottawa, ON: CAOT Publications ACE.

Pollock, N. (1993). Client-centered assessment. *Am J Occup Ther, 47,* 298-301.

Schkade, J. K., & Schultz, S., (1992). Occupational adaptation: Toward a holistic approach for contemporary practice, Part 1. *Am J Occup Ther, 46,* 829-837.

Simmons, D. C., Crepeau, E. B., & White, B. P. (2000). The predictive power of narrative data in occupational therapy evaluation. *Am J Occup Ther, 54,* 471-476.

Spencer, J., & Davidson, H. (1998). The community adaptive planning assessment: A clinical tool for documenting future planning with clients. *Am J Occup Ther, 52,* 19-30.

Spencer, J., Davidson, H., & White, V. (1996). Continuity and change: Past experience as adaptive repertoire in occupational adaptation *Am J Occup Ther, 50,* 526-534.

Spencer, J., Hersch, G., Eschenfelder, V., Fournet, J., & Murray-Gerzik, M. (1999). Outcomes of protocol-based and adaptation-based interventions for low-income elderly persons on a transitional unit. *Am J Occup Ther, 53,* 159-170.

Toth-Fejel, G. E., Toth-Fejel, G. F., & Hedricks, C. A. (1998). Occupation-centered practice in hand rehabilitation using the experience sampling method. *Am J Occup Ther, 52,* 381-385.

Upshur, R. E. G. (2001). The status of qualitative research as evidence. In J. Morse, J. Swanson, & A. Kuzel (Eds.), *The nature of qualitative evidence* (pp. 5-26). Thousand Oaks, CA: Sage.

World Health Organization (WHO). (2001). *International classification of function, disability and health.* Geneva: Author.

Home Modifications That Enable Occupational Performance

Susan Stark, PhD, OTR, FAOTA

CHAPTER OBJECTIVES

* Explain the environmental strategies that occupational therapists can use to improve occupational performance in the home.
* Discuss the theoretical underpinnings that support the role of occupational therapy in the home.
* Describe a clinical application that provides environmental support in the homes of individuals who have occupational performance dysfunction, based on models of person-environment-occupation (PEO) fit.
* Describe the role for occupational therapists interested in supporting occupational performance in the home setting.

INTRODUCTION

As described throughout this text, occupational performance is the result of complex interactions among person, environment, and occupation. The environment has been presented as a valuable intervention area for occupational therapists that hope to improve the occupational performance of their clients. The home environment is one of the most essential contexts within which this intervention can occur. It is within the home that individuals perform some of their most valued occupations. Home is the context of many occupations including activities of daily living (ADLs) and instrumental activities of daily living (IADLs) such as bathing, dressing, and preparing meals, leisure activities such as hobbies and productive pursuits such as managing a household or "working from home." The home is also the context of many activities related to valued roles, such as preparing and sharing holiday meals and caring for children. Home, as the setting of these important activities is frequently one of the most cherished environments in people's lives. Home has been described as a reflection of individual values (Marcus, Sarkissian, Wilson, & Perigut, 1986) and has been associated with personal identity (Dovey, 1985). The meaning of home becomes obvious to those who interact with individuals dealing with the onset of a new disability who often state that their primary rehabilitation goal is to return "home." When asked, older adults state that they prefer to live out their later years in their own home (American Association of Retired Persons, 2000). The goal of maintaining a home is often threatened by institutionalization due to the high number of barriers that are present in the homes of individuals who experience functional limitations (Trickey, Maltais, Gosselin, & Robitaille, 1993). Mismatches between the home environment and the physical capacity of individuals with disabilities pose real threats to maintaining independence. Common barriers in the home of individuals who have functional limitations include objects that are located out of reach, stairs, controls that are difficult to manage, and safety issues (Clemson, Roland, & Cumming, 1997; Mann, Hurren, Tomita, Bengali, & Steinfeld, 1994; Steinfeld & Shea, 1993).

An understanding of how to modify a home environment to enable clients to participate in their desired activities is an important skill for occupational therapists interested in helping their clients live at home safely and as independently as possible.

TRADITIONAL ROLES OF
OCCUPATIONAL THERAPISTS IN HOME MODIFICATIONS

Traditionally, occupational therapists have had the opportunity to implement home modifications in two different rehabilitation settings. One opportunity occurs if an individual is receiving services in an inpatient rehabilitation setting, the other if an individual is receiving home health care or home rehabilitation services. In the inpatient rehabilitation setting, therapists are typically encouraged to make a visit to a client's home prior to discharge from the hospital. The visit is carried out in order to ascertain the types of environmental barriers present in the home of an individual with a newly acquired disability. For example, a therapist could determine if a home would require the addition of a ramp or a bathroom modification. The home visit also allows the therapist to identify the type and size of durable medical equipment (e.g., tub bench) that would fit in the home to allow the client to perform ADLs independently. With recent cost containment measures by health care systems and demands for occupational therapists to increase productivity, the opportunity for occupational therapists to make home visits while a client is in an inpatient rehabilitation setting has decreased. Often, therapists rely on family or client reports for descriptions of homes in order to make recommendations or specify equipment. This practice often results in a poor fit between a person and his or her environment once they return home. One common problem in this circumstance occurs when the therapist makes equipment recommendations based on verbal descriptions of a space. This can result in equipment that does not fit in the home environment. The photo that is provided in Figure 14-1 illustrates this problem. The photo shows the different types of equipment that several successive therapists prescribed for this client who identified difficulty bathing as a problem. There are two types of tub benches, a grab bar, and a long-handled sponge in the photo. In this case the client received equip-

Figure 14-1. An example of equipment that was issued in a health care setting by a therapist who did not have the opportunity to see the actual layout of a bathroom. (Copyright © Polly Gray. Printed with permission.)

ment from rehabilitation providers who had not visited her home to observe the problems she was having while attempting to bathe. She was unable to get into the tub even with the help of the numerous devices that she had collected. She reported to the therapist that she was taking sponge baths at the time this photo was taken.

The other typical opportunity for home interventions occurs during home health or home rehabilitation services. Occupational therapists, as part of a home care team, see clients in their homes to improve their ability to perform self-care activities or IADLs. Generally, these therapists are able to observe the home environment and recommend equipment and modifications that are appropriate in the setting. The demands of this system limit the occupational therapist in different ways. With changing health care reimbursement, clients are often discharged from acute care hospitals very early and come home while still dealing with symptoms of their illness or injury. Therapists in the home rehabilitation system often do not have time to provide a comprehensive home evaluation and recommendations regarding removal of environmental barriers. Often, home care therapists find it difficult to provide an intervention that includes home modification beyond simple strategies such as providing the client with and training the client in the use of durable medical equipment.

The focus of these two therapy approaches has traditionally been the remediation of person factors (e.g., upper extremity strength) as a means to improve occupational performance. An approach that includes a systematic evaluation of environmental factors in the home and intervention strategies that include home modifications has not typically occurred in traditional health care systems due to the reasons indicated above.

The limitations of the current health system have resulted in a service gap. Because clients often do not receive support or advice from their health providers regarding how to remove the environmental barriers that exist in their homes, mismatch occurs between the person's capacity and the environment, which poses a threat to the ability of the clients to achieve their occupational goals.

ENABLING OCCUPATIONAL PERFORMANCE
THROUGH THE PROVISION OF HOME MODIFICATIONS

A lack of accessible, affordable housing for individuals with disabilities has long been identified as a primary concern by individuals with disabilities (Fawcett et al., 1988) and by professional service providers (Jones, Petty, Boles, & Mathews, 1986). Currently there are few private homes that have home modifications that support the performance of individuals who have disabilities. For example, fewer than ten percent of the 100 million existing homes in the United States contain modifications (Center for Universal Design, 1997). This is a startling figure when compared to the number of adults with chronic and disabling conditions, many of whom live in the community, suggesting that the restrictive living environments of individuals who have functional limitations could be contributing to occupational performance dysfunction.

There is evidence to suggest that limitations in the ability to perform tasks in the homes faced by these people occur as a result of a lack of environmental support (Connell, Sanford, Long, Archea, & Turner, 1993; Mann, et al., 1994). Although the home modifications literature is limited, current evidence suggests that modification of the home can enhance functional independence of older adults, improve occupational performance, improve safety, and reduce the stress placed on caregivers (Corcoran & Gitlin, 2001; Corcoran et al, 2002; Gitlin & Corcoran, 1993; Olsen, Ehrenkrantz, & Hutchings, 1993; Steinfeld & Shea, 1993). Further evidence demonstrates cost effectiveness of home modifications when compared to the costs of institutionalization and hospitalizations that result from injury (Mann, Ottenbacher, Fraas, Machiko, & Granger, 1999; Salkeld et al., 2000). The literature also points out that even when home barriers are identified, individuals do not take steps to having them removed or have a difficult time removing them (Cumming et al., 2001; Steinfeld & Shea, 1997).

Although the need for home modification services for people who experience functional limitations is clear and obvious, the outcome of improved performance after home modifications is still largely undocumented. This lack of evidence has allowed payers of health and human services to ignore the need. The shortage of accessible housing diminishes quality of life and ability to carry out daily activities, signaling to occupational therapists that there is a great need for interventions to reverse the current trend of medical professionals ignoring the problems caused by barriers in the home.

Current models of occupational therapy practice support maintaining occupational performance by providing environmental modifications. Strategies that improve safety and performance by removing barriers will decrease the likelihood of premature institutionalization (Mann et al., 1999). Assisting individuals with functional limitations to live their lives successfully in their homes by providing environmental modifications is an important aspect of the continuum of rehabilitation services.

It is often possible to compensate for functional limitations by modifying a person's home environment (Law et al., 1996). Occupational therapists can use their existing knowledge and skills, a PEO model of practice (Christiansen, 1991; Christiansen & Baum, 1997; Law et al, 1996), and a client-centred approach (Law, 1998) to achieve this goal.

A gap continues to exist for many consumers of traditional health care interventions. Current health systems do not allow occupational therapists to address the environmental barriers that are present in the homes of individuals who have disabilities. Ironically, the loss of function at home is not caused by the pathology itself, but is a result of other factors such as environmental barriers that affect performance. This excess disability can be prevented by the provision of home modifications.

A new clinical application framework based on PEO models was developed in response to the demands of community-dwelling individuals with disabilities who were having difficulty maintaining independence in the community due to the environmental barriers that they faced in their homes. In many cases the individuals were unable to leave their homes and had become virtually "trapped." These individuals reported disruptions in their occupational performance and requested help in achieving independence in their daily activities.

The Structure of the Clinical Application Framework

Occupational therapists can provide interventions to improve the occupational performance of people who have occupational problems caused by medical conditions or environmental barriers. The unique contribution of occupational therapy is to maximize the fit between what it is a person wants to do and his or her capability to do it (Christiansen & Baum, 1997). In order to accomplish this, individuals should be seen in their homes, the true context of their performance, to determine if the environment in which the person lives supports his or her occupational performance. If environmental barriers are present, the occupational therapist can employ a comprehensive client-centred approach to removing the barriers and build on environmental supports. Intervention strategies include caregiver education, architectural modification, equipment recommendations/training, and program referral. Interventions should be planned with consideration of the cultural, economic, institutional, political, and social context of the client and the family in light of the impairment and functional limitations associated with the acute or chronic condition.

This clinical application framework is different than traditional "home care or home rehabilitation." It is not a continuation of typical medical treatment with a focus on remediation of physical capacity; rather, the outcomes include enhanced ability to perform occupational roles and activities (such as bathing, preparing meals, etc.). The framework is not based on a medicalized definition of disability; the classification of disability as stated in the International Classification of Functioning, Disability and Health (ICF), published by the World Health Organization (WHO) (2001), is used. This conceptualization of disability appreciates the interaction between health conditions and contextual factors, and accounts for the personal as well as the environmental factors that influence the person's performance.

This framework is also organized around sustaining independence, community reintegration, and enhancement of well-being through client-centred interventions that improve the function of the individual within their community. This approach would be relevant for individuals who experience functional performance loss due to developmental disability, acquired disability, illness, or the process of aging. Table 14-1 describes the goals of such a program.

THE PROCESS

The PEO model provides the theoretical basis, or underlying concepts, for the assessment and for the intervention strategies of this clinical application framework. The process for this intervention includes the basic steps of traditional occupational therapy services

Table 14-1

Goals of a Clinical Application Framework That Support the Provision of Home Modifications in the Homes of Persons Who Have Disabilities

- Improve the occupational performance of persons with disabilities in their home.
- Improve the quality of life of individuals who have disabilities and their family members.
- Build family's and client's skills and knowledge of resources to support independent living.
- Decrease the risk of premature institutionalization.
- Ensure that environmental modifications address individual's and family's goals, skills, and needs to decrease the chance that the intervention will not be used.
- Train the person with a disability and the family to use equipment or environmental modifications safely within his or her everyday life.
- Increase the capacity of family members or caregivers to provide supportive care.
- Provide assessment and intervention in the every day context of the person's life.

(American Occupational Therapy Association, 2002; Fearing, Law, & Clark, 1997) with a shift in intervention strategies from person-centred to environment-centred approaches. As with traditional occupational therapy interventions, through the use of interview, observation and clinical assessment procedures, occupational therapists identify occupational performance problems including the functional performance deficits and environmental factors that are contributing to the occupational performance dysfunction. For the intervention to be relevant and meaningful, the assessment must be comprehensive.

After completion of the initial evaluation, the occupational therapist and client identify the client's goals for performance. A plan is developed that is based on interpretation of the assessment data (including the occupational factors, contextual factors, and personal factors), in addition to the goals of the clients/caregivers and the resources available.

Occupational therapists have the skills in occupational and task/activity analysis necessary to determine the types of environmental support that may be necessary to improve occupational performance. There are three major types of home modifications that are successful in improving the person-environment fit in the home and within the purview of occupational therapy practice. The three major types of home modifications are described below. They are also represented in Table 14-2.

Modifications can include changes of the physical environment (both spaces and objects within the environment), changes or modification of the activity and education about the environment and how to use it in a different way, or changing the social support in the environment. Examples of each type of modification will be described.

Physical Environment Changes

First, in the area of modification of the physical environment, there are several tools and techniques that can be used to improve person-environment fit. The existing space can be rearranged to accommodate the needs of the individual. For example moving most often

---

Table 14-2

Types of Environmental Modifications

Changes to the Physical Environment

- Modify the layout (remove a door to make the opening wider)
- Provide adaptive equipment (a tub bench)
- Architectural modifications (provide a ramp, bathroom modifications)

Modification of the Occupation

- Education about how to use the environment in a different way (always turn on lights before entering a room for an individual who has low vision)
- Use everyday items to achieve goals (use of a portable phone for safety)

Supports From People

- Caregiver education (proper transfer techniques, how to use a lift)
- Engage social services (home delivered meals)

---

used items within reach would be a modification that would fit in this category. Rearranging furniture to allow space for an individual's wheelchair is another example. A second strategy commonly used by occupational therapists is the provision of specialized adaptive equipment or durable medical equipment. Examples of these types of modifications can include providing a tub bench and handheld shower for someone who is unable to transfer into a bathtub to take a bath. A third example of change to the physical environment is the provision of architectural modifications. These can include simple modifications, such as adding rails to stairs or adding grab bars in the bathroom to assist in transfers to the toilet. More complicated architectural modifications can include the provision of ramps to enter the home or major remodeling of a home. More involved changes can include providing a roll-in shower where there was previously a bathtub, adding an elevator or providing an accessible addition that includes a bedroom and bathroom for a member of the family. Depending on their knowledge and experience, occupational therapists may request assistance in determining the best architectural modification for the client. In this case, the occupational therapist would bring an understanding of the occupational goal and the ability of the client and work together with a building professional to determine the best modification for the situation.

Modifying the Occupation

The next major category, modifying how one performs an activity, includes providing education about safety in the home and task modification. For example, instructing someone with low vision to turn on lights before going up and down stairs or reminding them to take their portable phone with them when they go to the basement to do laundry would fit in this category. Individuals are instructed to use their existing environments in ways that make it safer or easier for them to achieve their occupational performance goals. An additional strategy that can be used in this category is teaching individuals who have disabilities how to use everyday items to achieve their goals. For example, teaching an individual who has fine motor problems how to use a large button, universal remote control for the television may give him independence in pursuing his leisure interest of watching television.

Supports From People

The final type of modification strategy, supports or assistance from people, involves using personal assistance to compensate for functional loss. For example, an individual who is interested in eating, but is unable to independently prepare meals using a wheelchair because of the barriers in her kitchen, may benefit from home-delivered meals to overcome the barrier of not being able to access the kitchen appliances. A second strategy is providing family or friends with education regarding the type of help that an individual may need. Teaching a caregiver proper transfer techniques or the proper use of durable medical equipment are examples of this approach.

Usually, occupational therapists do not complete the actual home modifications, rather they work as a member of a team that may include the client, a contractor, and an architect. The role of the occupational therapist is to determine the types of modifications that are necessary to improve the occupational performance of the individual with a disability and their caregivers. They also work with the team to see that the modifications are completed in a manner that is consistent with the needs of the family. For example, the occupational therapist may determine the optimal placement of a grab bar in a bathroom, but a qualified contractor would install the grab bar. An occupational therapist may work with an architect to determine the space needs and layout of an accessible room for a client who recently acquired a spinal cord injury and uses a wheelchair. The role of the occupational therapist might include educating the architect about the types of spaces needed by the client, the types of fixtures that a client could independently use in the bathroom or the reach heights or space needs of a client that would assist in the development of a set of accessible plans.

The process of obtaining funding for modifications is often the largest hurdle for individuals with disabilities attempting to fund expensive modifications (Duncan, 1988). Occupational therapists may work in conjunction with a social worker or a case manager to help their clients obtain funding for home modifications. There are several methods to do so. In some cases clients refinance their home using traditional mortgages. In some cases grants may be obtained by local independent living centres, social service agencies, or even religious institutions. In some rare instances, insurance companies will fund some home modifications, though this is not the norm.

The decisions that occupational therapists make regarding which type of modification to use is dependent on several things. First, a comprehensive assessment of the capacity of the person, the environment, and the occupations of the individual must be completed. Second, the occupational performance goals of the client must be considered. There are several other aspects of the situation that must be clarified by the therapist prior to recommending modifications. These include the resources of the individual and family, the cultural values of the individual and family, the willingness of the individual to make changes in or to the home, and the level of independence that the individual wishes to achieve.

In determining the intervention type, it is imperative that the therapist act in a client-centred manner. There are often several methods to solve one problem. For example, the occupational performance problem of obtaining a meal for a person who has limited use of her left upper extremity can be solved by modification of the kitchen or the use of adaptive equipment, and task modification or the provision of a meal delivery service. The level of the intervention needs to be based on the wishes of the individual. If the person wishes to cook for herself alone in her home, a kitchen modification may be necessary, whereas if the person lives with a caregiver, she may be satisfied with adaptive equipment and "set-up" at the kitchen table. Determining how to implement the plan must be client-centred or the risk of abandoning expensive intervention strategies exists.

CASE SCENARIO: THE STORY OF MATT

Matt is 16 years old. He lives with his mother and father in a two-story house. Matt has cerebral palsy that results in dystonic movements leaving him with minimal function of one arm and moderate function of his right hand. He uses a power wheelchair to move from place to place. He transfers with the maximum assistance of his mother who lifts him on her own. Matt is very bright and loves to spend time with his family and friends. His mother has cared for Matt since he was an infant. She has performed his self-care activities and attended to all of his personal care needs including all transfers, toileting, dressing, and preparing his meals. Recently, his family circumstances have changed and Matt's mother will be returning to work, which means that Matt will be spending time alone in his home for the first time in his life. An occupational therapist was contacted to help Matt and his mother make some modifications to his home so he can be safe while she is away. The occupational therapist met with Matt and his family to identify the goals that they had for Matt in the home. Using a client-centred approach to understanding the occupational performance issues that were affected by home accessibility, the therapist used the Canadian occupational performance measure (COPM) (Law et al., 1998), a semi-structured interview procedure, to identify the occupational performance areas that Matt wanted to address in the home. Matt identified the need to be able to discreetly, with as much independence as possible, care for his bathing and toileting needs. Matt has also identified a concern that he has with mobility within his home. Currently, Matt is unable to move from the second floor of his home to the first floor. He is also unable to enter the house independently because of steps to enter the front door. Previously, Matt's parents carried him up the stairs to his bedroom and to enter the home. His entire family has stated a concern with Matt's safety getting into and out of the home and going from one level to another within the home.

Matt prioritized his goals as: 1) entering the home independently, 2) moving between first and second floor independently, and 3) bathing and toileting with least amount of personal assistance possible and maximum support from the physical environment. The occupational therapist then determined Matt's personal capacity, identifying both his strengths and limitations cause by his impairments by using standardized instruments. After the occupational therapist determined the client's occupational goals, and his capacity, she next observed how Matt and his family performed the activities of using the bathroom, getting in and out of the house and getting up and down the stairs to the second floor. The therapist then identified the environmental barriers that existed in the home. The findings indicated that there were several architectural barriers present. These included stairs to the entry door (Figure 14-2), and stairs within the house (Figure 14-3). The home also had an existing bathroom that did not have enough room to maneuver, had a tub Matt could not access without maximum assistance, had a mirror that Matt could not see, and had a sink/vanity area that Matt could not approach because of a lack of knee clearance (Figure 14-4).

The therapist then assessed the personal resources of the family, including financial support, tolerance for change of their home, and services that Matt was receiving. Matt's family had the ability to obtain the financial resources to make major renovations to their home and were willing to make physical changes to their home. Matt and his family were more interested in technology and environmental support than securing services, but they agreed to have a personal care attendant assist Matt during his self-care. The entire family made it clear that it was important that Matt have access to as much of the home as possible.

Once the therapist identified these barriers, she developed a plan for intervention that took into consideration the goals of Matt and his family, the resources the family had, the barriers that were present, the types of modifications that were possible in the existing home,

Figure 14-2. The entrance to the home; Matt's family carried him up to enter the home. (Copyright © Polly Gray. Printed with permission.)

Figure 14-3. The interior stairs that Matt's family had to address. (Copyright © Polly Gray. Printed with permission.)

Figure 14-4. The sink that Matt could not use independently. (Copyright © Susan Stark.)

and the local laws that guide home renovations. The occupational therapist provided the family with the written plan to remove the barriers, provided advice on how to choose a contractor and architect, as well as provided resources for seeking additional funding. The plan for renovation is provided in Table 14-3.

The family decided to fund most of the modifications by refinancing the home through a local bank and by requesting funds from their local housing authority. After the family secured the services of the architect and contractor, the occupational therapist reviewed plans and visited the home during construction to be sure that her recommendations were carried out. After the modifications were complete (Figures 14-5, 14-6, 14-7, 14-8), she met with Matt and his family to ensure that they knew how to use each piece of equipment safely and that Matt had received training from the agency that provided the attendant in how to manage his employee. She also repeated the COPM to determine Matt's perception of how he was able to perform his daily activities with the modifications in place.

Matt reported an improvement in his performance and satisfaction scores in each of the areas that were identified as problems. In this instance, the therapist provided only interventions that changed the environment, not the level of personal ability. This is illustrated in Figure 14-9, which depicts Lawton and Nahemow's (1973) ecological model, previously presented in Chapter 2. The white dot on the model is estimation of how Matt fit in his environment. As is indicated by the model, prior to the modifications, Matt experienced more environmental press than he was able to manage due to the nature and extent of his disabling condition. The shaded dot on the model is an estimation of how Matt currently fits into his environment. As is suggested by the arrow, Matt's abilities did not change during the inter-

Table 14-3
Matt's Renovation Plan

Matt's Problems (Identified Using the COPM)	Environmental Barriers Affecting His Performance	Solutions
Enter the home independently	• Stairs are present at the front (10) and rear (9) entrances.	• Provide a ramp to the rear entrance to the home. Ramp to have slope of 1:12 minimally and comply with standard safety regulations (including handrail and no-skid surface). Consider drainage and snow removal and grade appropriately.
Move between the first and second floors	• Stairs	• Provide an elevator with controls operable by one hand. Provide large-button controls.
Bathing and toileting as independently and with as little personal help as possible	• Layout of bathroom does not have adequate space. • Tub configuration and height does not permit easy transfer. • Toilet position and height does not permit easy transfer. • Mirror is located too high for Matt to see himself. • Sink has no knee clearance. • Faucet is not operable by one hand. • Light controls are too high.	• Addition of a new bathroom with the following inclusions: • Space for a 5-foot turning radius. • Overhead lift transfer system. • Roll in shower. • Sink with knee clearance (and pipes configured so they do not encroach in the space). • Single lever handle on faucet. • 19-inch toilet with fold down grab bars (to help Matt balance). • Bidet style toilet with washing/drying system.

Figure 14-5. The new elevator. (Copyright © Polly Gray. Printed with permission.)

Figure 14-6. The new entrance. (Copyright © Polly Gray. Printed with permission.)

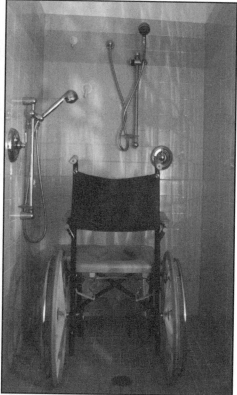

Figure 14-7. The new shower. (Copyright © Polly Gray. Printed with permission.)

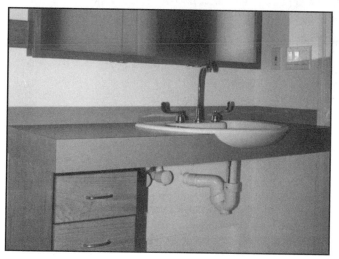

Figure 14-8. The new sink. (Copyright © Polly Gray. Printed with permission.)

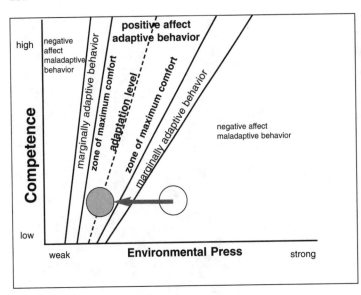

Figure 14-9. Lawton's model—Matt's abilities before and after home modifications. (Adapted with permission from Lawton, M. P., & Nahemow, L. [1973]. Ecology and the aging process. In C. Eisdorfer & M.P. Lawton (Eds.). *Psychology of adult development and aging.* Copyright © 1973 by the American Psychological Association.)

vention process. It was assumed that Matt was not likely to change in terms of his strength or range of motion, but by modifying his environment, the team was able to improve Matt's fit and improve his occupational performance.

CONCLUSION

This chapter presents a clinical application framework that will allow occupational therapists to expand their current practice to include environmental interventions as strategies to improve occupational performance. Although home modifications are not a new area to be considered by occupational therapists, current health care systems and diminished health care dollars have nudged occupational therapists away from an area of practice that has the potential to greatly influence occupational performance in the home and community. The framework's central focus on environmental changes provides a means for occupational therapists to expand their practice to include home modifications. This innovative application framework fills a gap in the health care system. Individuals who are left with residual disability after the rehabilitation process is complete now have the opportunity to focus on environmental modification as a means of improving their occupational performance.

This clinical application framework is based on existing models of occupational therapy practice such as PEO models (Christiansen, 1991; Christiansen & Baum, 1997; Law et al, 1996) and treatment using the occupational performance process model (Fearing, Law, & Clark, 1997), while acknowledging the emerging model of disability offered by the WHO (2001). The model also relies on a body of evidence from various disciplines including environment and behavior science and occupational therapy that provides justification for and guides the provision of home modifications.

Client-centred practice is an approach to providing therapy that demands respect for, and partnership with, people receiving services. It recognizes the autonomy of individuals, the strengths clients bring to a therapeutic relationship, the need for client choice, and the benefits of client-therapist collaboration (Law, 1998). A client-centred approach becomes extremely important when considering changes to the home. As discussed, home is a very important place. Out of respect for an individual's home, prior to making modifications to a home, it is important that the entire family participate in the planning process. For this reason, the application framework is driven by a client-centred approach. The client and family determine the intervention areas and are involved in decision making and managing the home modification process.

The application is generalizable to multiple populations. Although the case study presented in the chapter dealt with private, single family homes, the process has worked equally well for individuals who have developmental disabilities and live in supported apartments, older adults who experience changes in vision and need home modifications to stay at home safely and independently in their apartment, and young adults who are returning to a dorm room after a spinal cord injury. Other successful applications of the framework have supported older caregivers interested in relieving some of the physical work related to caring for a spouse, or caregivers of individuals who have dementia.

Although the homes may be different for people who have physical disabilities, the procedures to determine environmental barriers are consistent. This practice framework encourages occupational therapists to explore the possibility of eliminating excess disability, which results from person-environment misfit in the home. While not a new area of practice, this presents occupational therapists with an opportunity to expand existing practice to include environmental modifications or to develop a new clinical service that fills a gap on the existing continuum of care.

Study Questions

1. Discuss the three types of strategies that occupational therapists can use to improve occupational performance in the home. Give an example of each.

2. Describe how environmental modifications in the home support a practice guided by the PEO model.

3. How is the clinical application framework described in this chapter different than traditional practice areas for occupational therapy? How is it the same? How would an occupational therapist in traditional practice include these principles in her practice?

4. Who are the members of the "team" in a home modification practice? What is the role of the occupational therapist on the team? How is this different than traditional practice?

REFERENCES

American Occupational Therapy Association. (2002). Occupational therapy practice framework: Domain and process. *Am J Occup Ther, 56*, 609-639.

American Association of Retired Persons. (2000). *Fixing to stay: A national survey on housing and home modification issues—Executive summary.* Washington, DC: Author.

Center for Universal Design. (1997). *A blueprint for action: A resource for promoting home modifications.* Raleigh, NC: North Carolina State University, School of Design.

Christiansen, C. (1991). Occupational therapy intervention for life performance. In C. Christiansen & C. Baum (Eds.), *Occupational therapy overcoming human performance deficits* (pp. 3-45). Thorofare, NJ: SLACK Incorporated.

234   Chapter 14

Christiansen, C., & Baum, C. (1997). Person-environment occupational performance: A conceptual model for practice. In C. Christiansen & C. Baum (Eds.), *Occupational therapy: Enabling function and well-being* (pp. 47-70). Thorofare, NJ: SLACK Incorporated.

Clemson, L., Roland, M., & Cumming, R. (1997). Types and hazards in the homes of elderly people. *Occupational Therapy Journal of Research, 17,* 200-213.

Connell, B. R., Sanford, J. A, Long, R. L., Archea, C. K., & Turner, C. (1993). Home modifications and performance of routine household activities by individuals with varying levels of mobility impairments. *Technology and Disability, 2*(4), 9-18.

Corcoran, M. A., & Gitlin, L. N. (2001). Family caregiver acceptance and use of environmental strategies provided in an occupational therapy intervention. *Physical & Occupational Therapy in Geriatrics, 19,* 1-2.

Corcoran, M. A., Gitlin, L. N., Levy, L, Eckhardt, S., Earland, T. V., Shaw, G., & Kearney, P. (2002). An occupational therapy home-based intervention to address dementia-related problems identified by family caregivers. *Alzheimer's Care Quarterly, 3*(1), 82-90.

Cumming, R. G., Thomas, M., Szonyi, G., Frampton G., Salkeld, G., & Clemson, L. (2001). Adherence to occupational therapist recommendations for home modifications for falls prevention. *Am J Occup Ther, 55,* 641-8.

Dovey, K. (1985). Home and homelessness. In I. Altman, & C. M. Werner (Eds.), *Home environments, human behavior and environment: Advances in theory and research* (pp. 33-63). New York: Plenum.

Duncan, R. (1988). Funding, financing and other resources for home modifications. *Technology and Disability, 8,* 37-50.

Fawcett, S. B., Suarez de Balcazar, Y., Whange-Ramos, P., Seekins, T., Bradford, B., & Mathews, R. M. (1988). The concerns report: Involving consumers in planning for rehabilitation and independent living services. *American Rehabilitation, 14,* 17-19.

Fearing V. G., Law, M., & Clark, J. (1997). An occupational performance process model: Fostering client and therapist alliances. *Can J Occup Ther, 64,* 7-15.

Gitlin, L., & Corcoran, M. (1993). Expanding caregiver ability to use environmental solutions for problems of bathing and incontinence in the elderly with dementia. *Technology and Disability, 2*(1), 12-21.

Jones, M. L., Petty, C. R., Boles, C., & Mathews, R. M. (1986). Independent living: A survey of programs and service needs. *Rehabilitation Counseling Bulletin, 29,* 278-283.

Law, M. (Ed.). (1998). *Client-centered occupational therapy.* Thorofare, NJ: SLACK Incorporated.

Law, M., Baptiste, S., Carswell, A., McColl, M., Polatajko, H., & Pollock, N. (1998). *Canadian occupational performance measure* (3rd ed.). Ottawa, ON: CAOT Publications ACE.

Law, M., Cooper, B. A., Strong, S., Stewart, D., Rigby, P., & Letts L. (1996). The person-environment-occupation model: A transactive approach to occupational performance. *Can J Occup Ther, 63,* 9-23.

Lawton, M. P., & Nahemow, L. (1973). Ecology and the aging process. In C. L. Eisdorfer & M. P. Lawton (Eds.), *Psychology of adult development and aging* (pp. 619-674). Washington, DC: American Psychological Association.

Mann, W. C., Hurren, D., Tomita, M., Bengali, M., & Steinfeld, E. (1994). Environmental problems in homes of elders with disabilities. *Student Occupational Therapy Journal of Research, 14*(3), 191-211.

Mann, W., Ottenbacher, K., Fraas, L., Machiko, T., & Granger, C. (1999). Effectiveness of assistive technology and environmental interventions in maintaining independence and reducing home care costs for the frail elderly: A randomized controlled trial. *Archives of Family Medicine, 8,* 210-217.

Marcus, C. C., Sarkissian, W., Wilson, S., & Perigut, D. (1986). *Housing as if people mattered: Site designing guidelines for medium density family housing.* Berkley, CA: University of California.

Olsen, R. V., Ehrenkrantz, E., & Hutchings, B. (1993). Creating supportive environments for people with dementia and their caregivers through home modifications. *Technology and Disability, 2*(4), 47-57.

Salkeld G., Cumming, R. G., O'Neill, E., Thomas, M., Szonyi, G., & Westbury, C. (2000). The cost effectiveness of a home hazard reduction program to reduce falls among older persons. *Aus N Z Public Health, 24,* 265-71

Steinfeld, E., & Shea, S. M. (1993). Enabling home environments: Identifying barriers to independence. *Technology and Disability, 2*(4), 69-79.

Steinfeld, E., & Shea, S. M. (1997). *Enabling home environments: Strategies for aging in place.* Buffalo, NY: IDeA.

Trickey, F., Maltais, D., Gosselin, C., & Robitaille, Y. (1993). Adapting older persons homes to promote independence. *Physical & Occupational Therapy in Geriatrics, 12,* 1-14

World Health Organization. (WHO) (2001). *International classification of functioning, disability and health.* Geneva: Author.

Accessible Transportation: Novel Occupational Therapy Perspectives

Susanne Iwarsson, PhD, Reg. OT; Agneta Ståhl, PhD; and
Gunilla Carlsson, PhD, Reg. OT

CHAPTER OBJECTIVES

* To describe how occupational therapy can be applied to societal planning, especially targeting transportation.
* To define accessibility to the physical environment as well as related concepts, applied to urban transportation from occupational therapy, traffic planning, and engineering perspectives.
* To summarize experiences and knowledge acquired through collaborative research between occupational therapy and traffic planning and engineering.

INTRODUCTION

Occupational therapists have a declared interest in barrier-free design, for intervention with individual clients as well as for preventive influences on the public planning process. Still, based on the traditional individual client-therapist relationship, focus on environmental factors in practice remains on the environments most intimate to the person, predominantly indoor housing environments. Most published research studies address housing and its close surroundings. However, many occupations occur in neighborhoods and communities (Iwarsson, 1997), thereby implying greater demands on mobility. In order to enable occupational therapists to advocate accessibility at the societal level, the occupational therapy process must be applied to societal planning. Competence generated by experience gained from individual client-therapist relationships in rehabilitation contexts should be transferred to another kind of relationship, the population-health agent relationship. This novel approach challenges traditional occupational therapy practice; new methods and collaborative work must be developed.

Transportation is one prerequisite for participation in a wide range of other daily occupations, but it is also an important occupation in itself (Carlsson, 2002a). In the International Classification of Functioning, Disability and Health (ICF), the World Health Organization (WHO) (2001) has organized transportation as an area of activity and participation that is categorized at the same level as communication, self-care, and major life areas. Accessibility problems in the public environment have an impact on occupational performance in individuals with functional limitations, implying activity limitations and participation restrictions, but knowledge about how these factors are related to each other is scarce. An important aspect is the fact that in modern societies, many of the services previously concentrated in city centres are now frequently located in peripheral or suburban areas, and they are often designed for access by private automobiles (Holmberg & Hyden, 1996). Such changes have a negative impact on older people's abilities to lead active and independent lives, ultimately threatening health and participation. Even if public transportation is increasingly more accessible (e.g., by the use of low-floor buses in ordinary mainstream traffic), research indicates that vehicle design is not enough. In order to achieve the goal of accessibility for all, solutions between ordinary mainstream traffic and special transportation services are imperative (Ståhl, 1998, 1999). In addition, a generally accessible public outdoor environment is a prerequisite in order to make "the entire travel" possible (Ståhl, 1998). Such a goal challenges the entire societal planning process, and different stakeholders must collaborate more intensively. Occupational therapy competence could be an important part in this work, but we need to adapt our traditional approach and methods to work at the societal level.

The increased average length of life in the Western world implies that more people live more years with some ill health. In combination with the natural process of aging, even minor ill health may lead to functional limitations (e.g., limitations in stamina or restricted sight). Preserved walking or wheeled mobility is important for independence in activities of daily living (ADLs), and necessary in order to be able to go shopping or to the bus stop. The design of the public outdoor environment (e.g., high curbs or sloping walking surfaces) often causes accessibility problems. Efforts to increase public environment accessibility (to housing, public facilities, vehicles, public outdoor environments) give older people the opportunity to live independently, with preserved quality of life (Ståhl, 1998). Environments supporting activity and participation support health and quality of life as well. Accessible public outdoor environments and public transportation are important prerequisites for older people's travel and independence in daily activities (Ståhl, 1987).

The innovative ideas and strategies outlined in this chapter are based on a methodological approach, focusing on the challenges and opportunities in:

* Collaborative work between occupational therapy and traffic planning and engineering.
* Transferring knowledge (including theoretical perspectives) and experiences derived from environmental occupational therapy interventions implemented as part of individual rehabilitation processes to planning processes at a societal level.

According to current policy and legislation, all housing, official buildings, and public transportation should be accessible for all citizens (United Nations, 1993), but many problems still occur. Accessibility for all is one of the three target areas pointed out in a recent Swedish governmental proposition "From patient to citizen..." posing specific demands on knowledge generation and practical solutions, as well as advocating universal design (Regeringens proposition 1999/2000:79). Accessibility of the physical environment is an important prerequisite for activity, participation, and health, but still, this aspect of environmental design is widely neglected (Iwarsson & Ståhl, 2003).

In this context, the concept of universal design should be introduced. Universal design focuses on the mainstreaming process in design (i.e., buildings, tools, etc. should be usable for all individuals, regardless of functional limitations or special needs) instead of adapting environments to individual problems (Connell & Sanford, 1999; Preiser & Ostroff, 2001). One of the most important differentiations between accessibility and universal design is that accessibility is related to compliance with norms and standards, while universal design is process oriented. Universal design is based on the notion that the built environment and manmade products should be usable for as many members of the population as possible, regardless of age, functional capacity, etc. (Preiser & Ostroff, 2001). In itself, the concept of universal design is more health promoting than the concept of accessibility, because it targets the population at large, not only users with special needs (see Chapter 7 for further description of universal design). According to the WHO (1986), creating environments supportive of health is among the most vital priorities. The WHO has also incorporated environmental factors into the ICF conceptual framework (WHO, 2001), describing activity, participation, and health. A more specific aspect is the fact that inaccessible environments imply risks for injuries (e.g., injuries resulting from falls or traffic accidents (Berndtman, Brundell-Freij, Hyden, & Ståhl, 1998). Foresighted planning of future city environments, involving universal function and design requires novel approaches based on the needs of user groups (e.g., children, persons with functional limitations, older people).

While public environment accessibility and universal design issues are gaining increased attention, there still is a lack of systematic procedures for making efficient priorities. Measures undertaken to solve accessibility problems are often made ad hoc, rather than based on the results produced by use of systematic assessments (Fänge, Iwarsson, & Persson, 2002). Current legislation and international intentions specify requirements for accessibility to the public environment, but the regulations are in large measure subject to local interpretations. Only with valid and reliable data on local accessibility problems at hand, including users' opinions (Fänge & Iwarsson, 1999) as well as objective, norm-based assessments, can efficient discussions with the stakeholders involved take place (Fänge & Iwarsson, 2003). Among all stakeholders involved in the societal planning process, occupational therapy knowledge and experience represent one important basis for the development and use of methodology (Carlsson, 2002a). For example, accessibility assessments using a travel chain perspective (Carlsson, 2002b) (i.e., moving from origin to destination [Iwarsson & Ståhl, 1999]), target solid solutions for universal design and increased accessibility.

TRADITIONAL PERSPECTIVES AND CURRENT KNOWLEDGE

The very early writings of occupational therapy suggested the relationship between ability to perform everyday activities and life satisfaction. Current occupational therapy practice is based on the notion that daily occupation is an important determinant of health and well-being (Law, Steinwender, & Leclair, 1998; Wilcock, 1998). In 1966, Reilly described the health of human beings in terms of level of adaptation to their environment, rather than in freedom of pathology. One important issue that relates to occupational performance in older adults and individuals with impairments is the need to adapt the physical environment throughout society (e.g., related to housing, transportation), thereby increasing general accessibility. Even if most occupational therapy models explicitly target environmental aspects, the concept of physical accessibility has not been problematized within occupational therapy theories, especially regarding transportation.

The overarching goal of traffic policy in Sweden and many other countries is to provide safe and environmentally friendly transportation resources at the lowest possible cost, while at the same time integrating people with impairments and functional limitations into all parts of society. The aim is to offer, by means of accessible public transportation, all citizens equal freedom of choice regarding travelling. Still, the ordinary solutions for public transportation are marred with barriers (Ståhl, Brundell-Freij, & Makri, 1993), and to a great extent, individuals with impairments have to use special transportation services (STS). One crucial task for the scientific discipline of traffic planning and engineering is the development of knowledge concerning the planning and design of an accessible traffic transportation environment for individuals with impairments (Holmberg & Hyden, 1996). Key to research and educational activities in the field is the person-environment relationship. The design of the traffic transportation system must take into account that persons with different, individual abilities are supposed to move in the environment, being able to live and develop in society.

In accessibility research (Iwarsson, 1997; Iwarsson, & Slaug, 2001; Iwarsson & Ståhl; 1999; Steinfeld & Danford, 1999), the environmental docility hypothesis (Lawton & Simon, 1968), and the ecological model (Lawton & Nahemow, 1973) have been used as the main theoretical foundation (Carlsson, 2002a; Carlsson, Iwarsson, & Ståhl, 2002a). The environmental docility hypothesis proposes that the less competent the individual, the greater the impact of environmental factors on that individual (Lawton & Simon, 1968). The ecological model also deals with the interaction between individual competence and environmental press (see Figure 2-2). Individual competence refers to basic domains such as biological health, sensorimotor functioning, cognitive skills, and ego strength, and there may be differences between objective competence and perceived competence. Environmental press refers to the demands that environments place on people as they live their lives. Some environments pose great demands on people while others do not. There may also be a difference between the subjectively experienced environment and objective environment (Lawton, 1986). For example, a large proportion of older adults expresses fear of moving around in the public outdoor environment, in particular in the traffic system. At the same time it is a rather low percentage that actually has been involved in any accidents or incidents (Ståhl, 1987). When it comes to methodology, in objective terms accessibility problems can be analyzed only after reliable and valid professional environmental assessment, in relation to existing norms and guidelines, and related to functional capacity in individuals or groups of individuals (Iwarsson & Slaug, 2001). In contrast to this, stakeholders in the disability movement argue that the only experts on accessibility issues are users themselves—i.e., they stress the very subjective character of the word (Iwarsson & Ståhl, 2003). In ongoing research on housing accessibility, the results indicate differences between professional, objective accessibility assessments and users' subjective ratings (Fänge & Iwarsson, 2003). On the contrary, there are differences in the extent to which subjective ratings co-vary with objective accessibility assessments between different sub-samples of subjects, presumably depending on varying personal experiences when it comes to really performing activities in different sections of the housing environment. In another study, focusing on public environment accessibility, similar tendencies were identified (Fänge, Iwarsson, & Persson, 2002).

In order to understand the prerequisites for developing occupational therapy strategies targeting societal planning in this field, mobility and transportation must be understood from an occupational perspective or applying current ICF terminology (WHO, 2001) from an activity and participation perspective. Knowledge about the daily activities of certain social groups and factors behind different travel patterns are needed in addition to basic knowledge in traffic planning. The everyday life deals with the organization of occupations in time and space (CAOT, 1997). A trip is mostly motivated by the need to participate in an activity at the destination (Zemke & Clark, 1996). Consequently, travel during an average day is most commonly described in relation to other activities (i.e., the purposes of the trips). Salomon and co-workers examined the out-of-home activities per capita in Europe, and they reported that the mean number of activities ranged from 1.5 to 2.5 per day (Salomon, Bovy, & Orfeuil, 1993). This is only a rough score of social interaction, and as the authors declared, the activity analysis is of more interest when home and out-of home activities are examined together. Walking is included in mobility, but usually ignored in transportation research if another mode is used for the trip. Studies among adults and teenagers in the United States showed that driving and transportation took up 6 to 9% of daytime activities, which can be compared with productive activities, such as working or studying (24 to 60%), or leisure activities (20 to 43%). Driving and transportation were described as parts of maintenance activities, taking up 20 to 42% of daytime (Csikszentmihalyi, 1998). It can be seen, then, that transportation plays a significant role in daily activities, and is therefore important to be addressed in research and practice.

Four important factors affecting patterns of daily occupations, including mobility patterns are: a) having no car, b) acquiring a disability, c) being out of work, and d) having a low income (Krantz, 1999). In addition, age is an important determinant for mobility patterns. For example, a recent Swedish study identified that the main method of transportation for some teenagers with low vision was by car driven by parents, while other teenagers walked and went by public transportation (Kroksmark & Nordell, 2001). In another study carried out in Germany, Italy, and Finland, age, health situation, and driving ability were reported as significant factors for outdoor mobility. Further factors of importance for mobility among older people were their social situations and their networks of family and friends (Mollenkopf et al., 1997). To a substantial extent the physical environment and access to transport systems, not age and functional limitations, affect mobility in old age (Ståhl, 1987).

Travel is a link between different activities, but it may also be an occupation in itself (Ståhl, 1987). A 76-year-old woman told how she met the same "gang" in the morning at the bus (Carlsson, 2002b). The people on the bus talked to each other during the trip; they went to different destinations in the city but met again during the return trip and talked about their experiences. Sometimes they did not actually have any errands downtown, but they traveled in order to meet with each other. That is, every activity has a relation to everything else we do and activities often have additional qualities, such as making new social contacts, decreasing feelings of loneliness, and promoting participation. The content of these experiences determines quality of life (Csikszentmihalyi, 1998). Mobility is a prerequisite for participating in daily activities at different places; mobility is important in our orchestration of daily lives. Restricted mobility influences activity patterns and well-being (Carlson, Clark, & Young, 1998) as well as participation.

Traditional rehabilitation and occupational therapy assessments of ADLs often comprise mobility activities, but they are frequently restricted to physical movement indoors (Carlsson, 2002a). When it comes to transportation, a few ADL instruments include this activity but the operationalizations used are very general and tend to have low validity depending on the substantial environmental component inherent (Iwarsson, 1998; Iwarsson & Isacsson, 1997). An important occupational therapy contribution to health promotion is to convey knowledge about how physical environmental design affects the prevalence and incidence of activity limitations and participation restriction in the community, but also about differences in mobility patterns between persons with and without functional limitations (Kroksmark & Nordell, 2001). Of importance are also such questions as Ståhl (1999) called attention to, namely if the number of trips increases with more accessible travel chains. Occupational therapy represents knowledge and expertise—e.g., in how to assess individual functional limitations in relation to environmental barriers—while traffic planning and engineering represents the knowledge of how to plan and design the traffic environment and the transportation system. The design of public transportation must proceed from a holistic perspective and presupposes that people have different needs and desires when they travel (Iwarsson & Ståhl, 1999).

FUTURE STRATEGIES

To be able to successfully influence societal planning to promote accessibility to the physical environment—e.g., in transportation, occupational therapists must acknowledge that definitions differ between professions (Iwarsson & Ståhl, 2003). For example, the terms of functioning and disability as defined in the ICF (WHO, 2001) are known among rehabilitation professionals but are scarcely known among other professions. Furthermore, our comprehension of accessibility problems is different from how technicians and engineers perceive them. A highway engineer might not consciously reflect upon what is obvious and self-evident to an occupational therapist—i.e., that accessibility in public transportation is a prerequisite for activity and participation. If different stakeholders have different goals for their interventions in mind, their communication is not likely to be successful. Unless occupational therapists make crucial discrepancies very clear and communicate them, they will not succeed in efficient interventions at the societal level. Thus, one strategy is to transfer tacit occupational therapy knowledge into conceptual definitions, discuss them at an interdisciplinary level to reach consensus, and communicate them to society at large. Next, based on these definitions, occupational therapists should develop methods and strategies applicable to societal planning, use them in collaborative research and practice, and in concrete terms present their results to stakeholders in the societal planning process.

It is obvious that we need to develop reliable and valid methods to survey accessibility and evaluate interventions. Ten years ago, methodological development based on Steinfeld's enabler concept (Steinfeld et al., 1979) started in Sweden, based on the traditional patient-therapist perspective. Derived from practical experiences from community-based occupational therapy in relation to inefficient housing adaptation processes, an instrument for housing accessibility assessment was needed. The instrument, housing enabler (Iwarsson & Slaug, 2001), was used in health care science research targeting the older adult population, subsequently generating knowledge on accessibility problems at group and population levels (Iwarsson, 1997). In 1996, fruitful cooperation between occupational therapy and traffic planning and engineering saw its start (Iwarsson & Ståhl, 1999), and the enabler concept has since then been used as the basis for further methodological development and research (Iwarsson, Jensen, & Ståhl, 2000). This novel interdisciplinary collaborative approach has generated knowledge in several different respects.

Conceptual Definitions

An analysis and critique of definitions used in accessibility research and practice from the perspective of person-environment theory is necessary. A basic problem regarding accessibility is the fact that different societal stakeholders, including people in general and persons with functional limitations themselves, use different definitions of the concept. A first prerequisite for successful implementation of accessibility throughout society is to define what accessibility to the physical environment means, and to make sure that different actors in the planning process share the same conceptual definitions.

Like many other words, accessibility has a common, everyday meaning as well as specific meanings in different contexts (Iwarsson & Ståhl, 2003). According to the *Oxford Popular Dictionary and Thesaurus* (1998, p. 2), "accessible" is an adjective synonymous with "approachable, at hand, attainable, available, close, convenient, handy, and within reach." A definition sometimes used in engineering and planning contexts is the simplicity with which activities in the society can be reached, including needs of citizens, trade, industries, and public services—i.e., accessibility is defined primarily in terms of distances and time and not related to human capacity (Pirie, 1979). According to Pirie, the ideas expressed in the publication referred to are borrowed from travel-behavior modeling and time-geography. In its specific meaning in this chapter, accessibility is a relative concept because it expresses the relationship between the capacity of the individual and the demands from the environment (Iwarsson, 1997). In objective terms, accessibility can be assessed reliably and validly by professionals, in relation to existing norms and guidelines (Iwarsson & Slaug, 2001; Iwarsson & Ståhl, 2003; Preiser & Ostroff, 2001).

A word often used in parallel with accessibility is usability—perceptions of how well the design of the environment enables functioning and well-being, mainly from the user's perspective (Fänge & Iwarsson, 1999; Steinfeld & Danford, 1999). For example, Swedish building and planning legislation has for many years required that all housing, work premises, or other premises open to the public must be accessible and usable for persons with restricted mobility or restricted sense of locality. In this legal framework, usability was interpreted and defined as follows: the built environment has to allow any individual, in spite of impairments, to be able to perform daily activities within it (Didon, Magnusson, Millgard, & Molander, 1987). When it comes to the word in itself, "usable" is an adjective synonymous with "fit to use, functioning, operational, serviceable, valid, and working" (*Oxford Popular Dictionary and Thesaurus*, 1998, p. 354)—i.e., it is not synonymous with accessibility. Although the words accessibility and usability have different definitions, some authors, for example Steinfeld and Danford (1999), use them in parallel, stating that they both usually are defined in terms of observed task performance. Defined in this way, accessibility as well as usability represent the concept of person-environment fit, but the distinction Steinfeld and Danford make is that usability is based on individual interpretations.

Turning to occupational therapy theories, for example the Canadian model of occupational performance (CMOP) (CAOT, 1997), they all comprise person as well as environmental components, but they also includes the component occupation. As specifically defined in the CMOP, occupational performance represents the transaction between person, environment, and occupation (CAOT, 1997), and thus it seems closely related to usability (Carlsson, Iwarsson, & Ståhl, 2002b; Iwarsson & Ståhl, 2003). In a recent paper, Iwarsson and Ståhl (2003) suggested the following definitions of accessibility and usability:

* ✳ *Accessibility* is a relative concept, implying that accessibility problems should be expressed as a person-environment relationship. In other words, accessibility is the encounter between the person's or group's functional capacity and the design and

demands of the physical environment. Accessibility refers to compliance with official norms and standards, thus being mainly objective in nature.

* The concept of *usability* implies that a person should be able to use (i.e., to move around) and be in the environment on equal terms with other citizens. Accessibility is a necessary precondition for usability, implying that information on the person-environment encounter is imperative. However, usability is not only based on compliance with official norms and standards; it is mainly subjective in nature, taking into account user evaluations and subjective expressions. Most important, there is a third component distinguishing usability from accessibility: occupation.

Specific Methodological Considerations

In order to operationalize the concept of accessibility for physical planning, information about the person as well as the environment is imperative to assess person-environment fit. First, the person component of accessibility should be expressed and assessed in terms of functional capacity (Carlsson, Iwarsson, & Ståhl, 2002a, 2002b) or in negative valence, functional limitations (Nagi, 1991). Functional limitations are restrictions in a person's ability to perform fundamental physical and mental actions in daily life (Verbrugge & Jette, 1994). That is, functional limitations refer to the person component of the concept of accessibility with minimum influence of the environmental component. In relation to current ICF terminology (WHO, 2001), the definition of functional capacity we apply (Carlsson et al., 2002b) is included in the "activities and participation" component. The fact that this ICF component is very comprehensive makes the ICF categorization unsuitable for the stringent operationalization needed in accessibility research. The "activities and participation" component includes very different aspects of human functioning; activities close to "body functions and structure" as well as participation aspects of great diversity. Even if the ICF details a separate "environmental factors" component, the "activities and participation" component includes aspects influenced by environmental components as well, for example, instrumental activities of daily living (IADLs) (Iwarsson & Isacsson, 1997). This argument is the main reason for keeping the more specific definition "functional capacity" in order to operationalize the personal component of the concept of accessibility. A profile of functional limitations is a description of all of the functional limitations reported in a person (Carlsson et al., 2002b; Iwarsson, 1999; Iwarsson & Slaug, 2001). This approach can be applied to the individual as well as the group or population levels, but using an epidemiological perspective with profiles of functional limitations at the population level has only recently been suggested, and the lack of knowledge is substantial (Carlsson et al., 2002b).

Second, it is crucial to be aware of the characteristics of the environmental component of the concept of accessibility (Jensen, Iwarsson, & Ståhl, 2002). Objective environmental assessment should be based on norms and guidelines, but research so far raises questions about the range of anthropometric and biomechanical values in the population and what values the norm should be based upon (Connell & Sanford, 1999; Jensen et al., 2002). Of vital importance is the degree to which anthropometric data are derived from a sample representative of the entire population (i.e., whether all persons with impairments are included or whether some are excluded) when defining norms and standards. Despite this questioning of our existing norms, norms are necessary. It is impossible to make objective and reliable assessments of environmental details and discuss design issues without norms, because society sets up its ambition of general accessibility from norms and legislation. Another problem is the dynamic nature of the outdoor environment (e.g., daylight and dusk or seasonal variations), which poses specific demands on assessment methodology. Still another kind of challenge to be dealt with in environmental assessments is the repetition of barriers. Different environmental design features or barriers sometimes occur once along a route, others occur many times. The problems of dynamic features and repetition of environmental barriers threaten the reliability of accessibility assessments (Iwarsson et al., 2000), and are more obvious when the outdoor environment at societal level is at target than when the issue is individual housing.

Not until information on the person as well as the environmental components has been gathered is it possible to analyze and judge the degree of accessibility, which is understood as person-environment fit. This third step, rendering a quantification of the problems anticipated, is the final result of accessibility assessments applying the enabler concept (Iwarsson & Slaug, 2001).

When it comes to accessibility in public transportation targeting mobility and travel as activities, further methodological steps are required. The enabler concept yields a description of the profile of functional limitations in the person or target group as well as a detailed description of the physical environment, and it has potential to generate a quantification of the accessibility problems anticipated (Iwarsson & Slaug, 2001). This is needed, but it is not sufficient for describing the complexity of a travel chain (i.e., moving from origin to destination) (Carlsson, 2002b). This outlook emphasizes the fact that travel as such is an occupation, and the degree of accessibility should not only be based on a vast number of problematic design features described in detail. To really understand the complexity of travelling in relation to accessibility problems, the importance of contextualizing travelling is obvious. A trip should not be seen separately from its context (Carlsson, 2002a, 2002b). On the contrary, travelling is only one of many occupations required for active participation in society. That is, the whole is more than the sum of the parts (Ståhl, 1998). In order to approach accessibility problems from the perspective of occupational therapy, methods incorporating the occupation component are imperative. In Iwarsson and Ståhl's definition (2003), we need information on usability as well.

Development of Novel Research Methods

Physical accessibility as one of the complex areas in transportation calls for systematic procedures of assessment. The person and the environmental components of the concept of accessibility need to be focused and investigated separately as well as together. However, the results always need to be analyzed and interpreted from the theoretical understanding of person-environment-occupation (PEO) interactions. With the person component, a challenge is to operationalize functional capacity in a way that could be utilized in more efficient analyses of accessibility problems. The first steps to classify profiles of functional limitations have been taken (Carlsson et al., 2002b) but substantial research efforts must be accomplished in the exploration of the personal component at group and population levels.

Examples of complementary methods that give fruitful results when they are used in parallel in the investigation of the environmental component of accessibility were observations of critical incidents (Flanagan, 1954) and independent, objective and norm-based environmental assessments based on the enabler concept (Jensen et al., 2002). During observation of critical incidents, the goal is to make the participants use their upper limits of functional capacity. By experience, occupational therapists are used to handling situations when the environmental demands are on the verge of exceeding a person's capacity (Mattingly & Flemming, 1994). In the study by Jensen et al. (2002), the subjective individual experience of environmental barriers was highlighted during real trips by bus, while a societal, norm-based perspective was highlighted by means of the objective environmental assessment.

When exploring people's experiences of accessibility problems, focus-group interviews in combination with participant observations have also been successfully used (Carlsson, 2002b). Focus-group interviews provide a breadth of information about environmental demands causing accessibility problems, while further in-depth participant observation studies are required as well. We expect this kind of research to advance knowledge about the transactional process comprising PEO (CAOT, 1997), in a travel chain perspective elucidating aspects of usability as well (Carlsson, 2002a; Carlsson et al., 2002a).

Occupational Therapy in Relation to Traffic and Societal Planning

Many persons, in particular older persons, suffer from several different diseases at the same time. Thus, a person's functional capacity is a result of several diagnoses in combination with the consequences of the normal process of aging (Femia, Zarit, & Johansson, 2001; Lieberman et al., 1999). Thanks to traditional medical care and research, present knowledge on functional limitations in relation to specific diagnoses is sufficient, while knowledge about the prevalence of different combinations of functional limitations in the population is still scarce (Carlsson et al., 2002b). Occupational therapists hold important knowledge and experience on such combinations of functional limitations among their clients, and it should be used in accessibility planning at the society level. Because decisions about physical accessibility planning at the societal level target the population at large, information based on individual data cannot be used. Instead, an epidemiological approach must be adopted. As a result of the traditional patient-therapist relationship, occupational therapists have not yet presented such data on an aggregated level, but novel studies are underway (Carlsson et al., 2002b). From a transport planning perspective, data is usually collected on an individual level, based on large and solid studies. However, limited awareness of the need for more adequate knowledge and understanding of functional capacity and its complexity prevails in transport planning, and thus there is a risk that results might lack depth. Furthermore, in order to understand complex PEO transactions in a travel chain perspective, there is a need for introducing the importance of acknowledging the component of occupation as well (Carlsson, 2002a, 2002b). In this respect, occupational therapists have unique competency, but the value of explicitly articulating and introducing the concept of usability into practical work remains still to be proven. Nevertheless, combining occupational therapy knowledge with knowledge from technical planning should promote more efficient accessibility planning.

CASE SCENARIO: NANCY

In order to describe how occupational therapy can be applied to societal planning, especially targeting transportation, a case scenario is presented, applying our experiences so far. Most occupational therapists are still used to starting their intervention from the individual perspective, and the organizations they work in expect them to do so. Thus, an initial challenge is how to bring about an arena allowing for novel approaches targeting the societal level. To stimulate occupational therapists to try the suggested kind of interventions, we have chosen a vignette starting with a traditional occupational therapy client but ending with the occupational therapist being invited to take part in the societal planning process.

Exploration of Accessibility and Usability Problems in Urban Public Bus Transportation in a Travel Chain Perspective

Nancy lived alone in an apartment in a block of flats in the suburban area of a town in south Sweden. She was 82 years old and had recently experienced declining functional capacity. For the last 2 years, Nancy had weekly assistance from home care services, and she had discussed her situation several times with the persons attending her. Several of the problems Nancy was concerned about were related to everyday activities, and therefore the home care assistant suggested that an occupational therapist should be consulted. Within a few days, Mary, the district occupational therapist, called Nancy to pay her a home visit. During their first talk in Nancy's kitchen, Mary realized that Nancy really missed her outdoor activities, especially the opportunity to independently use public transportation.

To determine if reduced functional capacity was influencing Nancy's problem, Mary interviewed and observed Nancy while moving around. Nancy confirmed that she had poor balance, difficulties in reaching with her arms, and difficulties in bending and kneeling. Indoors, she did not use walking aid but she used a wheeled walker outdoors. Mary then explained that in order to identify environmental problems when travelling by public transportation, they together should make a bus trip. Nancy was very hesitant, because she had not done this for a long time. However, Mary encouraged her, and Nancy then decided that they together could go by bus to visit her friend in another part of the town. They made an appointment a few days later. When Mary arrived she explained to Nancy that she intended to take detailed notes during the trip, and encouraged Nancy to comment on any problems she experienced.

Holding on to the handrail and her cane, Nancy walked the six steps down to reach the wheeled walker standing inside the entrance door. To be able to pass the front door, Nancy had to perform several movements at the same time. The heavy door had to be simultaneously unlocked and pulled toward her. Leaning on the wheeled walker, Nancy struggled to keep the door open and left the house. On the way to the bus stop she reported pain in her arms, due to the uneven paving stone surface covering the pathway. She took advantage of the curb cuts to cross the street. Bicycles were parked on the pathway and she had to change direction to avoid walking into them. At the bus stop Nancy needed help from Mary to lift the wheeled walker onto the bus, even though the driver activated the kneeling function, which lowers the bus so that the first step is closer to the ground (Figure 15-1). Nancy had no problem in the bus aisle because the bus was rather empty. She was able to find a vacant seat where she could have the wheeled walker in front of her. As the bus approached her stop, Nancy needed a few moments to put the wheeled walker in the right direction toward the door. The driver kneeled the bus but Nancy needed Mary's assistance to lift the wheeled walker out of the bus, because there was no raised platform outside and therefore the step down was too high. Nancy walked from the bus stop toward her destination and had difficulties overcoming small differences in levels with the wheeled walker. Unfortunately, her friend's home was too far away for her and Nancy chose to give up and return to the bus stop. Many problems were repeated on Nancy's way back home (e.g., the difficulties due to the low platform at the bus stop). In addition, there was a handrail only at one side in the bus entrance door, which further increased the difficulties for Nancy when entering. When reaching the bus stop near her home, she had difficulties getting up from the seat. She needed help to lift the wheeled walker out of the bus and got Mary's assistance to step down and out. On her way back home she had some difficulties managing the curb cuts with the wheeled walker due to the uneven pavement surfaces. Subsequently, the difficulties accessing the entrance door at home were repeated.

Figure 15-1. Bus stop.

The following day, Mary traveled Nancy's route by herself, assessing environmental barriers objectively using a norm-based instrument based on the enabler concept. With the two data sets at hand—one reflecting the subjective client perspective, the other reflecting the societal, norm-based perspective—she then made a comparative analysis. In order to document her findings she wrote a summary of the results, detailing the most apparent accessibility problems identified along Nancy's route, explaining why the problems occurred and their negative consequences in terms of restricted activity and participation. Even though Mary had been working in her municipality for quite a long time, she had not before had any contacts with the local traffic/highway department. She contacted their office and presented her findings briefly over the phone, and was immediately invited to come and give an oral presentation to a group of highway engineers and planners. The engineers understood the importance of adding occupational therapy competence to their group. Mary's results were taken seriously and she was invited to further collaborate with them.

As a result of several meetings and discussions at the traffic/highway department involving representatives from the local public transport authority, a raised platform was built at the bus stop near Nancy's home. One argument Mary used for this was the fact that many older adults used that particular bus stop. In this case, no more modifications were made immediately. Instead, Mary instructed Nancy in how to overcome her problems by alternative ways of performing the activity (e.g., using her wheeled walker for several rests during the trip). After only a few trips together with Mary, Nancy managed to make the trip on her own.

Besides the positive effects for Nancy, Mary's novel initiative also resulted in the start of more systematic collaborative work between the occupational therapists, the traffic/highway department, and the local public transport authority in the municipality in accessibility discussions. Within a few months, the traffic/highway department and the local public transport authority initiated a pilot project, involving Mary and a few of her colleagues. The two authorities jointly decided to reserve financial resources, covering the more research-oriented part of the project (methodology development) as well as the concrete physical measures to be taken. The goal of the pilot project was to start a process to develop methodology for implementing accessibility measures promoting older people's travel in urban areas. The pilot project was limited to a defined geographic district, representing a relatively high proportion of people over age 65, living in different kinds of housing. The first step was to identify the travel chains most frequently used by older people living in the district. A postal questionnaire was sent to all people in the targeted age group, asking for some descriptive data, information on their travel habits, and their most important destinations. In the questionnaire, they were also invited to participate in information meetings arranged by the occupational therapists. Quite a lot of people were interested and a number of such meetings were held. At these meetings, Mary and her colleagues provided information about the project, and asked for volunteers to participate more actively in the project in surveying and discussing accessibility in the district. Many of the older adults at the meetings wanted to take part, and during the next 3 months several group discussions were arranged, led by occupational therapists together with engineers and planners from the traffic/highway department and the local public transport authority. In parallel, some of the volunteers took part in participant observation sessions of the same kind as Nancy initially did, providing the occupational therapists with valuable additional information from real travel chains. With this participatory approach, information on actual needs and problems were collected, analyzed, and documented in cooperation between the older inhabitants and the occupational therapists.

Next, all the information gathered was handed over to the engineers at the traffic/highway department and the planners at the local public transport authority. The information was used as the basis for concrete suggestions to change the physical environment in the district as well as in public transport system in the area. The suggestions were presented and discussed at new meetings with the older inhabitants of the district and the occupational therapists involved in the project. Measures agreed upon in these meetings were then continually implemented in the project district. The pilot project was considered very successful, but in order to evaluate its long-term effects, Mary and her colleagues were asked to plan for a follow-up survey after one year. In the same manner as in earlier project phases, the traffic/highway department and the local public transport authority decided to jointly finance the survey.

CONCLUSION

The collaborative work so far between occupational therapy and traffic planning and engineering described in this chapter has resulted in a joint exploration and definition of accessibility to the physical environment applied to urban transportation. In addition, the concept of usability has been defined and introduced. This was not reached without difficulty. Different professional cultures establish their own languages and approaches and have different opinions on which measures are adequate and should be taken. However, in such intense professional meetings, each profession's tacit core knowledge must be obvious and identified. Awareness of what is unique in occupational therapy knowledge is a necessary prerequisite in order to apply it to new arenas (e.g., to societal planning). The collaboration described is only a beginning and real development requires much time and effort, including mutual openness for novel perspectives.

Summary of Environmental Intervention

The focus of this work is on accessibility to the physical environment, more specifically in public transportation in urban areas, defining accessibility as a relative construct. Its relativity is based upon the fact that accessibility problems occur as a result of the combination between the functional limitations in an individual or a group/population and the demands from the environment. This work is innovative, because it is based on intense interdisciplinary collaboration between occupational therapy and traffic planning and engineering. Input from both disciplines has generated a new research platform with novel scientific questions and a solid methodological base for further development.

The understanding of the importance of making the distinction between objective and subjective understandings of accessibility problems is a distinctive result of our work so far, and the utility of introducing usability as a specific concept remains to be proven. Occupational therapy interventions should always be client centred, but for efficient interventions at the societal level, professional as well as client perspectives must be integrated. The knowledge communicated in this chapter is based on empirical research, integrating experience from occupational therapy and traffic planning and engineering and client-centred perspectives.

Applications and Recommendations

How does this work link to current occupational therapy practice and theory? Occupational therapists working in community contexts have much practical experience and interest in accessibility efforts at societal levels (e.g., in planning the physical environment), but they lack strategies for systematic interventions. Our experiences and the results presented can assist occupational therapists in making their knowledge visible and more reliable and efficient at the societal level. The conceptual definitions and methods provided help to make occupational therapy knowledge more concrete and communicable, facilitating occupational therapists to engage in societal planning processes (Carlsson, 2002a). Further readings on occupational therapists in consultation roles (Jaffe & Epstein, 1992) might support occupational therapists to take on new challenges such as the ones discussed here.

In this chapter we have illustrated an occupational therapy process applied to societal planning targeting public transportation focusing on older adults. The same kind of procedure can of course be applied to other areas of society (e.g., accessibility to public facilities such as libraries and museums). Likewise, it can be applied to other user groups, for example with severely disabled teenagers as reported by Fänge et al. (2002). However, because activity and participation issues differ between population groups, experience gained from one specific setting and user group cannot automatically be generalized to other arenas. Each setting and user group requires its own specific considerations.

Study Questions

1. Why is transportation an issue of importance for occupational therapists?

2. Define accessibility.

3. Define usability.

4. What makes a transportation system accessible and usable?

5. Think about your local transportation system. If you had a visual field deficit, and needed a wheeled walker to ambulate, how would you get from your home to the closest department store? Would you be able to walk? Could you walk to the closest bus stop or subway station? Would you be able to board a bus? Would you need to transfer? What barriers and enablers might you face in the trip?

6. A small city (population 100,000) has a public bus system that operates daytime, a special transit system for people with disabilities, and three taxi companies. A group of older adults, all of whom live in a seniors' apartment building are feeling very frustrated by the problems they have had getting groceries and other necessities. The closest grocery shop is about 1 kilometre (~0.6 miles) away, and the travel chain to get there involves walking a fairly steep hill. The major shopping area is about 7 kilometres (~4 miles) away. Many of the residents no longer drive, especially during the winter months. You are an occupational therapist working for a community rehabilitation company. You have been seeing two of the residents of the building as clients, and by the elevator you read about an organizing meeting residents are planning to discuss their transportation difficulties. You decide to attend the meeting as an interested party (on a volunteer basis), and the residents are eager to have you involved. Who else would you recommend they involve? What strategies would you propose they try to address their concerns?

REFERENCES

Berndtman, M., Brundell-Freij, K., Hyden, C., & Ståhl, A. (1998). *Single accidents among unprotected road-users.* Department of Traffic Planning and Engineering. Sweden: Lund University.

Canadian Association of Occupational Therapists (CAOT). (1997). *Enabling occupation: An occupational therapy perspective.* Ottawa, ON: CAOT Publications ACE.

Carlson, M., Clark, F., & Young, B. (1998). Practical contributions of occupational science to the art of successful aging: How to sculpt a meaningful life in older adulthood. *Journal of Occupational Science, 5,* 107-118.

Carlsson, G. (2002a). *Catching the bus in old age. Methodological aspects on physical accessibility in public transport.* Unpublished doctoral dissertation, Department of Clinical Neuroscience. Sweden: Lund University.

Carlsson, G. (2002b). *Travelling by urban public transportation: Exploration of accessibility, and usability problems in a travel chain perspective.* Manuscript submitted for publication.

Carlsson, G., Iwarsson S., & Ståhl, A. (2002a). *Exploration of physical accessibility in society: Reflections on theory and methodology in a travel chain perspective.* Manuscript submitted for publication.

Carlsson, G., Iwarsson, S., & Ståhl, A. (2002b). The personal component of accessibility at group level: Exploring the complexity of functional capacity. *Scandinavian Journal of Occupational Therapy, 9,* 100-108.

Connell, B. R., & Sanford, J. A. (1999). Research implications of universal design. In E. Steinfeld, & G. S. Danford (Eds.), *Enabling environments: Measuring the impact of environment on disability and rehabilitation* (pp. 35-57). New York: Kluwer Academic/Plenum.

Csikszentmihalyi, M. (1998). *Living well: The psychology of everyday life.* London: Orion Books.

Didon, L. U., Magnusson, L., Millgard, O., & Molander, S. (1987). *Plan-och bygglagen, en kommentar [The Law on Building and Planning. A commentary].* Stockholm: Norstedts.

Fänge, A., & Iwarsson, S. (1999). Physical housing environment—development of a self-assessment instrument. *Can J Occup Ther, 66,* 250-260.

Fänge, A. & Iwarsson, S. (2003). *Accessibility and usability in housing: Construct validity and implications for research and practice.* Manuscript submitted for publication.

Fänge, A., Iwarsson, S., & Persson, A. (2002). Accessibility to the public environment in city centres as perceived by teenagers with functional limitations. *Disability and Rehabilitation, 24,* 318-326.

Femia, E. E., Zarit, S. H., & Johansson, B. (2001). The disablement process in very late life: A study of the oldest old in Sweden. *Journal of Gerontology: Psychological Sciences, 56B(1),* 12-23.

Flanagan, J. C. (1954). The critical incident technique. *Psychological Bulletin, 51,* 327-358.

Holmberg, B., & Hyden, C. (1996). *Trafiken i samhället. Grunder för planering och utformning [Traffic in society. Foundations for planning and design].* Lund: Studentlitteratur.

Iwarsson, S. (1997). Functional capacity and physical environmental demand. *Exploration of factors influencing everyday activity and health in the elderly population.* Unpublished doctoral dissertation, Department of Community Health Sciences. Sweden: Lund University.

Iwarsson, S. (1998). Environmental influences on the cumulative structure of instrumental ADL: An example in osteoporosis patients in a Swedish rural district. *Clin Rehabil, 12(3),* 221-227.

Iwarsson, S. (1999). The housing enabler: An objective tool for assessing accessibility. *British Journal of Occupational Therapy, 62,* 491-97.

Iwarsson, S., & Isacsson, A. (1997). On scaling methodology and environmental influences in disability assessments: The cumulative structure of personal and instrumental ADL among older adults in a Swedish rural district. *Can J Occup Ther, 64,* 240-251.

Iwarsson, S., Jensen, G., & Ståhl, A. (2000). Travel chain enabler: Development of a pilot instrument for assessment of urban public bus transportation accessibility. *Technology and Disability, 12,* 3-12.

Iwarsson, S., & Slaug, B. (2001). *The housing enabler. An instrument for assessing and analysing accessibility problems in housing.* Nävlinge och Staffanstorp: Veten & Skapen HB & Slaug Data Management.

Iwarsson, S., & Ståhl, A. (1999). Traffic engineering and occupational therapy: A collaborative approach for future directions. *Scandinavian Journal of Occupational Therapy, 6,* 21-28.

Iwarsson, S., & Ståhl, A. (2003). Accessibility, usability and universal design-Positioning and definition of concepts describing person-environment relationships. *Disabil Rehabil, 25,* 57-66.

Jaffe, E., & Epstein, C. F. (1992). *Occupational therapy consultation: Theory, principles, and practice.* St. Louis, MO: Mosby.

Jensen, G., Iwarsson, S., & Ståhl, A. (2002). Theoretical understanding and methodological challenges in accessibility assessments focusing the environmental component: An example from travel chains in urban public bus transport. *Disability and Rehabilitation, 24,* 231-242.

Krantz, L. G. (1999). *Rörlighetens mångfald och förändring. Befolkningens dagliga resande i Sverige 1978 och 1996. [Diversity and Change in Mobility. People's daily travel in Sweden 1978 and 1996. In Swedish, summary in English].* Gothenburg University, Sweden: Department of Human and Economic Geography, School of Economics and Commercial Law.

Kroksmark, U., & Nordell, K. (2001). Adolescence: The age of opportunities and obstacles. What is it like to be a teenager with low vision? *Journal of Visual Impairment and Blindness, 25,* 213-225.

Law, M., Steinwender, S., & Leclair, L. (1998). Occupation, health, and well-being. *Can J Occup Ther, 65,* 81-91.

Lawton, M. P., (1986). *Environment and aging* (2nd ed.). Albany, NY: Center for the Study of Aging.

Lawton, M. P., & Nahemow, L. (1973). Ecology and the aging process. In C. Eisdorfer & M. P. Lawton (Eds.), *The psychology of adult development and aging* (pp. 619-674). Washington, DC: American Psychological Association.

Lawton, M. P. & Simon, B. (1968). The ecology of social relationships in housing for the elderly. *Gerontologist, 8,* 108-115.

Lieberman, D., Friger, M., Fried, V., Grinshpun, Y., Mytlis, N., Tylis, R., et al. (1991). Characterization of elderly patients in rehabilitation: Stroke versus hip fracture. *Disability and Rehabilitation, 21,* 542-547.

Mattingly, C., & Fleming, M. H. (1994). *Clinical reasoning: Forms of inquiry in therapeutic practice.* Philadelphia: F. A. Davis.

Mollenkopf, H., Marcellini, F., Ruoppila, I., Flaschenträger, P., Gagliardi, C., & Spazzafumo, L. (1997). Outdoor mobility and social relationships of elderly people. *Archives of Geronotology and Geriatrics, 24,* 295-310.

Nagi, S. Z. (1991). Disability concepts revisited: Implications for prevention. In A.M. Pope, & A. R. Tarlov (Eds), *Disability in America: Toward a national agenda for prevention* (pp. 309-327). Washington DC: National Academy Press.

Oxford Popular Dictionary & Thesaurus (3rd ed). (1998). Oxford, UK: Oxford University.

Pirie, G. H. (1979). Measuring accessibility: A review and proposal. *Environment and Planning Architecture, 11,* 299-312.

Preiser, W. F. E., & Ostroff, E. (Eds.). (2001). *Universal design handbook.* New York: McGraw-Hill.

Regeringens proposition (1999/2000:79). *Från patient till medborgare—en nationell handlingsplan för handikappolitiken [Governmental proposition. From patient to citizen—An national agenda for disability policies].* Stockholm: Author.

Reilly, M. (1966). A psychiatric occupational therapy program as a teaching model. *Am J Occup Ther, 22,* 61-67.

Salomon, I., Bovy, P., & Orfeuil, J. P. (Eds.). (1993). *A billion trips a day. Tradition and transition in European travel patterns.* Dordrecht, Netherlands: Kluwer Academic Publishers.

Ståhl, A. (1987). Changing mobility patterns and the aging population in Sweden. *Transportation Research Record,* 1135.

Ståhl, A. (1998). Service routes or low-floor buses. Study of the travel behaviour among elderly and disabled people. *Proceedings for the TRANSED '98, 8th International Conference on Mobility and Transport for Elderly and Disabled People,* 595-602. Western Australia: Indomed.

Ståhl, A. (1999). Public transport or special service or a mix? *Proceedings for the 2nd KFB—Research Conference.* Sweden: Lund University.

Ståhl, A., Brundell-Freij, K., & Makri, M. (1993). *The adaptation of the Swedish public transportation system—Yesterday, today and tomorrow. An evaluation.* TFB-report 1993:14. Stockholm.

Steinfeld, E., & Danford, G. S. (1999). Theory as a basis for research on enabling environments. In E. Steinfeld, & G. S. Danford (Eds.), *Enabling environments: Measuring the impact of environment on disability and rehabilitation* (pp. 11-33). New York: Kluwer Academic/Plenum.

Steinfeld, E., Schroeder, S., Duncan, J., Faste, R., Chollet, B., Bishop, M., et al. (1979). *Access to the built environments: A review of the literature.* Washington, DC: Government Printing Office.

United Nations. (1993). *Standard rules on the equalization of opportunities for persons with disabilities.* New York: Author.

Verbrugge, L. M., & Jette, A. (1994). The disablement process. *Social Science and Medicine, 38*(1), 1-14.

Wilcock, A. (1998). *An occupational perspective of health.* Thorofare, NJ: SLACK Incorporated.

World Health Organization (WHO). (1986). *Ottawa charter for health promotion.* Geneva: Author.

World Health Organization (WHO). (2001). *International classification of functioning, disability and health.* Geneva: Author.

Zemke, R., & Clark, F. (Eds.). (1996). *Occupational Science. The evolving discipline.* Philadelphia: F. A. Davis.

Chapter 16

Therapeutic Design of Environments for People With Dementia

Barbara Acheson Cooper, PhD, DipP&OT; and Kristen Day, PhD

CHAPTER OBJECTIVES

* To discuss the current and expanded role of occupational therapists in the design and implementation of therapeutic environments.
* To present the evidence on the therapeutic impact of design in long-term care settings for people with dementia.
* To discuss the application by occupational therapists of therapeutic design principles in long-term care settings for people with dementia.

INTRODUCTION

The therapeutic effects of the environment are well acknowledged (Birren & Sloane, 1980; Halpern, 1995; Lawton, 1982). In long-term care facilities for people with dementia, the design of the built environment has tremendous potential to improve function and well-being. This chapter summarizes the evidence on the effectiveness of the application of therapeutic environmental designs in long-term care settings for people with dementia and discusses the current and potential role of occupational therapists as members of the design team.

For pragmatic reasons, the term environment is limited to considerations of the physical or built environment and, to a lesser extent, the organizational/institutional environment. The population chosen for this illustration is one that is highly sensitive and vulnerable to attributes of the built environment (Brawley, 1997; Cohen & Weisman, 1991).

CURRENT ROLE OF THE OCCUPATIONAL THERAPIST

Occupational therapy practice is client centred, focusing on the self-identified needs and priorities of people with disabilities, with therapists acting as facilitators for achievement (Law & Baum, 2000). Since the early 1990s, occupational therapists have become increasingly aware that the context or environment within which occupation occurs has a powerful influence on occupational performance outcomes (Law, 1991). Models of practice in occupational therapy now include the environment as a key variable (Dunn, Brown, & McGuigan, 1994; Law et al., 1996). Professional focus on the relationship between health and the built environment has fostered interaction between occupational therapy and the design professions, particularly architecture. This is well illustrated by a funded project under the auspices of the Aging Design Research Program (ADRP) involving the American Association of Occupational Therapists (AOTA), the American Institute of Architects (AIA), and the Association of Collegiate Schools of Architecture (ACSA). This resulted in the development of numerous educational tools, such as library resources (ADRP, 1993) and relevant monographs (e.g., *Design for Aging: Strategies for Collaboration between Architects and Occupational Therapists*) (Ware, 1993).

Nonetheless, in spite of these thrusts and the generally accepted view that occupational therapists can best assess the functional abilities and limitations of clients, their professional input is still seldom sought by design teams. Instead, their role is limited to providing direct service to clients. A case will be developed in this chapter for an expanded role for occupational therapists, to enable them to contribute more definitively to the design of environments that promote optimal occupational performance for people with dementia. The person-environment-occupation (PEO) model (Law et al., 1996) and the most current evidence on the effectiveness of therapeutic environments (Day, Carreon, & Stump, 2000) will be used to frame and develop this position.

THE PERSON-ENVIRONMENT-OCCUPATION MODEL

The PEO model (Law et al., 1996) provides a useful tool for examining the barriers to occupational performance and therapeutic potential of the environment, for determining client strengths and needs, and for establishing occupational (functional) priorities. An assessment of the relationship among the three components (person, environment, and occupation) and of the various interfaces existing among them extends the possibilities for intervention. The PEO model predicts that in situations where clients' functional abilities are not expected to improve and may in fact be deteriorating, increasing the therapeutic benefits of the environment will result in maintained or increased occupational performance. Research supports these assumptions of the model (Green & Cooper, 2000; Strong et al., 1999). For an expanded description of this conceptual model, refer to Chapter 2 of this book and Law et al. (1996).

STRATEGIES AND APPROACHES FOR CONSIDERING DESIGN ISSUES

The authors of this chapter argue that occupational therapists can increase their involvement in the design of long-term care facilities for people with dementia and, by so doing, can maximize the therapeutic potential of these facilities. To realize this potential, occupational therapists must first have some understanding about how long-term care facilities are designed. The design of dementia care environments typically follows one of two models. (The two models are over-generalized for the purpose of this chapter; in reality, many design projects fall somewhere along a continuum between these two extremes. The authors thank Dr. Margaret Calkins, Innovative Designs in Environments for an Aging Society [IDEAS], and Dr. Uriel Cohen, University of Wisconsin-Milwaukee, for their helpful comments on the design process.)

In the first, "progressive" model of long-term care design, the design project begins long before an architect or design team is hired. The Board of Directors and upper-level administration make initial decisions about the overall scope of the design project (type of facility, size, etc.). They work with a consultant to develop a market feasibility study and marketing plan. Early in the planning process, management assembles a project team that includes staff from numerous departments in the facility (nutrition, activities, nursing, maintenance, etc.), representing various levels within those departments (e.g., the Nursing Director, Registered Nurses, and Certified Nursing Assistants). Using the overall scope of the project as a guide, the project team carefully considers how they want the new facility to function: What services will be provided? What is the desired "character?" What will be the activity program? Which staffing model is best? How will dining take place?

The program (called an "operational" or functional program) is the guide or recipe that the architect follows. Too often, the program is disappointingly brief, detailing only square footage and adjacencies—what spaces should be included, how large they should be, and which spaces should be next to each other. A more detailed program would include elements such as desired qualities, activities to accommodate, relationships between activities, and so on. The latter requires that facility staff carefully plan the type of facility they desire. A more detailed program also gives the architect more information, minimizing opportunities for miscommunication or false assumptions.

Armed with this information, the facility hires an architect, who then develops design alternatives to meet the priorities of the program. The project team continues its involvement throughout the design process, giving feedback on the design at each stage of design development, as the design becomes increasingly detailed.

By contrast, in the "conventional" model of long-term care design, the Board of Directors and upper-level administration assume most of the responsibility for the design of the facility. An architectural team is engaged early in the process and is asked to design the facility based on general information such as the number of projected residents, budget, regulatory codes, etc. Without specific information on the vision for the facility (i.e., how staff and management want the facility to function, and what they want to happen in each space) the architect may rely heavily on his or her past projects to guide the design of this facility. Consequently, the design of the building heavily dictates how the facility can be used, instead of the reverse. In this model, other staff members are not brought into the design process until most of the important decisions have already been made, and their feedback is used to assess design alternatives that are already quite specific.

Certainly, the "progressive" model of design is better suited to considering the therapeutic impacts of design, and is more likely to accommodate the active participation of occupational therapists in the design process.

Table 16-1

Therapeutic Design Principles for Long-Term Care Facilities for People With Dementia

Planning Principles	General Attributes of the Environment	Specific Activity Area
• Provide a responsive continuum of care • Tap local resources • Consider the appropriate group size for units	• Provide non-institutional character • Eliminate environmental barriers • Incorporate "things from the past" into the facility • Provide sensory stimulation without stress • Cluster small activity areas together • Provide opportunities for meaningful wandering • Incorporate other living things into the facility • Provide a range of public to private spaces	• Include positive outdoor spaces • Provide safe, welcoming, and identifiable entry and transition areas • Provide non-institutional shared spaces in each household • Provide domestic kitchens for residents' use • Ensure that dining is intimate and familiar • Create alcoves • Ensure that resident rooms are private and homelike • Ensure that bathing spaces foster dignity and independence • Ensure that toileting areas increase function and independence • Provide places for visiting that make this a positive experience
adapted from Cohen & Weisman, 1991		• Provide retreat areas for staff

THE THERAPEUTIC IMPACT OF DESIGN ON PEOPLE WITH DEMENTIA

The impact of the physical environment as a therapeutic intervention is well substantiated. Numerous "design guides" have been written in recent years, to offer planning, architectural, and interior design recommendations to architects and facility administrators, with the goal of enhancing safety, homelikeness, and other therapeutic qualities of long-term care facilities (Brawley, 1997; Calkins, 1988; Cohen & Day, 1993; Cohen & Weisman, 1991; Coons, 1987; Hiatt, 1987; Hyde, 1989; Lawton, 1979; Pynoos, Cohen, & Lucas, 1988; Regnier, 1997; Zeisel, Hyde, & Levkoff, 1994). A wide range of design interventions has been identified as having potential therapeutic impact (Table 16-1).

To maximize therapeutic impact, one must consider design and planning at various levels, from design decisions that affect the entire long-term care facility, to decisions that address only specific locations within it (Cohen & Weisman, 1991). At the most general level, potential therapeutic impact should be considered when making overarching *planning decisions* about the initial development of a long-term care facility (e.g., the size of the facility location, type and number of residents). For example, by electing to provide a continuum of care to accommodate people in various stages of dementia, facilities may minimize the trauma for residents and families that accompanies relocation when dementia advances (Cohen & Weisman, 1991). Other design decisions address *general attributes of the environment*—those aspects of the overall long-term care environment that have a therapeutic impact throughout the facility, including building organization. For example, residents may have higher levels of concentration and reduced agitation if noise from pagers, alarms, and telephones is kept to a minimum throughout the facility (Evans, 1989; Hall, Kirschling, & Todd, 1986). Likewise, the organization of long-term care facilities into well-differentiated "households" with distinct identities may help to promote orientation and way-finding among residents (Elmstahl, Annerstedt, & Ahlund, 1997). Finally, and especially relevant for occupational therapists, decisions about the design of specific rooms and activity spaces within the facility should be considered in terms of their therapeutic potential. For example, if kitchens are made accessible for residents' use and if generous amounts of counter and sink space are incorporated, residents may be more likely to participate in food preparation and clean up, thereby maintaining motor skills and a positive self-image (Calkins, 1988). At the same time, safety concerns pertaining to potential burns and fire must be addressed in the design of kitchen areas.

Evidence of the Therapeutic Impact of Design on Dementia

A growing body of empirical research examines the relationships between the design of the physical environment and the well-being of people with dementia. This section summarizes what is known about the therapeutic benefits of design in long-term care facilities, based on findings from a recent review of over 70 English language research reports on design and dementia that were published since 1980 (Day, Carreon, & Stump, 2000). Research findings should, in many instances, be regarded as tentative or preliminary, because of the small sample sizes (often less than 30 participants) that are employed in many studies. See Day, Carreon, and Stump (2000) for more information on this literature review and its findings. The discussion follows the framework outlined above, moving from the therapeutic impacts of general facility planning decisions to the impacts of decisions regarding the design of specific rooms. In the subsequent section, the authors discuss the potential involvement of occupational therapists in enhancing the therapeutic design of long-term care facilities for people with dementia.

PLANNING DECISIONS

Little is known about how the general planning of long-term care facilities (concerning, for example, the size of a facility, its organization, its levels of care) impacts the well-being of people with dementia, beyond conjecture. It appears, however, that facilities that provide "segregated" facilities for people with dementia realize greater therapeutic benefits, as compared to facilities that serve both cognitively impaired and cognitively intact residents. Because special care units (SCUs) vary enormously in their design, staffing, philosophy, activities, etc., comparisons of SCUs and integrated units are difficult to conduct and harder to interpret. Recent trends in dementia care increasingly specify physically separate and programmatically distinct SCUs that accommodate only cognitively impaired individuals. Research findings mostly support the positive impacts of SCUs, including benefits such as improvements or slowed decline in communication skills, self-care skills, social function, and mobility for residents; reduced emotional strain for families; and greater competence and satisfaction for staff (Annerstedt, 1993; Bellelli, et al., 1998; Benson, Cameron, Humbacj, Servino, & Gambert, 1987; Greene, Asp, & Crane, 1985; McCracken & Fitzwater, 1989; Skea & Lindesay, 1996; Swanson, Mass, & Buckwalter, 1993; Wells & Jorm, 1987).

The division of long-term care facilities into smaller groups of residents (6 to 12 rather than 20 to 60) is also associated with positive outcomes for residents. Benefits associated with smaller units (ranging from 9 to 19 residents) include higher motor functions, lower agitation and aggressiveness, more mobility, and more social interaction (Annerstedt, 1993, 1997; McAllister & Silverman, 1999; McCracken & Fitzwater, 1989; Moore, 1999; Morgan & Stewart, 1998; Netten, 1993; Skea & Lindesay, 1996; Sloane, et al., 1998).

GENERAL ATTRIBUTES OF THE ENVIRONMENT

More is known about the therapeutic impacts of environmental features such as lighting, noise levels, non-institutional design, and so on. To begin, research findings associate positive benefits with long-term care environments that are designed to be less institutional, such as incorporating homelike furnishings and personal belongings. Because many people with dementia do not require the skilled level of nursing care that traditional nursing homes were designed to provide (Day & Calkins, 2002; Liebowitz, Lawton, & Waldman, 1979; Snyder-Hiatt, 1978; US Department of Health and Human Services, 1981), dementia care environments are free to focus instead on providing a supportive, less institutional setting that compensates for cognitive loss. The benefits of such non-institutional design include improved intellectual and emotional well-being, reduced agitation, reduced trespassing and exit-seeking, and improved function of older adults with dementia and other mental illnesses (Annerstedt, 1994, 1997; Cohen-Mansfield & Werner, 1998; Kihlgren et al., 1992; McAllister & Silverman, 1999; Sloane, et al., 1998).

Other environmental attributes that are associated with therapeutic benefits include: specialized signage to improve residents' orientation (Hanley, 1981; Lawton, Fulcomer, & Kleban, 1984; Namazi & Johnson, 1991b; Namazi, Rosner, & Rechlin, 1991), high levels of lighting to improve way-finding and reduce agitation (Elmstahl, et al., 1997; Netten, 1989; Sloane, et al., 1998), and reduced noise and sensory stimulation (e.g., neutral color scheme, eliminating pagers and alarms) to improve concentration and reduce behavioral disturbances and use of physical and chemical restraints (Bianchetti, Benvenuti, Ghisla, Frisoni, & Trabucchi, 1997; Cleary, Clamon, Proce, & Shullaw, 1988; Namazi & Johnson, 1992b; Netten, 1993; Swanson, et al., 1993). Design has also been harnessed as a tool to improve safety in long-term care environments. For example, design interventions can discourage residents' attempts to wander away from a facility (a major safety concern for staff and family members). Design solutions to prevent unwanted exiting often do so by applying camouflaging techniques that are not interpreted as such by cognitively impaired residents, such as by using cloth panels or mini-blinds to disguise doors, windows, or door handles (Chafetz, 1990; Dickenson, McLain-Kark, & Marshall-Baker, 1995; Hussian & Brown, 1987; Mayer & Darby, 1991, Morgan & Stewart, 1999; Namazi, Rosner, & Calkins, 1989). Other design interventions successfully reduce falls (e.g., low furnishings, raised toilet seats, railings) (Pynoos & Ohta, 1991; Scandura, 1995).

SPECIFIC ACTIVITY SPACES

The design of individual rooms in the long-term care facility, such as outdoor spaces, dinning rooms, toilet rooms, and so on, has further therapeutic impact. Often, such rooms are designed to be more homelike and familiar. Researchers hypothesize, for example, that outdoor spaces make dementia care environments less institutional, while also allowing residents to get more sunlight and more exercise. The provision in dementia care environments of outdoor spaces for residents' use is associated with reductions in violent episodes (Mooney & Nicell, 1992).

Like outdoor spaces, kitchens and dining rooms are often characterized as essential for a more homelike atmosphere. Less institutional dining arrangements are linked to increased social interaction and improved eating behavior among residents (Götestam & Melin, 1987; Melin & Götestam, 1981). Dining on the dementia unit itself rather than in a centralized location in the facility is associated with reduced aggression among residents, because dining on the unit alleviates the need to crowd residents into elevators (a frequent site of aggression) to reach a centralized dining room (Negley & Manley, 1990). In yet another example of non-institutional design, residents who were provided with single rather than group bedrooms spent less time in their rooms and more time in motion, but engaged in less interaction (perhaps by choice) (Lawton, Liebowitz, & Charon, 1970).

Specific activity spaces are also designed to increase function and independence among residents. For example, specially-designed clothes closets have been found to increase autonomy in dressing for those with middle stage dementia. These closets have a section that contains only the clothes to be worn that day, presented systematically in the order that they would usually be donned (Namazi & Johnson, 1992a). Likewise, toilet rooms have been designed to increase independent use, such as by providing special signs to help residents locate public toilets and by replacing doors around toilets with curtains, so that toilets are more visible when not in use (Namazi & Johnson, 1991a; 1991b). Bathing rooms have been successfully designed to reduce over-stimulation and stress in bathing (Whall, et al., 1997; Kovach & Meyer-Arnold, 1996; Lawton, et al., 1984; Namazi & Johnson, 1996; Sloane, Lindeman, Phillips, Moritz, & Koch, 1995).

THE ROLE OF OCCUPATIONAL THERAPIST: CURRENT AND EXPANDED

If included in the planning phase of design, the occupational therapist could use the PEO model *broadly* to identify and prioritize key factors of the three components (person, environment, and occupation, respectively), and particularly, the environmental needs of residents that would hinder or maximize their occupational performance. The PEO model clearly demonstrates the utility to occupational therapists of the physical environment as one among numerous therapeutic resources.

Occupational therapists are infrequently involved in the early stages of designing long-term care facilities when overarching planning decisions are made. Their absence is not surprising, because these planning decisions too often focus on issues such as market, cost, and "look" of the facility, without serious consideration of its therapeutic potential (which would be the domain of occupational therapists). Further, as noted earlier, the planning stage of facility design is too often restricted to a management-centred team that omits most facility staff, including occupational therapists and others (staff in nutrition, activities) who could contribute valuable information. Finally, facility administrators and architects may have a limited view of the potential role of the occupational therapist, namely, that the therapist is there to provide individual therapy to residents of the long-term care facility. Administrators and architects may also lack knowledge as to occupational therapists' holistic and client-centred approach to fostering occupational performance, including their ability to assess functional abilities and recommend modifications to the built environment that enable function and compensate for losses.

Future potential (albeit more limited) exists for involving occupational therapists early in the design of long-term care facilities, where they could articulate the need to consider therapeutic impacts of planning decisions. The involvement of occupational therapists at the planning stage might take at least two forms. First, occupational therapists (and other facility staff, such as nutrition, nursing, activities, and so on) should participate on the design team from early in the design process, to ensure that the design reflects their vision, experience, and clients' functional needs. Second, occupational therapists trained in design and dementia could serve as consultants to the architectural design team, where they would bring special expertise on the therapeutic role of design in long-term care environments (the inclusion of an occupational therapist could also help to differentiate design firms as they compete for projects).

Occupational therapists can play an even greater role in later stages of the design process. For instance, occupational therapists could provide valuable assistance to the design team in identifying the appropriate general attributes of long-term care environments. Occupational therapists could readily participate in selecting appropriate furnishings, materials, and equipment that would maximize residents' function and occupational performance. Occupational therapists might also increase their involvement in design decision-making by partnering with interior designers who have primary responsibility for decisions about lighting, furnishings, and so on.

Interior designers' expertise in gerontology and in occupational performance varies widely. Consequently, occupational therapists could work with them to evaluate equipment, furnishings, etc. in terms of their potential therapeutic impact. In this case, occupational therapists would likely serve as consultants and members of the design team, or as representatives of the specific facility under design or renovation. These roles are highly consistent with trends in the field of occupational therapy that demonstrate that occupational therapists are beginning to serve more as professional, organizational-level consultants and supervisors, rather than only as therapists to individual clients. While these roles appear to fall outside the health care system, in fact they are dependent on the health care knowledge links that occupational therapists are able to provide.

Clearly, occupational therapists can participate in the design of specific activity spaces to increase therapeutic benefits and minimize detrimental effects. This role is consistent with the function that occupational therapists currently serve, namely, to identify environmental as well as individual and social interventions to improve client function. Again, occupational therapists can be especially helpful in establishing design priorities (e.g., that safety must supersede function as a design consideration). Because of their knowledge of dementia and of the corresponding resident population, occupational therapists can also suggest where particular design decisions would be inappropriate, such as with the excessive and non-homelike use of technology that may be confusing and frightening to people with dementia (e.g., the use of recorded devices to warn patients against exiting or to guide them to activity areas).

Occupational therapists occupy a unique position that can increase their effectiveness in design decision-making. As staff members in the facility, they are trusted and are knowledgeable about how the facility actually operates. This can provide them with information to make appropriate recommendations about design that will promote optimal function for residents and staff. The occupational therapist can also help to anticipate and overcome resistance (both covert and expressed) that often accompanies change. Frequently, resistance is associated with a lack of knowledge of the traditional skills of the professional that are being applied in a manner perceived as unusual.

When helping to address general attributes of the environment and to design specific rooms and activity areas that would promote occupational performance, the occupational therapists would primarily use the PEO model to consider the relationship between the abilities and needs of the residents (P) of the facility, as well as their desired function (O). This step, added to knowledge of the evidence on the impact of therapeutic design, will allow the occupational therapist or team to develop priorities for desired characteristics of the physical environment (E) that would most improve the "fit" of the three components. Nonetheless, tradeoffs may be required when determining priorities—for example, making toilets more visible to encourage residents' independent toileting may simultaneously decrease the residents' privacy.

The knowledgeable occupational therapist will have most to offer when the director of the facility is supportive of a therapeutic approach to design. In reality, the extent to which a facility will adopt a therapeutic design approach depends largely on the leadership of the facility director or owner. A true commitment to using the environment as a therapeutic resource requires both knowledge of this potential and leadership to promote a culture change in the facility. Occupational therapists may need to promote their expertise to the directors of facilities through advocacy and marketing to ensure that their contributions are considered.

CASE EXAMPLE

Author Barbara Acheson Cooper, an occupational therapist holding a PhD in architecture (environment and behavior), was asked to consult to a private facility for seniors. The invitation was made after members of the local Alzheimer's Society heard BC speak on the importance of the environment for the well-being of people with dementia at National and International Alzheimer's Conferences. She was also listed as a consultant in an Aging Design Research Program (1993) monograph.

The complex in question was located in a small southern Ontario town. It consisted of several apartment clusters housing a continuum of care services ranging from independent living units for retirees to full long-term care. A significant number of residents in the independent living units who cared for partners with dementia had indicated that they required short-term relief services for themselves and their partners. In response, the owners were considering the conversion of the eighth and ninth floors of an unfilled building into closed units dedicated to this purpose.

Cooper conducted a post-occupancy evaluation of the complex and proposed conversion areas. This systematic process sets objectives for the evaluation, allows the assessor to determine the fit between the facility and the desired use and to make written recommendations for meeting the stated objectives. For a full description of the post-occupancy evaluation (POE) process see Cooper, Ahrentzen, and Hasselkus (1991).

During the first visit, the mission of the institution and objectives for the assessment were determined with the administrator. The main goal was to provide a report for the board on the overall viability of the proposed changes. Subject to reaffirming the proposal, the second goal was to identify the specific areas of focus for the renovations.

Cooper used the PEO model to structure the post-occupancy evaluation. She first identified the strengths and weaknesses of the proposal as these pertained to the institution's mandate, making note of major limiting factors of person, environment, and occupation (e.g., the progressive nature of Alzheimer's Disease, the need to use upper floors in the apartment building to house this facility, and the balance between safety concerns and functional independence for residents). Following this, Cooper identified strengths of the three components and the transactions required among them in order to maximize function and quality of life for residents and their families. An example of this step is recognizing the interdependence of positive institutional interest in promoting independent function for residents with cognitive disabilities and the provision of environmental amenities such as well-signed, easily-accessible washrooms.

The POE of the existing facility identified the following factors as organizational strengths:

* Flexible services and resources providing a continuum of care for residents
* Focus on a non-institutional environment
* Emphasis on independent living
* Recognition of the need for privacy
* Provision of frequent opportunities for resident-community interaction

Following this, an assessment of the two floors to be converted was carried out. Strengths of the existing built environment included:

* The location of the building, which allowed expansion of the existing continuum of care into similar adjacent units thereby reducing relocation stress for the residents
* The size of the apartments, which allowed easy conversion into one- or two-bed units
* An elliptical corridor that could easily be developed as a wandering path for residents

Four major weaknesses or problem areas were identified (safety, self-care/washrooms, wayfinding, therapeutic gardens) and possible solutions were suggested. Each of these issues represents an occupational performance issue (the transaction between person, environment, and occupation), where the overlap of the anticipated person skills of the residents was not adequate considering the challenges of the physical environment and the demands of the tasks themselves within the environment. Recommendations could then be generated with a focus on environmental modifications to better optimize the PEO relationship.

* Safety
    1. *Issue*—Sensory ambiguity: large single-paned windows extended to within a foot from the floor, thereby creating the illusion that there were no boundaries between the room and the outdoors

       Solution—Application of ornamental grillwork to the inside of the windows from their base to a height of 3.5 feet
    2. *Issue*—Fear of unwanted resident exiting

       Solution—The use of inobtrusive technology such as alarm bracelets, specialized locks for the unit entry and specialized call buttons for the elevator
    3. *Issue*—Fear of burns to residents and fire hazards from inappropriate use of kitchen stove

       Solution—Disengagement of equipment not in use, specialized locks and controls for equipment in use
    4. *Issue*—Fear of burns from hot water taps

       Solution—Installation of temperature controls for hot water taps (mixers)
* Self Care/Washrooms
    1. *Issue*—Inappropriate or lack of use of specialized areas such as bathrooms and activity rooms

Solution—The use of attention cues, such as bright colors or graphics to encourage entry into the area(s)

2. *Issue*—Inappropriate equipment in bathroom

   Solution—Age-relevant modifications to washroom equipment to facilitate safe use including: the use of lever handles on sinks, raised toilet seats, and relocation of toilet paper dispensers to allow for the installation of side rails to the toilets

* Way finding

   1. *Issue*—Facilitating family and professional visits

      Solution—Improving signage, such as the use of a large number across from the elevator doors to identify floor level

   2. *Issue*—Facilitating resident identification of bedroom

      Solution—Enhancement of room identification cues (i.e., larger numbers, name plates, areas for personal photographs)

   3. *Issue*—Enhancement of leisure activities

      Solution—Encouraging resident use of the wandering path and activity room through the use of devices such as activity stations containing brightly colored graphics or relevant memorabilia

* Therapeutic gardens

   1. *Issue*—The board wanted to consider information on therapeutic gardens as a possible future project

   2. *Solutions*—A resource file, articles, and information on the development and use of therapeutic gardens were made available. The conversion floors looked out over a school playground and in the absence of gardens, ways to facilitate resident observation of these outdoor activities were developed

The consultant used the POE to address all phases of design, focusing in particular on general attributes of the environment and specific spaces that would promote residents' function and quality of life. A final report making a general recommendation to proceed with the conversions and delineating the specific recommendations was forwarded to the board along with reprints of relevant research information. Typically, a consultant has no control over the degree to which recommendations are implemented. This was the case here; however, members of the local Alzheimer's Society later told Cooper informally that all recommendations had been implemented.

CONCLUSION

This chapter described the design process used for building or renovating long-term care facilities for people with dementia and explored the ways in which occupational therapists might collaborate. Additionally, it summarized the research on the therapeutic benefits for this population that can be provided by the built environment. The PEO model was used to support the contention that occupational therapists' specific abilities to assess client function and their familiarity with carrying out environmental assessments and modifications allows them to be valuable members of the design team, providing key information on the relationship between residents' function and the environment that might otherwise be missed. Notwithstanding the logic of this argument, the establishment of customary links for occupational therapists with design teams will require the combined efforts of management, members of traditional design professions, and the occupational therapy profession itself.

Summary of Environmental Interventions

The case described focused on two major components of the environment: the built and organizational environment. Specifically, it reported on the role played by the organizational environment in recognizing and seeking the expertise of the occupational therapy consultant. In this case, Cooper's knowledge of the characteristics and needs of the user population and of the therapeutic design literature as well as her ability to assess the physical environment allowed her to make practical, evidence-based suggestions for changes that would improve the fit between the environment and the residents. By placing an emphasis on therapeutic design, she also emphasized the importance of a client-centred approach. The occupational therapist was able to support her recommendations to the board with key articles from the literature, thereby providing unbiased evidence for the proposed changes. In this example, the focus was on environmental modifications that would maximize the functional potential and quality of life of residents, meet the needs of their family caregivers, and assist the organization responsible to provide best care.

Application and Recommendations

While the application of occupational therapy knowledge as described in the case currently rests with a few consultants, it need not be so limited in the future. The key lies with promoting mutual knowledge of the interdigitation of roles and abilities of the many professions responsible for therapeutic design. Educational approaches have already been used to foster the relationship between occupational therapists and architects. As previously described, the American Association of Architects' and the AOTA's collaboration to develop monographs and library resources that would foster mutual knowledge of both professions provides a precedent (ADRP, 1993; Ware, 1993). Extending this thrust to include interior designers and managers of long-term care facilities is highly recommended. This could take the form of combined projects or assignments for students in occupational therapy, architecture, interior design, and health administration that would result in increased knowledge of one another's expertise, and an increased likelihood that each would seek the other out for projects in the workplace. Professional conferences, courses, journals, books, newsletters, and invited talks also could provide important fora for dissemination of these ideas. Equally powerful however, is word-of-mouth promotion. In this regard, all practitioners can encourage awareness of how they could contribute to the development of designs that promote the occupational performance of their clients. Occupational therapists interested in acting as consultants in this area can hone and develop their expertise in many ways: through informal and formal studies, through affiliations with groups representing clients with dementia, and by working with architects and builders interested in this topic.

Study Questions

1. Name three ways in which environmental designs can be used to enhance the occupational performance of people with dementia; specify the expected effects.

2. Name and discuss three ways in which environmental designs could impede the occupational performance of people with dementia.

3. Identify two or more examples of design principles that are in conflict with each other. How can these conflicts be addressed?

4. Discuss areas where the design team's roles are clearly defined and areas where they may overlap. How can role conflict be avoided?

5. Describe ways by which occupational therapists can inform facility owners and managers of their ability to contribute to the design process.

REFERENCES

Aging Design Research Program. (1993). *Design for aging network directory.* Washington, DC: Author.

Annerstedt, L. (1993). Development and consequences of group living in Sweden. *Social Science and Medicine, 37,* 1529-1538.

Annerstedt, L. (1994). An attempt to determine the impact of group living care in comparison to traditional long-term care on demented elderly patients. *Aging Clin Exp Res, 6,* 372-380.

Annerstedt, L. (1997). Group-living care: An alternative for the demented elderly. *Dement Geriatr Cogn Disord, 8,* 136-142.

Bellelli, G., Frisoni, G. B., Bianchetti, A., Boffelli, S., Guerrini, G. B., Scoutuzzi, A., et al. (1998). Special Care Units for demented patients: A multicenter study. *Gerontologist, 38,* 456-462.

Benson, D. M., Cameron, D., Humbacj, E., Servino, L., & Gambert, S. R. (1987). Establishment and impact of a dementia unit within the nursing home. *J Am Geriatr Soc, 35,* 319-323.

Bianchetti, A., Benvenuti, P., Ghisla, K. M., Frisoni, G. B., & Trabucchi, M. (1997). An Italian model of dementia special care unit: Results of a pilot study. *Alzheimer Dis Assoc Disord, 11,* 53-56

Birren, J. E., & Sloane, B. (Eds.). (1980). *Handbook of mental health and aging.* Englewood Cliffs, NJ: Prentice-Hall.

Brawley, E. C. (1997). *Designing for Alzheimer's disease. Strategies for creating better care environments.* New York: John Wiley and Sons.

Calkins, M. P. (1988). *Design for dementia: Planning environments for the elderly and the confused.* Owing Mills, MD: National Health.

Chafetz, P. K. (1990). Two-dimensional grid is ineffective against demented patients exiting through glass doors. *Psychol and Aging, 5,* 146-147.

Cleary, T. A., Clamon, C., Proce, M., & Shullaw, G. (1998). A reduced stimulation unit: Effects on patients with Alzheimer's disease and related disorders. *Gerontologist, 28,* 511-514.

Cohen, U., & Day, K. (1993). *Contemporary environments for people with dementia.* Baltimore, MD: Johns Hopkins University.

Cohen, U., & Weisman, G. D. (1991). *Holding on to home: Designing environments for people with dementia.* Baltimore, MD: Johns Hopkins University.

Cohen-Mansfield, J., & Werner, P. (1998). The effects of an enhanced environment on nursing home residents who pace. *Gerontologist, 38,* 199-208.

Coons, D. H. (1987). Overcoming problems in modifying the environment. In H. J. Altman (Ed.), *Alzheimer's disease: Problems, prospects, and perspectives* (pp. 321-328). New York: Plenum.

Cooper, B., Ahrentzen, S., & Hasselkus, B. (1991). Post-occupancy evaluation: An environment-behaviour technique for assessing the built environment. *Can J Occup Ther, 58,* 181-188.

Day, K., & Calkins, M. (2002). Design and dementia. In R. B. Bechtel & A. Churchman (Eds.), *The new environmental psychology handbook* (pp. 374-393). New York: John Wiley & Sons.

Day, K., Carreon, D., & Stump, C. (2000). The therapeutic design of environments for people with dementia: A review of the empirical research. *Gerontologist, 40*, 397-421.

Dickinson, J. I., McLain-Kark, J., & Marshall-Baker, A. (1995). The effects of visual barriers on exiting behavior in demented care unit. *Gerontologist, 35*, 127-130.

Dunn, W., Brown, C., & McGuigan, A. (1994). Ecology of human performance: A framework for thought and action. *Am J Occup Ther, 48*, 595-607

Elmstahl, S., Annerstedt, L., & Ahlund, O. (1997). How should a group living unit for demented elderly be designed to decrease psychiatric symptoms? *Alzheimer Dis Assoc Disord, 11*, 47-52.

Evans, B. (1989). *Managing from day to day: Creating a safe and workable environment.* Minneapolis, MN: Department of Veterans Affairs Medical Center.

Gotestam, K. G., & Melin, L. (1987). Improving well-being for patients with senile dementia by minor changes in the ward environment. In L. Levi (Ed.), *Society, stress, and disease* (Vol. 5: Old Age, pp. 295-297). Oxford University.

Green, S., & Cooper, B. (2002). Occupation as a quality of life constituent: A nursing home perspective. *British Journal of Occupational Therapy, 63*, 17-24.

Greene, J. A., Asp, J., & Crane, N. (1985). Specialized management of the Alzheimer's disease patient: Does it make a difference? *Journal of the Tennessee Medical Association, 78*, 559-563.

Hall, G., Kirschling, M. V., & Todd, S. (1986). Sheltered freedom—An Alzheimer's unit in a ICF. *Geriatric Nursing, 7*, 132-137.

Halpern, D. (1995). *Mental health and the built environment: More than bricks and mortar?* Bristol, PA: Taylor & Francis.

Hanley, I. G. (1981). The use of signposts and active training to modify ward disorientation in elderly patients. *J Behav Ther Exp Psychiatry, 12*, 241-247.

Hiatt, L. G. (1987). Environmental design and mentally impaired older people. In H. J. Altman (Ed.), *Alzheimer's disease: Problems, prospects, and perspectives* (pp. 309-390). New York: Plenum.

Hussian, R. A., & Brown, D. C. (1987). Use of two-dimensional grid to limit hazardous ambulation in demented patients. *Journal of Gerontology, 42*, 558-560.

Hyde, J. (1989). The physical environment and the care of Alzheimer's patients: An experiential survey of Massachusetts' Alzheimer's units. *American Journal of Alzheimer's Care and Related Disorders and Research, 4*, 36- 43.

Kihlgren, M., Brane, G., Karlsson, I., Kuremyr, D., Leissner, P., & Norberg, A. (1992). Long-term influences on demented patients in different caring mileus, a collective living unit and a nursing home: A descriptive study. *Dementia, 3*, 342-349

Kovach, C. R., & Meyer-Arnold, E. A. (1996). Coping with conflicting agendas: The bathing experience of cognitively impaired older adults. *Scholarly Inquiry for Nursing Practice: An International Journal, 10*(1), 23-36.

Law, M. (1991). The environment: A focus for occupational therapy. *Can J Occup Ther, 58*, 171-179.

Law, M., & Baum, C. (2000). Measurement in occupational therapy. In M. Law, C. Baum, & W. Dunn, (Eds.), *Measuring occupational performance* (pp. 3-29). Thorofare, NJ: SLACK Incorporated.

Law, M., Cooper, B., Strong, S., Stewart, D., Rigby, P., & Letts, L. (1996). The person-environment-occupation Model: A transactive approach to occupational therapy. *Can J Occup Ther, 63*, 9-23.

Lawton, M. P. (1979). Therapeutic environments for the ages. In D. Canter & S. Canter (Eds.), *Designing for therapeutic environments: A review of research* (pp. 233-276). New York: John Wiley and Sons.

Lawton, M. P. (1982). Competence, environmental press, and the adaptation of older people. In M. P. Lawton, P. G. Windley, & T. Byerts (Eds.), *Aging and the environment: Theoretical approaches* (pp. 33-59). New York: Springer.

Lawton, M. P., Fulcomer, M., & Kleban, M. (1984). Architecture for the mentally impaired elderly. *Environment and Behavior, 16*, 730-757.

Lawton, P., Liebowtiz, B., & Charon, H. (1970). Physical structure and the behavior of senile patients following ward remodeling. *Aging and Human Development, 1*, 231-239.

Liebowitz, B., Lawton, M. P., & Waldman, A. (1979). Evaluation: Designing for confused elderly people. *American Institute of Architects Journal, 68*, 59-61.

Mayer, R., & Darby, S. J. (1991). Does a mirror deter wandering in demented older people? *Int J Geriatr Psychiatry, 6*, 607-609.

McAllister, C. L., & Silverman, M. A. (1999). Community formation and community roles among persons with Alzheimer's disease: A comparative study of experiences in a residential Alzheimer's facility and a traditional nursing home. *Qualitative Health Research, 9*, 65-85.

McCracken, A. L., & Fitzwater, E. (1989). The right environment for Alzheimer's: Which is better, open versus closed units? Here's how to tailor the answer to the patient. *Geriatric Nursing, 10*, 293-294.

Melin, L., & Gotestam, K. G. (1981). The effects of rearranging ward routines on communication and eating behaviors of psychogeriatric patients. *J Appl Behav Anal, 14*, 47-51.

Mooney, P., & Nicell, P. L. (1992). The importance of exterior environment for Alzheimer residents: Effective care and risk management. *Healthcare Management Forum, 5*, 23-29

Moore, K. D. (1999). Dissonance in the dining room: A study of social interaction in a special care unit. *Qualitative Health Research, 9*, 133-155.

Morgan, D. G., & Stewart, N. J. (1998). High versus low density special care units: Impact on the behavior of elderly residents with dementia. *Canadian Journal on Aging, 17*, 143-165.

Morgan, D. G., & Stewart, N. J. (1999). The physical environment of special care units: Needs of residents with dementia from the perspective of staff and care givers. *Qualitative Health Research, 9*, 105-118.

Namazi, K. H., & Johnson, B. D. (1991a). Environmental effects on incontinence problems in Alzheimer's patients. *American Journal of Alzheimer's Care and Related Disorders and Research, 6*, 16-21.

Namazi, K. H., & Johnson, B. D. (1991b). Physical environmental cues to reduce the problems of incontinence in Alzheimer's disease units. *American Journal of Alzheimer's Care and Related Disorders and Research, 6*, 22-29.

Namazi, K. H., & Johnson, B. D. (1992a). Dressing independently: A closet modification model for Alzheimer's disease patients. *American Journal of Alzheimer's Care and Related Disorders and Research, 7*, 22-28.

Namazi, K. H., & Johnson, B. D. (1992b). Pertinent autonomy for residents with dementias: Modification of the physical environment to enhance independence. *American Journal of Alzheimer's Care and Related Disorders and Research, 7*, 16-21.

Namazi, K. H., & Johnson, B. D. (1996). Issues related to behavior and the physical environment: Bathing cognitively impaired patients. *Geriatric Nursing, 17*, 234-239.

Namazi, K. H., Rosner, T. T., & Calkins, M. P. (1989). Visual barriers to prevent ambulatory Alzheimer's patients form exiting through an emergency door. *Gerontologist, 2*, 699-702.

Namazi, K. H., Rosner, T. T., & Rechlin, L. (1991). Long-term memory cuing to reduce visuo-spatial disorientation in Alzheimer's disease patients in a special care unit. *American Journal of Alzheimer's Care and Related Care and Related Disorders and Research, 6*, 10-15.

Negley, E. N., & Manley, J. T. (1990). Environmental interventions in assaultive behavior. *Journal of Gerontological Nursing, 16*, 29-33.

Netten, A. (1989). The effect of design of residential homes in creating dependency among confused elderly residents: A study of elderly demented residents and their ability to find their way around homes for the elderly. *Int J Geriatr Psychiatry, 4*, 143-153.

Netten, A. (1993). *A positive environment? Physical and social influence on people with senile dementia in residential care.* Aldershot, England: Ashgate.

Pynoos, J., Cohen, E., & Lucas, C. (1988). *The caring home booklet: Environmental coping strategies for Alzheimer's care givers.* Los Angeles: Long-Term Care National Resource Center at UCLA/USC.

Pynoos, J., & Otha, R. J. (1991). In-home interventions for persons with Alzheimer's disease and their care givers. *Physical & Occupational Therapy in Geriatrics, 9*, 83-92.

Regnier, V. (1997). Design for assisted living. *Contemporary Long-Term Care, 20*(2), 50-52.

Scandura, D. A. (1995). Freedom and safety: A Colorado center cares for Alzheimer's patients. *Health Progress, 76*, 44-46.

Skea, D., & Lindesay, J. (1996). An evaluation of two models of long-term residential care for elderly people with dementia. *Int J Geriatr Psychiatry, 11*, 233-241.

Sloane, P. D., Lindeman, D. A., Phillips, C., Moritz, D. J., & Koch, G. (1995). Evaluating Alzheimer's special care units: Reviewing the evidence and identifying potential sources of study bias. *Gerontologist, 35*, 103-111.

Sloane, P. D., Mitchell, C. M., Preisser, J. S., Phillips, C., Commander, C., & Burker, E. (1998). Environmental correlates of resident agitation in Alzheimer's disease special care units. *J Am Geriatr Soc, 46*, 862-869.

Snyder-Hiatt, L. (1978). Environmental changes for socialization. *Journal of Nursing Administration, 18*(1), 44-55.

Strong, S., Rigby, P., Stewart, D., Law, M., Letts, L., & Cooper, B. (1999). Application of the person-environment-occupation model: A practical tool. *Can J Occup Ther, 66*, 122-133.

Swanson, E. A., Maas, M. L., & Buckwalter, K. C. (1993). Catastrophic reactions and other behaviors of Alzheimer's residents: Special unit compared with traditional units. *Arch Psychiatr Nurs, 7*, 292-299

US Department of Health and Human Services. (1981). *Progress report on senile dementia of the Alzheimer's type* (81-2342). Washington, DC: National Institute of Health.

Ware, C. (Ed). (1993). *Strategies for collaboration between architects and occupational therapists*. Available from AIA/ACSA Council for Architectural Research, 1735 New York Avenue, NW, Washington DC 20006.

Wells, Y., & Jorm, A. F. (1987). Evaluation of special nursing home unit for dementia sufferers: A randomized controlled comparison with community care. *Aust N Z J Psychiatry, 21*, 524-531.

Whall, A. L., Black, M. E., Groh, C. J., Yankou, D. J., Kuperschmid, B. J., & Foster, N. L. (1997). The effect of natural environments upon agitation and aggression in late stage dementia patients. *American Journal of Alzheimer's Disease, 12*, 216-220.

Zeisel, J., Hyde, J., & Levkoff, S. (1994). Best practices: An environment-behavior (E-B) model for Alzheimer special care units. *American Journal of Alzheimer's Care and Related Disorders and Research, 9*, 4-21.

Chapter 17

Expanding Environments Through Technology

Linda Petty, BSc(OT), OT Reg. (Ont.)

Chapter Objectives

* To demonstrate how assistive technology can make environments more accessible and supportive of occupational performance.
* To describe developments in assistive technology and improved accessibility of commercial technology.
* To encourage readers to utilize technology to enable clients to overcome activity limitations and participation restrictions in their community, at home, work, school, or in virtual environments.
* To encourage readers to advocate for funding for technology for individuals, facilities and institutions, as well as advocate for the accessibility of technology by all people, including those with disabilities.

Introduction

Occupational therapy has traditionally focused on rehabilitating or habilitating individuals with disabilities by increasing the skills of the person, or perhaps by modifying the occupation they are unable to fulfill with aids. Technology offers numerous tools to further achieve these goals with the provision of communication aids for people with speech impairments, wheeled mobility aids for those clients unable to walk effectively, and computer access tools for clients unable to write or read due to motor, learning, or visual impairments. Occupational therapists are now involved in the assessment, fitting or training, and provision of technical aids for individuals with a wide range of impairments.

Technology is now so powerful, all-pervasive, and relatively affordable that its usefulness extends beyond compensating for individual needs. The environment can be expanded for people with disabilities or modified to such an extent that it can support individuals in the workplace or most school settings, as well as to meet their leisure needs. With the aid of technology, environments can be shaped, which offers opportunities to interact with others, acquire information, obtain employment, shop, play, and conduct financial transactions. This chapter highlights a range of new options offered by advances in technology, which occupational therapists can harness to enhance and expand environments to improve occupational performance of their clients. This chapter describes an expanding role for occupational therapists in which technology can be used to provide alternate ways for clients to accomplish activities, and to enable clients to overcome limitations in participation and improve their quality of life.

New Advances in Technology to Improve Access to Environments

In recent years, there have been great advances in technology while prices for technology have plummeted. This has resulted in a proliferation of technical tools for communication, writing, and work throughout the developed world. The vast majority of the population has benefited from improved communication and expanded options for leisure and work offered by telecommunications and the Internet. Assistive technology, in the form of adapted computer access, synthesized, and then digitized speech and specialized software began to be developed in the early 1980's.

Developments in Assistive Technology and Improved Accessibility of Commercial Technology

There are number of key factors that have influenced the growth and development of the assistive technology industry since its onset in the days of the Apple IIe, the IBM compatible XT system and mechanical drive "electric" wheelchairs. As with any industry, the quantity of consumers and size of market is a major influence on development. There has been an increase in the number of people requiring assistive technology due, in part, to increases in the total population, and because the proportion of the United States and other western populations with disabilities has risen markedly during the past 25 years. Two distinct trends have contributed to the increasing overall prevalence of disability. First, there has been a gradual rise, due largely to demographic shifts associated with an aging population. Second, there has been a rapid increase that took place during the early 1990's of the number of children and young adults reported as having disabilities (Kaye, LaPlante, Carlson, & Wenger, 1996).

To provide a sampling of both the magnitude of the needs and the effect of aging on disability, statistics are offered in Table 17-1, from the United States Census brief of 1997, *Disabilities Affect One-Fifth of All Americans*.

As the baby boom generation becomes senior citizens, the increasing market share of aging consumers is providing an incentive to the developers of assistive technology. Similarly, mainstream manufacturers are increasingly aware of the need to provide information, appliances, and other products in formats that will be more accommodating to varia-

Table 17-1

Prevalence of Disability by Selected Characteristics

Out of a total population of 267.7 million non-institutional individuals, 52.6 million (or 19.7 percent) had some type of disability. Among those with a disability, 33.0 million (or 12.3 percent of the total population) had a severe disability and 10.1 million (or 3.8 percent of the total population) needed personal assistance with one or more ADLs or instrumental activities of daily living (IADLs).

The likelihood of having a disability increased with age. Among those 45 to 54 years old, 22.6 percent had some form of disability, 13.9 percent had a severe disability, and 3.6 percent needed personal assistance. For those 65 to 69 years old, the comparable estimates were 44.9 percent, 30.7 percent, and 8.1 percent. For the oldest age group shown in Table 1, 80 years old and over, the estimates were 73.6 percent, 57.6 percent, and 34.9 percent.

(U.S. Census, 1997)

tions in mobility, vision, hearing, cognition, and manual dexterity. An increase in numbers of individuals with visual impairments is widely predicted, as many eye conditions are age-related, as indicated in the following excerpt from the *Statistics on Vision Impairment: A Resource Manual* (Leonard, 2002):

"*The following projections are based on estimates of self-reported vision impairment from The Lighthouse National Survey on Vision Loss (The Lighthouse Inc., 1995) and applied to US Census population projections:*

* *17% of persons age 45 and older report some form of vision impairment, representing 16.5 million persons. By the year 2010, when all baby boomers are age 45 and older, this number will increase to 20 million.*

* *9% of persons age 45 and older report a severe vision impairment, representing 8.7 million persons. By the year 2010, when all baby boomers are age 45 and older, this number will increase to 10.7 million.*

* *About 7.3 million, or 21% of persons age 65 and over, report some form of vision impairment. As baby boomers age, this number will reach 8.3 million in the year 2010, 11.3 million in 2020, and in 2030, 14.8 million persons age 65 and older will report some form of vision loss.*

* *About 3.8 million or 11% of persons age 65 and over report a severe vision impairment. As baby boomers age, this number will reach 4.3 million in the year 2010, 5.9 million in 2020, and in 2030, 7.7 million persons age 65 and older will report a severe vision impairment*" (Leonard, 2002, p. 13).

Universal design is an approach to creating everyday environments and products that are usable by all people, regardless of age or ability without the need for adaptation or specialized design (Trachtman, Mace, Young, & Pace, 1999). More manufacturers of standard products are moving toward universal design, while others have improved and increased the range of specialized assistive technology available to people with disabilities. (Universal design is discussed in greater detail in Chapter 7.)

Finally, advocacy on the part of people with disabilities and their support networks has resulted in improved integration into the mainstream and in changes to legislation that now enforce equality in the school and work place. The legislative requirements, in turn, support and escalate the development of adaptations and technology required by people with disabilities, as it assures their adoption in a wider range of employment and school settings. The Americans With Disabilities Act (ADA) (US Government Access Board, n.d.) was a landmark civil rights law for individuals with disabilities in the United States. It enables them to seek and obtain gainful employment on a level playing field with non-disabled peers. Employers are required to provide reasonable accommodations so that persons with disabilities can perform the essential functions of the job. Technology solutions are one of the reasonable accommodations available to assure these individuals an opportunity to use their skills in gainful employment (Struck, 1999).

Provision of assistive technology in the school environment has been improving in both funding and service delivery, which again widens the market for commercial development of devices. While funding varies by province in Canada, access was implemented nationally in the United States with the passing of the Individuals With Disabilities Education Act (IDEA) in 1990 and the reauthorized IDEA '97, which included the explicit direction to meet the student's need for assistive devices and services (Zabala, J., & Quality Indicators Assistive Technology Consortium [QIAT] Leadership Team, 2000).

The Rehabilitation Act Amendments of 1998 in the United States was the next major landmark in legislation and has improved the accessibility of technology used around the world. The law strengthens Section 508 of the Rehabilitation Act and requires that all electronic and information technology provided by the federal government be made accessible to people with disabilities. The Access Board is responsible for developing accessibility standards for such technology for incorporation into regulations that govern federal procurement practices (US Government, 1998). As the US government is a major buyer of technology, all of the major technology firms, such as Microsoft, Hewlett Packard, IBM, and Adobe have implemented the standards set by the Access Board. As new interfaces and software are being developed, accessibility for people with disabilities is being built into the products.

"Section 508," as it has come to be known, has forcibly brought the need for accessibility to the attention of companies who develop software for Web design. Extensive work had been done on developing standards for Web site design and access by the Web Accessibility Initiative of the W3C, the standard-setting body for the World Wide Web, and several software accessibility-checking engines are available free of charge (World Wide Web Consortium, n.d.). Prior to the enforcement of Section 508, Web site accessibility was mainly a priority only to groups already aware of the needs of people with disabilities. Now, major Web design software players, such as Macromedia, a developer of software for Web content and design, offer support such as accessibility templates, solution kits, and Web site testing to promote accessibility.

USING TECHNOLOGY TO ENABLE
OCCUPATIONAL PERFORMANCE WITHIN VARIOUS ENVIRONMENTS

Technology in the School Environment

Generally, technology has been applied to individual needs, to improve occupational performance at the individual level for written and spoken communication, mobility, reading, signaling, etc. For school-age children, the equipment needs to be available in their classroom or needs to be transported with them to various classrooms. It can be made available for use at home, for communication, or to accomplish school tasks. Because everyone has a need for personal and leisure activities, the equipment can usually be used to facilitate these activities without major additional expenditures. For example, an elementary school student with cerebral palsy, requiring mobility aids and both face-to-face and written communication devices will need a range of technology. The portable communication and writing aid at school can also be used to communicate and write at home. With the addition of the game Wheel of Fortune (Hasbro, Providence, RI), it can serve as a leisure tool as well. While some communication devices can also meet writing needs, there is often the need for an additional stationary computer system dedicated to writing, projects, and Internet access, with the most effective and speedy access method for that dedicated task. The stationary system should also include a flatbed scanner and optical character recognition (OCR) software if the individual has difficulties either with manipulating books and paper, or has a learning disability or visual impairment that prevents grade/age appropriate reading. It is very important that training in the needed technologies be provided for both home and school environments and that the same or compatible hardware and software be used in both environments to minimize new learning (Treviranus & Petty, 2002).

For students with low vision or blindness, technology supports their need to take notes, read, write, and view educational learning media. Accessing closed circuit televisions (CCTVs), computer equipment and learning keyboarding skills is critical at an early age, to support development of reading and writing skills. The assistive technology is a fundamental work tool, the equivalent to paper and pencil for the non-disabled students (Texas School for the Blind and Visually Impaired, n.d.). Academic and vocational skills are heavily dependent on computer-based technology for this client group. The student will need early access to screen enlargement software with speech support for writing, reading, and using educational software, and a CCTV for reading, seeing toys, coloring, etc. If the student has no functional vision, access to scanning/reading technology and a screen reader will support reading and writing, with substantial training. For information on screen reading, screen enlargement, OCR and CCTV technology, see the technical glossary at the University of Toronto's Adaptive Technology Resource Center at www.utoronto.ca/atrc/reference/tech/techgloss.html.

It is important to assess tactile skills and ensure additional supports are available within the educational and home environment to learn Braille (Texas School for the Blind and Visually Impaired, n.d.). Braille can then become an important option for reading, reading presentation notes, making personal notes, labeling articles, and reading information from a computer screen with a Braille display. It is critical to ensure periodic re-evaluations of the student's functional vision and academic accommodations to ensure that the student has materials adapted to accommodate his or her visual needs throughout the curriculum. Teamwork is vital, as education, therapy, and vision professionals often work independently, instead of together, in providing service. This is particularly critical for the student with both visual and physical disabilities, as access to the assistive technology to accommodate for vision may need modification for physical needs, with sticky keys (sticky keys, filter keys, and toggle keys are accessibility options to facilitate use of the regular keyboard by those with motor impairments. They are found in the control panel of Macintosh or Windows operating systems), alternative keyboards, or other physical access methods (Petty & Buskin, n.d.). For example, a young teenager with severe ataxia affecting her speech, vision, and motor control was able to use an expanded keyboard with a custom overlay to input text and commands to a computer system with screen magnification and speech feedback (Figure 17-1). This same individual found a great deal of satisfaction from reading, particularly Nancy Drew mysteries. With her ataxia she was only able to read one or two paragraphs in 15 min-

Figure 17-1. Computer system using 20-inch monitor and screen magnification and reading software to support vision, with high contrast, programmed keyboard overlay to operate screen magnification software and type.

utes. However, with a scanned-in book or stories from the Nancy Drew Web site, she was able to read independently at an effective rate by listening to the computer software and focusing her vision on the text as the software scrolled it across the screen.

Children with learning disabilities need psycho-educational testing to determine the type and extent of learning disability, and an assessment for assistive technology to support them in academic pursuits. Assistive technology and computer use can help students work around or bypass their learning disabilities, enabling them to function at a level more in line with their intellectual abilities, by compensating for difficulties in areas such as reading, writing, spelling, math, organization, and memory (Frostig Center, n.d.). Voice recognition software has been demonstrated to actually remediate for learning disabilities, improving an individual's writing skills over a prolonged trial (Higgins and Raskind, 2000).

For vision, physical, and learning impairments, the use of textbook material in a digital format is recommended. New digital technology allows text and audio to be combined and incorporates sophisticated searching and indexing facilities. New standards have been adopted that will replace audio tape "talking books" with digital versions. The searchable aspect of this technology, higher quality audio and more convenient format will be a great advantage to individuals with low vision, blindness, or learning disabilities in navigating their way around text (Recording for the Blind and Dyslexic, n.d.). Internet and Intranet options allow individuals the access to electronic documents that their motor impairment would normally limit or restrict them from acquiring.

Web-based educational programs and Web-based libraries have also influenced the availability of educational options for people with disabilities. While many of the Web-based learning software packages are not particularly user-friendly to those with poor vision or motor impairments, efforts are underway to improve the accessibility of the software (University of Toronto Resource Centre for Academic Technology, 1999).

Technology for children has to grow to accommodate their developing skills and changing educational programs. Likewise, computer equipment is usually functional for only 4 to 5 years due to changes in system requirements by newer software and peripherals (Treviranus & Petty, 2002). For those at high school and post-secondary levels, technology needs to keep pace with the academic needs by adding advanced level math functions, foreign languages, geography, and science capacity. Access to advanced-level research is important, through scanned in reference material, on-line material, and electronic reference material on compact discs. Accommodations in the classroom can be very important for such tasks as note taking, as those with physical, visual, or learning disabilities will usually not be able to copy information from lecture or blackboards at an effective rate. Accommodation can take the form of a note-taking device, tape recording, and photocopies or electronic files of the instructor's notes.

Technology in the Work Environment

Whether employment equity is mandated by legislation or not, employers can access a vital pool of trained individuals by either accommodating employees who have acquired disabilities or by hiring individuals who have disabilities. Assessment and modification of the work environment should include a review of all computer hardware and software in use, reading and writing materials, and communication mechanisms in place. Recommendations can then incorporate modification to the technology to improve the accessibility for the individual and ways of ensuring future accessibility of the workplace technology. In addition to technology, recommendations can encompass any needed changes to methods of accomplishing tasks and other aspects of the occupation that the technology cannot accommodate. Environmental modifications may include changes in door opening mechanisms and controls for lighting and heating for access by an individual using a wheelchair. Use of email and electronic and online documents can facilitate accommodation for individuals with visual or learning disabilities, as the text can be enlarged or efficiently read out loud. CCTVs can be integrated into computer systems to allow an individual with low vision to view hard copy material on the same computer monitor as their electronic documents. For example, this can facilitate the copying of invoice information into budget spreadsheets (Figure 17-2).

Office equipment designed for postural support and flexible configuration is now readily available. While the term "ergonomic" is much over-used, the wide range of variable height desks and keyboard trays, footrests, armrests, and chair configurations facilitate accommo-

Figure 17-2. CCTV camera integrated with computer system to allow viewing of paper documents (on left side of monitor) while inputting figures into spreadsheet document (on right side of monitor). Use of "false colors" for document viewed with CCTV assists in distinguishing between work areas.

dating individuals with orthopedic or repetitive strain injuries. Accommodating the individual who has held a manual labor position after he acquires a visual or physical disability is not always possible. However, retraining to use computer technology can be a viable rehabilitation option.

Portable digital assistants, such as Palm Pilots, are being harnessed to improve the success of people with cognitive impairments in work environments. Individuals with learning disabilities or head injuries can use the commercial devices to support organizational skills and short/long term memory (Zaraska, 2001). For severe cognitive impairments, customization creates one or two picture interfaces that cue the individual verbally, in the same way a job coach would, when activated (Pocket Coach, n.d.). These devices are relatively low cost, look modern and attractive, and offer supports for independence and competency.

Working from home has become almost conventional, and can greatly benefit people with disabilities. One can connect to the workplace using videoconferencing, email, and FTP or an Intranet for sharing files. Work from home can be suitable for those who suffer from disorders strongly affected by fatigue or fragile medical conditions, or who require specialized assistive technology that is not easily supported in their work environment. For example, one professor at a major urban university who experienced progressive loss of vision found it most effective to have his assistive technology set up in his home office, and receive student papers by email and fax. He used screen magnification in conjunction with screen reading to read his work and that of his students in electronic form. He also used an OCR reading program to read research materials that he or a graduate student would scan into the system.

Technology in the Home Environment

Technology can maintain people in their environment of choice or enable them to move to more diverse, private, and individualized settings without risk to their safety or health, thus maximizing occupational performance. Many older adults or individuals with fragile health may require emergency assistance. They would not be able to stay alone in their environment of choice without emergency radio-based signaling aids (Steggles & Leslie, 2001). For example, a senior who had several strokes injured her leg in a fall. She was a widow who lived alone in a one-level home in a small village, with her busy family members living in neighboring cities. This individual was able to use a wheeled walker and community agency support for transportation, but was at risk for further falls. Based on assessment and trials she was able to choose a personal response system that gave her a remote button to call an emergency monitoring service, even from outdoors in the garden. She was able to continue living in her home, while recovering from the leg injury, and kept the system afterward as it ensured her safety and allowed her to remain independent in her home. Similarly, electronic aids to daily living (EADLs), also known as environmental control units, allow individuals with very limited physical skills to live in the community with attendant care or family support, as they can control aspects of their home, school, or work environments (Bain, 1997). Single and multiple output transmitters are used to activate and operate a range of personal entertainment equipment, home security systems, including video surveillance and door openers, telephones, lights, thermostats, and appliances. The transmissions can be activated by voice, switch access, or wheelchair controls and can also be set up for access from bed. EADLs in a home can enable occupational performance by active participation in educational, vocational, or leisure activities such as computer and Internet-based educational programs or vocational tasks and games. Independent access to the community is facilitated through telephone and door controls, allowing the individual to exit and enter and arrange transportation at will. Many individuals with severe physical disabilities are otherwise dependent on passive recreation, such as watching TV or listening to the radio, and dependent on caregivers for controlling lights, room temperature, window blinds, and the entrance or exit of visitors (Rigby et al., 2003).

For example, one individual with C4 level quadriplegia uses a combination of direct mouthstick selection to a multifunction transmission remote and computer keyboard and trackball, voice recognition software, integrated power chair and door opener controls and a speaker phone as EADLs to support his occupational performance (Figure 17-3). His parent's home was renovated to provide a wheelchair accessible entrance, elevator, and a mini apartment. He has attendant care and family support to assist with activities of daily living (ADL) needs. Initially, as a student at a local university, he was able to arrange accessible transportation or drive his wheelchair to the nearby campus and pursue his studies with support from adapted computer systems at home and on-campus. He then pursued his interest in Web design by developing several Web sites and participating in beta testing Web site creation software. He now does contract work from his home computer as the technical editor for major reference works on Web development software. He has been either the technical editor or contributing author for six major publications over the last 3 years and finds the adaptive technology allows him to work from home at his own pace to complete contracts in a timely manner, even in the fast-paced world of Web development. Leisure interests are also supported through email communication, computer graphics, and computer-based photography with a remote-controlled video camera and a digital camera. He is able to operate the window blinds, TV, video, and stereo equipment with the same remote control with a mouthstick, which is also used to access the adapted light switches and a speaker phone (Figure 17-4). A switch activated call system is in place at night. These supports

Figure 17-3. Client with workstation accessed with voice recognition and mouth stick for trackball and keyboard operation, used for web design and technical editing, leisure, and communication.

enable him to be productive and independent in adjusting aspects of his environment, and to pursue his leisure interests.

Occupational therapists are moving into a leading role in consulting to institutions and community agencies on implementing assistive technology such as door openers, call systems, and accessible computer and information kiosks, beyond the traditional realm of wheelchair ramps and accessible washrooms. EADLs can make a considerable difference in supporting the educational, vocational, and leisure interests of individuals with disabilities, as well as ensuring safety and enhancing communication and community access. Occupational therapists can support clients as advocates for equipment, accessibility, and funding, and can be a strong proactive voice in public arenas. We can enable clients to grow in occupational performance and prevent restrictions to community participation.

Figure 17-4. The entertainment unit, adjacent drapery controls, and up to 12 different devices can be accessed with one multi-purpose remote control.

Accessing Virtual Environments Through Technology

New communities are being created on the World Wide Web that can enhance support networks for people with disabilities, and thus expand the environments in which they live. Technology thus enables people to join and contribute to groups that are geographically diverse, but share the same needs created by their disability. These interactive Web sites create a supportive virtual environment for the exchange of ideas, jointly working on projects and advocacy, sharing best practices, and more. Examples of these include the Learning Disabilities Resource Community (LDRC) project for learning disabilities (http://www.ldrc.ca), SNOW Web site for special education (http://snow.utoronto.ca) and Ability On-line BBS Web site for children with disabilities and their families (www.ablelink.org). One of the primary goals of the LDRC, for example, is to bring together the four major groups that make up the learning disabilities community. These groups consist of individuals with learning disabilities (and their families), community organizations, educational institutions, and research centres. The LDRC initiates an active communication loop between members of these four groups with the goal of developing relationships that are beneficial to all individuals. LDRC members may communicate in "real time" with others using the site, bringing together a virtual community of groups across the nation who study, work, or live with special needs learners, or are affected themselves by a learning disability. Methods for doing this include a newsletter, a chat area on the Web site, forums for parent and teacher discussion or general discussion, posting of relevant articles, and project information (Learning Disabilities Resource Community, n.d.). "Ability Online," on the other hand, is focused on sharing and support rather than factual communication and information exchange. It offers a friendly and safe computer friendship network where children and youth with disabilities or chronic illnesses connect to each other as well as to their friends, family members, caregivers, and supporters. Networking in this way removes the social and physical barriers that can come with having a disability or illness, and provides opportunities for children and adolescents to exchange information, build self-confidence, share hope and encouragement, and make valued friends (Ability On-Line, n.d.).

There are a range of Web sites created by people with disabilities to share their experiences and connect with others via the Internet. Some are personal Web pages prepared by one person about his or her experiences with life and disability. Others take the form of online magazines or "e-zines," which are similar to conventional magazines in their format. Reference sites, often called "jump sites," will list and catalogue online disability culture resources. Some Web sites are hybrids of all three types mentioned. Beyond these differences in format is the commonality that all disability culture Web sites provide a perspective upon the experience of living with a disability (Heard, 1999). Steven E. Brown, co-founder of the Institute on Disability Culture, offers the following definition of disability culture, while acknowledging that there is much controversy over the meaning of the term:

> "People with disabilities have forged a group identity. We share a common history of oppression and a common bond of resilience. We generate art, music, literature, and other expressions of our lives and our culture, infused from our experience of disability. Most importantly, we are proud of ourselves as people with disabilities. We claim our disabilities with pride as part of our identity. We are who we are: we are people with disabilities." (Brown, 1996, p. 2)

Using technology to change the interface to the World Wide Web to suit individuals' needs and disabilities can then enable access to information and job searches, and email communication and Web surfing for personal and productivity needs. This can be accomplished both individually with home computer access equipment and for larger groups of people to access community or government systems. One interface, called Web-4-All (W4A) package was developed for Industry Canada to ensure seniors, new Canadians, and people with disabilities or literacy challenges have computer access to services and information on the Internet. It accomplishes this by creating a customizable computer interface based on universal design principles. Each user can select, save, and deploy their own interface preferences using smart card technology that automatically configures a suite of assistive software and hardware (University of Toronto, Resource Centre for Academic Technology, n.d.).

CONCLUSION

Major developments in commercial technology and assistive technology, influenced by growth in populations of consumers and legislation, have greatly expanded the options available to improve occupational performance and environments to support people with disabilities. Technology can improve the support for individuals, their families, and support staff working with them, decrease isolation, and improve the quality of education, employment, and life. It can also greatly expand the local, personal environment to provide social, emotional, and spiritual support from individuals across the Internet, improving the occupational performance of those who move from relative isolation into a broader culture.

Evidence on the Use of Technology to Improve Occupational Performance

Technology has been an intervention of choice with the growth of computers in the workforce over the last 20 years, and occupational therapists are taking an increasing role in recommending hardware and software, as well as ergonomic accommodations. With the explosive growth in mainstream industry, and the wide-ranging development of the assistive technology industry, the establishment of standards of practice, protocols, and outcome measures has lagged far behind. Rather, the focus has been to give people with disabilities access to the ever-increasing range of commercial tools and devices as quickly as possible. Standardized assessments for accessing technology such as switch interfaces or computers are not widely available, and what is in use has rarely been tested for reliability and validity, nor published (Bromley, 2001). The bulk of documentation on changing occupational performance with technology is found in multi-disciplinary literature publications, such as *Technology and Disability, Assistive Technology, Journal of Augmentative and Alternative Communication*, or in even less readily available sources, such as conference proceedings from technology-focused conferences. Evidence upon which to base practice and clinical recommendations tends to be found locally, or on list serves or bulletin board services, by drawing on the experience of others. The disadvantage of this is obvious, as many individual therapists do not have access to experts, and the waiting list for cross referrals can be extensive, or services are unavailable. The other disadvantage is the unfortunate potential to repeat mistakes, if the general practice is perhaps not the most effective, but widely accepted. There is a strong need for outcome-based studies in the field of technology, and for publication in peer-reviewed journals that are indexed for electronic access. The growing international effort to obtain outcome measures in this field can be seen on the Assistive Technology Outcome Measures Web site at http://www.utoronto.ca/atrc/reference/atoutcomes/. A few commonly used measures for evaluation of outcomes of assistive technology services are listed in Table 17-2.

The Canadian occupational performance measure (COPM) (Law, et al., 1998) has been successfully applied to assistive technology (Reid, Rigby, & Ryan, 1999), and is incorporated as a needs review and follow-up measure in both research studies and clinical practice in a growing number of assistive devices service delivery centres (Lim & Lenker, 2000; Petty, Treviranus, Lee, & Minevich, 2001).

Conference proceedings demonstrate that many outcome measures are now being implemented in athletic training practice in some areas (Brandt & Iwarsson, 2001; Lim & Lenker, 2000; Petty, Treviranus, Lee, & Minevich, 2001). Several major studies on the psychological impact of the use of EADLs and consumer satisfaction have been published (Jutai, Rigby, Ryan, & Stickel, 2000; Stickel, Ryan, Rigby, & Jutai, 2002). These studies, and others of similar nature, will add considerable credibility to advocates seeking funding for what can be expensive yet essential equipment to enable people with disabilities to live safely, independently, and higher quality lives in the community.

Summary of Environmental Intervention

As described, technology can have a major impact on the environment, making it more suitable, safe, and enabling for the individual. The process of assessing individuals' needs and choosing technology that effectively supports them keeps the focus client-centred, preventing the focus on changing to technology for the sake of technology. The use of technology

Table 17-2

Outcome Measures That Focus on Assistive Technology

Outcome Measures	Acronym	Relevant Publication
Cost-Effective Rehabilitation Technology Through Appropriate Indicators	CERTAIN	Andrich, Ferrario, & Moi, 1998
Efficiency of Assistive Technology and Services	EATS	Brodin & Persson, 1995
The Matching Person and Technology	MPT	Scherer, 1996
Psychosocial Impact of Assistive Devices Scale	PIADS	Day & Jutai, 1996
The Quebec User Evaluation of Satisfaction With Assistive Technology	QUEST	Demers, Weiss-Lambrou, & Ska, 1996

to create new environments or improve occupational performance within environments is innovative as it draws on the exponentially growing wealth of products and Web sites. Applications of technology rapidly change, with each intervention offering more effective and improved solutions with changes in the field. The process of considering technology to accommodate the individual more effectively in the home, school, or work environment is only practiced in relatively small pockets of occupational therapy services, so much so that there is not a large body of evidence from which to base practice. The above approach is drawn from the growing body of literature available and expert clinical interventions as illustrated by international conferences and list serves.

Prevalence of Technology in Occupational Therapy Practice

Technology is not widely discussed in occupational therapy literature as a treatment modality to improve occupational performance, yet it is evolving into an obvious option of environmental accommodation in today's increasingly technological world. The statistic that more than 49 million Americans demonstrate an ongoing disabling condition, and of these only just over 13 million use assistive technology suggests that technology is not universally accessible (Olson & DeRuyter, 2002). Occupational therapists need to expand their current practice, theories, and models to incorporate applying technological accommodations to clients' environments. Knowledge is the key factor needed to enable practicing occupational therapists to make the shift to using technology. Many in the profession have never had formal training in technology, and may only use computers for basic word processing to support their practice. Technology brings with it a high learning curve with its ongoing rapid change and chronic breakdowns. Occupational therapists do not want to feel incompetent or unable to properly support clients, thus many may avoid the area all together. As new, techno-savvy generations enter the field, therapists will be better equipped to embrace assistive technology and implement it with their clients.

Advocacy is also a critical role for therapists: for funding for technology, for accessibility of mainstream Web sites and technical devices, and for accessibility of software and hardware for use in school and work environments. One study of American occupational therapists found that therapists' lacked knowledge of ADA Title III and consequently reported inaction with regards to this important act (Redick, McClain, & Brown, 2000). They concluded that this lack of knowledge and action could affect the accessibility of the environment, and the independence and empowerment of clients who are wheelchair mobile, which would impede progress toward fully inclusive communities. Legislation outside of the United States is frequently inadequate in assuring full access to needed equipment, services, and information for citizens with disabilities. Occupational therapists need to be aware of existing legislation, advocate for client rights and education, and encourage their clients to self-advocate.

Implementation of Technology

The implementation of technology to improve the fit with the environment crosses age groups and disability types. Seniors use computers with screen magnification or CCTVs to maintain their visualization of leisure and self-care materials. Students and workers manage reading, communication, and writing tasks with accessible computer technology. The World Wide Web has become a place of information, communication, shopping, work, and play for an escalating number of individuals, from home, work, school, libraries, and an increasing range of access sources. Technology can be a support for mental illness, providing support through on-line communication and information. It can also greatly expand the local, personal environment to provide social, emotional, and spiritual support from individuals across the Internet, improving the occupational performance of those who move from relative isolation into a broader culture. Regardless of with which client group an occupational therapist works, applications exist for expanding environments and supporting occupational performance with technology.

Recommendations

Critical areas of current and future need are in advocacy around legislation and funding needs and in systematic collection and publication of outcome data on use of technology. Occupational therapy students need to be equipped to act in these areas; educational institutions and health care organizations need to make these areas a priority for research and publication. Similarly, courses, papers, and columns on assistive technology ought to be widespread among national occupational therapy organizations and educational programs, not limited to specialty conferences and publications. Resources can be focused on the development of these areas to further increase the base of evidence from which to practice occupational therapy in assistive technology.

Study Questions

1. How can technology within the school environment enable occupational performance for:

    a. The primary student with mobility and speech impairments?

    b. The grade-eight student with visual impairments?

    c. The high school student with learning disabilities?

2. How can technology be used to maintain people with severe or progressive impairments in their environment of choice?

3. What are the major outcome measures that focus on assistive technology?

4. How could the COPM be used to evaluate the provision of assistive technologies?

5. Suggest how technology could be utilized to improve occupational performance in the following case study involving the work environment. Delineate how you could evaluate the outcome of the intervention.

CASE STUDY: JOE

Joe was hired as an Employment Specialist by XYZ, an agency that facilitated return to work by women on social assistance programs. Joe had acquired quadriplegia at the C4, 5, and 6 level 21 years ago and had some grasp movement in his right upper extremity, less in his left. Joe could type with a typing stick consisting of a pencil in a universal cuff around his right hand, and drives a power chair for mobility. Joe has worked in employment services prior to being hired and has been a computer user for a number of years. XYZ was in the process of planning renovations to a large workshop room to increase office space, so the assessment request was to include recommendations for construction of an accessible office space and environment, as well as computer access methods.

Joe's major job tasks involve:

* Intake interviews with clients, recording employment data into Contact 4 database software. This software consists of specific data fields with drop-down lists or self-formatting boxes. It uses the tab key to move between fields, or the mouse to select menus or new sections of the client's file. Job counseling with clients also requires use of the Contact 4 software.

* Job development with employers and service agencies via letters and telephone contact. The office uses MS Office 2000, especially MS Word, with which Joe is familiar.

* Telephone contact with employers and clients.

* Preparing and presenting workshops to clients using MS PowerPoint.

Problems identified:

* Accessing office environment from the wheelchair with limited hand function. Concerns were raised regarding the access to light switches, door handles, and access to the workshop room, door widths, office heat controls, and computer set-up and access.

* Joe fatigues quickly in making notes with a pen, so is unable to document interviews with handwriting.

* Joe types with one typing stick in a universal cuff, the equivalent of one finger typing; however, this requires movement from the shoulder and can be fatiguing for prolonged input needs.

* Joe requested information to improve his safety in his home environment, an attendant care-supported apartment, as he had no method to contact the attendants or use the phone from bed.

REFERENCES

Ability On-Line. (n.d.). *What is Ability On-Line?* Retrieved January 4, 2002, from www.ablelink. org/public/about.htm.

Andrich R., Ferrario, M., & Moi, M. (1998). A model of cost-outcome analysis for assistive technology. *Disabil Rehabil, 20*, 1-24.

Bain, B. K. (1997). Environmental control systems. In B. K. Bain & D. Leger (Eds.), *Assistive Technology: An interdisciplinary approach* (pp. 119-139). New York: Churchill Livingstone.

Brandt, A., & Iwarsson, S. (2001). Do certain groups of older people benefit the most from the use of powered wheelchairs? *Proceedings of the RESNA 2001 Annual Conference*, Arlington VA: RESNA Press, 212-214.

Brodin, H., & Persson, J. (1995). Cost-utility analysis of assistive technologies in the European Commission's TIDE Program. *Int J Technol Assess Health Care, 11*, 276-83.

Bromley, B. (2001). Assistive technology assessment: A comparative analysis of five models. *Proceedings of the 2001 CSUN Conference*. Retrieved on January 6, 2002 from http://www.csun.edu/cod/conf2001/proceedings/0193bromley.html.

Brown, S. (1996). *Disability culture: A fact sheet.* Las Cruces, NM: Institute on Disability Culture. Retrieved on January 5, 2002 from the Institute on Independent Living Web site: http://www.independentliving.org/docs3/brown96a.html.

Day, H., & Jutai, J. (1996). Measuring the psychosocial impact of assistive devices: The PIADS. *Canadian Journal of Psychology, 9*(2), 159-168.

Demers L., Weiss-Lambrou, R., & Ska, B. (1996). Development of the Quebec user evaluation of satisfaction with assistive technology (QUEST). *Assistive Technology, 8*, 3-13.

Frostig Centre. (n.d.). *Trafford center for technology and learning disabilities.* Retrieved January 5, 2002, from the Frostig Center Web site: http://www.frostig.org/technology/tctld.htm.

Heard, C. P. (1999). Disability culture on-line: Consumers informing occupational therapists: Overview of disability culture and relevance to occupational therapy practice. *Occupational Therapy NOW, 1*(4), 14-16.

Higgins, E., & Raskind, M. (2000). Speaking to read: The effects of continuous vs. discrete speech recognition systems on the reading and spelling of children with learning disabilities. *Journal of Special Education and Technology, 15*, 1. [Electronic version] Retrieved September 22, 2002, from http://jset.unlv.edu/15.1/higgins/first.html.

Jutai, J., Rigby, P., Ryan, S., & Stickel, S. (2000). Psychosocial impact of electronic aids to daily living. *Assistive Technology, 12*, 123-131.

Kaye, H., LaPlante, M. Carlson, D. & Wenger, B. (1996). Trends in disability rates in the United States, 1970-1994. *Disability Statistics Abstract, 17*, 1- 4.

Law, M., Baptiste, S., Carswell, A., McColl, M., Polatajko, H., & Pollock, N. (1998). *Canadian occupational performance measure* (3rd ed.). Toronto, ON: CAOT Publications ACE.

Learning Disabilities Resource Community. (n.d.) *Community.* Retrieved January 5, 2002 from http://www.ldrc.ca/.

Leonard, R. (2002). *Statistics on vision impairment: A resource manual* (5th ed.). New York: Lighthouse International. Retrieved September 17, 2002, from http://www.lighthouse.org/prodpub_publications.htm#surveys.

Lim, S. F., & Lenker, J. A. (2000). Evaluating the clinical utility of the PIADS for computer accommodation. *Proceedings of the RESNA 2000 Annual Conference* (pp. 110-112). Arlington VA: RESNA.

Olson, D., & DeRuyter, F. (2002). Introduction. *Clinician's guide to assistive technology.* St. Louis, MO: Mosby.

Petty, L., & Buskin L. (n.d.). *Identifying and adapting for visual impairment in the client with physical disabilities.* Retrieved January 5, 2002, from University of Toronto, Resource Centre for Academic Technology Web site: http://www.utoronto.ca/atrc/rd/library/papers/visualdisabled.html.

Petty, L. S., Treviranus, J., Lee, I., & Minevich, A. (2001). Study on the severity of vision loss and client satisfaction and performance. *Proceedings of the RESNA 2001 Annual Conference* (pp. 242-244). Arlington VA: RESNA.

Pocket Coach. (n.d.). *Pocket coach research history.* Retrieved January 3, 2002 from http://www.ablelinktech.com/ResearchPage.asp?SelectedProduct=Pocket%20Coach.

Recording for the Blind and Dyslexic. (n.d.). *RFB&D's AudioPlus™.* Retrieved January 3, 2002, from http://www.rfbd.org/digitalaudio.htm.

Redick, A. G., McClain, L., & Brown, C. (2000). Consumer empowerment through occupational therapy: The Americans with Disabilities Act Title III. *Am J Occup Ther, 54*, 207-213.

Reid, D., Rigby, P., & Ryan, S. (1999). Evaluation of the rigid pelvic stabilizer on occupational performance of children with cerebral palsy. *Pediatr Rehabil, 2*, 37-55.

Rigby, P., Ryan, S., Joos, S., Cooper, B., Jutai, J., & Steggles, E. (2003). *The impact of electronic aids to daily living upon the lives of persons with cervical spinal cord injuries.* Manuscript submitted for publication.

Scherer, M. J. (1996). Outcomes of assistive technology use on quality of life. *Disabil Rehabil, 18*, 439-448.

Steggles, E., & Leslie, J. (2001). Simple technology to enhance independent living. *Proceedings of the International Conference on Technology and Aging* (pp. 150-153). Toronto, ON: ICTA.

Stickel, S., Ryan, S., Rigby, P., & Jutai, J. (2002). Toward a comprehensive evaluation of the impact of electronic aids to daily living: Evaluation of consumer satisfaction. *Disability and Rehabilitation, 24*, 115-125.

Struck, M. (1999). Technology solutions for ADA compliance. *Occupational Therapy in Health Care, 11*, 23-28.

Texas School for the Blind and Visually Impaired. (n.d.). *Principles of assistive technology for students with visual impairments.* Retrieved September 15, 2002, from http://www.tsbvi.edu/technology/principles.htm.

Texas School for the Blind and Visually Impaired. (n.d.). *Texas School for the Blind and Visually Impaired Pre-Braille Assessment.* Retrieved January 5, 2002 from website: www.tsbvi.edu/education/index.htm#Assessment.

Trachtman, L. H., Mace, R. L., Young, L. C., & Pace, R. J. (1999). The universal design home: Are we ready for it? *Phys Occup Ther Geriatr. 16*, 1-18.

Treviranus J., & Petty L. (2002). Computer access. In D. Olson & F. DeRuyter, (Eds.) *Clinicians' guide to assistive technology* (pp. 91-113). St. Louis, MO: Mosby.

United States Government Access Board. (n.d.). *The Americans With Disabilities Act (ADA) of 1990.* Retrieved September 21, 2002, from http://www.access-board.gov/about/ADA.htm.

United States Government. (1998). *Section 508 of the Rehabilitation Act: Electronic and information technology accessibility standards.* Retrieved January 3, 2002, from http://www.access-board.gov/508.htm

United States Census. (1997). *Disabilities affect one-fifth of all Americans.* Retrieved Sept 20, 2002, from http://www.census.gov/hhes/www/disability.html.

University of Toronto Resource Centre for Academic Technology. (1999). *SNOW courseware evaluation 1999—summary.* Retrieved January 4, 2002, from http://snow.utoronto.ca/initiatives/crseval/summary.html.

University of Toronto, Resource Centre for Academic Technology. (n.d.). *Web-4-all.* Retrieved January 3, 2002, from http://www.utoronto.ca/atrc/rd/smartcard/index.html.

World Wide Web Consortium. (n.d.). *References on web accessibility.* Retrieved January 3, 2002 from http://www.w3.org/WAI/References/.

Zabala, J., & Quality Indicators Assistive Technology Consortium (QIAT) Leadership Team. (2000). Quality indicators for assistive technology services in school settings [Electronic version], *Journal of Special Education Technology, 15*, 4. Retrieved October 5, 2002, from http://jset.unlv.edu/15.4/Zabala/first.html.

Zaraska, N. (2001). Personal digital assistants (PDAs): Putting technology in the palm of your hands. *Occupational Therapy NOW, 3*, 10.

SECTION FOUR

Conclusion

In this last section of the book, the editors synthesize the paradigms, theories, models, and applications presented by the authors for using environments in occupational therapy interventions. The challenges facing occupational therapists as they continue to advance their thinking and daily practice are discussed. Future directions of the profession are reflected upon as occupational therapists continue to develop knowledge, skills, and abilities in using environments to enable occupational performance.

Using Environments in Occupational Therapy: Challenges, Opportunities, and Future Directions

Lori Letts, PhD, OT Reg. (Ont.); Patty Rigby, MHSc, OT Reg. (Ont.); and Debra Stewart, MSc, OT Reg. (Ont.)

CHAPTER OBJECTIVES

* To synthesize key insights gained from the previous chapters.
* To summarize major challenges and opportunities facing occupational therapists using environments to enable occupational performance.
* To describe possible future roles for occupational therapists in using environments to enable occupational performance.
* To share strategies occupational therapists can implement to move practice toward that future.

LESSONS LEARNED ABOUT USING ENVIRONMENTS IN OCCUPATIONAL THERAPY INTERVENTION

Previous chapters of this text have demonstrated that occupational therapists use environments in many different ways with varying populations to promote occupational performance. Examples have confirmed the value of modifying social, institutional, cultural, and physical environments at societal, community, group, and individual levels. Scenarios have been shared to demonstrate working with children, adults with physical disabilities and mental illness, and older adults who are dealing with challenges of typical aging and issues such as dementia and stroke. From these examples, a number of lessons can be culled that are themes across the book. These include:

* The need to focus on occupational performance first.
* The tremendous potential of the environment to enable occupational performance.
* The links between other disciplines and occupational therapy in valuing environments.
* The potential to expand occupational therapy practice to new settings.

Begin With a Focus on Occupational Performance

Although the focus of this text is on the use of environments in intervention, it is essential that the starting point for any occupational therapy interaction be occupational performance. The central importance of occupational performance as the focus of occupational therapy has become evident in theory development, as described in Chapter 2. For example, the person-environment-occupation (PEO) model of occupational performance (Law et al., 1996), the ecology of human performance (EHP) model (Dunn, Brown, & McGuigan, 1994), the model of human occupation (MOHO) (Kielhofner, 2002), the person-environment model of occupational performance (PEOP) (Christiansen & Baum, 1997), and the model of competent occupational performance in the environment (Hagedorn, 2000) all incorporate occupational performance or occupation as a central concept.

The application of these models in practice means that first, occupational performance issues need to be identified. This identification is then followed by an analysis to identify what person, environment, and occupation factors contribute to the problem, and what might be changed to address the issue. By beginning with a focus on occupational performance issues, therapy is client-centred from the outset. The process has been demonstrated clearly in Chapters 9, 11, 12, 13, and 17. In each of these chapters, explicit analysis of an occupational performance issue is described to show the link between occupational performance and the environmental intervention. This same link has also been emphasized in the occupational therapy practice framework (American Occupational Therapy Association (AOTA), 2002), and in the occupational performance process model (Canadian Association of Occupational Therapists [CAOT], 1997; Fearing, Law, & Clark, 1997). The theoretical ideas that have been evolving in occupational therapy in recent years are translated into practice in these documents, further demonstrating the theory-practice link.

Environments Are Powerful Sources of Change to Improve Occupational Performance

Once occupational performance issues are identified, and intervention approaches are considered, it becomes clear that environments are an important focus for change to improve occupational performance. Occupational therapy models explicitly incorporate environment as a factor that can influence occupational performance and can be modified to enhance performance. Although different words are sometimes used to describe environments, such as context or setting, the principle of using the environment to enhance or improve occupational performance is a central theme of this book.

The chapters in this text demonstrate that there are many ways that environments can be created and modified to enhance occupational performance. At the societal level, Chapter 6 describes how cultural environments can be used to develop rapport and identify interventions that are appropriate for clients. Chapter 7, on universal design, illustrates how physical environments can be created and modified so that they better meet the needs of all users, while Chapter 16 demonstrates how therapeutic design principles can be used to promote occupational performance for people with dementia living in long-term care facilities. At the individual level, Chapter 14 uses a PEO model to consider modifications to the physical environment as well as interventions focused on the social environment to enable people to continue living in their own homes. Chapter 15 demonstrates the value in focusing on a travel chain as an approach to consider when environmental modifications are needed to improve the accessibility of public transportation systems. Chapter 17 describes how technology can be used to promote occupational performance through helping people negotiate existing environments and create new environments.

There is tremendous diversity in the ways in which environments can be used to enable occupational performance. The chapters in this text have provided a number of illustrations of that diversity.

Occupational Therapy Is Not Alone

The contributing authors in this text have drawn on diverse sources to demonstrate the value of using environments to enable occupational performance. Many areas of study recognize the importance of the environment. Occupational therapists can draw upon the evidence that is mounting about the environment and its influence on health and well-being from numerous sources, including:

* The International Classification of Functioning, Disability and Health (ICF) (WHO, 2001)
* The Independent Living Movement (ILM)
* Universal design
* Person-environment studies

The ILM, for example, has emphasized the importance of the social and cultural environment in having the potential to disable but also enable people to live, play, and work in community settings. The WHO's (2001) ICF presents a model that acknowledges the interaction of environmental factors with functioning at the level of the body, the person, and society. Throughout the developed world, trends toward de-institutionalization have been recognized for people with developmental disabilities and mental illness. In education, the principle of inclusive environments has become accepted.

All of these examples demonstrate that many sectors recognize the influence that environments have on health and well-being. Occupational therapists who cross boundaries of health, education, and many other sectors are one of many groups that see the importance of the environment in promoting health as well as occupational performance.

Occupational Therapy Practice Has Potential to Expand

Although the future roles for occupational therapists with environment as a focus are discussed in a later section of this chapter, that potential has already been demonstrated through the scenarios shared by the contributors to this book. With the focus of occupational therapy on occupational performance, it becomes possible to think about enhancing performance through a variety of approaches and systems. Environments at their broadest structural and institutional levels need to be supportive of optimal occupational performance for all of its citizens. This means that occupational therapists have a role to play in health promotion (as demonstrated by Chapters 4 and 5), and in designing public spaces

(Chapter 7) and transportation systems (Chapter 15) that meet the needs of all of its users. Contributions can by made by occupational therapists in a number of systems in which few currently work, such as public health, transportation, building, and public space design.

CURRENT CHALLENGES AND OPPORTUNITIES IN USING ENVIRONMENTS TO ENABLE OCCUPATIONAL PERFORMANCE

While it is clear that occupational therapists can enable occupational performance through environmental interventions, the use of environments is not common in all areas of practice. Occupational therapists face a number of challenges in attempting to work in ways that focus on occupational performance, and in using the environment as a means to modify occupational performance. While there are challenges that need to be addressed, there are also opportunities for occupational therapists to work in new ways that mesh with theoretical and practice perspectives.

Funding to Support Practice Focused on Occupational Performance

One of the major challenges faced by occupational therapists is related to remuneration. In countries such as the United States, it may be related to the private remuneration system, in which occupational therapists are limited in what they are able to do and for which they will receive payment. In countries such as Canada, where there is a public payment system for occupational therapists, challenges still remain in trying to branch out into areas not traditionally part of the health system, such as industry, housing, social welfare, and health promotion.

No matter where the location, occupational therapists may need support from their own organizations to practice in new ways. For example, support may be needed so that occupational therapy practice can evolve to become more community-based than centre-based. It may be particularly challenging for occupational therapists to receive funding to provide services in systems in which there has not traditionally been a role. It may require that occupational therapists use examples from the work of others to demonstrate that occupational therapists can make a contribution. Funding for a pilot project might be available, with continued funding contingent on positive evaluation results. Strategies that might be implemented include searching for non-traditional sources of funding, and presenting occupational therapy in terms of its conceptual model of occupational performance, excluding examples of occupational therapy service delivery in health care.

Enabling Transitions in Occupational Therapy Practice

Change can be difficult for everyone, even when people are motivated to make the change. Occupational therapists interested in changing their practice to focus on occupational performance and to consider environments more explicitly as a potential source of enhancing occupational performance may find it challenging to make the transition. Change can be difficult because like all people, occupational therapists develop habits in the ways they think about and approach tasks. Making changes to practice means that it is necessary to pause and re-think things. Sometimes this takes a bit more time, and sometimes it just means having to be more aware of what is being done and why.

That is not to say, however, that occupational therapists are the biggest barriers to making the transition in practice. Other factors also play significant roles. For example, occupational therapists frequently receive referrals that reflect the roles that referral sources believe they fulfill. For example, on an acute psychiatry unit, the referral might suggest occupational therapy provide assertiveness training. While assertiveness training is an appropriate intervention in occupational therapy, there are other interventions (as described in this book, Chapters 8 and 9) that focus more on environments that might be just as effective, or in some situations even more effective.

The same sorts of expectations can be considered from the perspective of clients. Clients and their families may have preconceived ideas about what an occupational therapist can do to assist them. For children and their families, it may be that families expect the therapist to concentrate on changing the skills of the child, rather than enhancing the environment to improve occupational performance (as described by Chapters 10 and 11).

Further education about the evolving practice of occupational therapists may be needed to help clients and referral sources understand the occupational therapy process as it continues to evolve. This can also involve marketing occupational therapy in a way that is more readily understood by the people who are the focus of influence. Team members, referral sources, and clients might be better informed about the potential contributions that an occupational therapist could make when a model like the PEO model is used. The model can demonstrate how an occupational therapist might work through and address issues with a client. This approach might help others see occupational therapy as broader than what they had originally thought.

Curricula in Occupational Therapy Programs

In order to support the continued transitions in occupational therapy practice to reflect current thinking in occupational therapy and areas that influence occupational therapy practice, there are wonderful opportunities to promote that transition through occupational therapy curricula.

Student occupational therapists are exposed to the most recent developments in theoretical contributions to practice. They are being trained to focus on occupational performance issues first, and to consider environments as potential sources of change to promote occupational performance. They are often encouraged to think broadly about the contributions occupational therapists can make in a variety of settings. Many evaluations of student performance reflect the changing models of practice. In Canada for example, the CAOT certification examination for new graduates has been modified over the years to incorporate issues of environment (personal communication, B. Cooper, October 23, 2002).

Many occupational therapists who supervise students on clinical placements view the experience as one that is mutually beneficial. Students gain opportunities to apply the skills and knowledge they have gained through academic preparation with clients. Supervising therapists have an opportunity to engage with students in discussions about the latest research evidence and theory development that can help them to ensure their practice fits with current models of practice. This benefits the therapists and the students and sets the stage for new graduates to find work that reflects the thinking and practice frameworks they have learned about in their programs.

However, there continue to be challenges in the area of occupational therapy curricula. It can be frustrating for students and faculty alike when the ideas discussed in the classroom setting are not carried out in practice, for many of the reasons that have already been discussed. An important step in ensuring that occupational therapists adopt new practice models is to teach them to new graduates, who can then carry them into their practice.

FUTURE VISIONS FOR USING ENVIRONMENTS IN OCCUPATIONAL THERAPY

If the ideas shared in this text demonstrate a different type of occupational therapy practice, in which occupational performance is the focus and environments are considered important sources of change to promote occupational performance, what then is the future for occupational therapy? What will occupational therapy look like once these transitions in practice occur? The ideas shared in this book certainly allude to some of the ways that occupational therapy will change into the future, including:

* *More collaboration with consumers.* Occupational performance is the starting point of all interactions in occupational therapy, and is something that is experienced by clients. Therefore, client-centred practice is central to the occupational therapy process. As occupational therapy practice continues to expand beyond working with individual clients to working with families, groups, communities, and society at large, collaboration with consumers of services will continue to expand.

* *Expanding scope and settings for practice.* Occupational therapists already assume roles that are outside of the health care system. Even within health systems, there is some expansion outside of the traditional work with individual clients with disabilities through efforts in health promotion and community development. When our practice is consistently described in light of enabling occupational performance, the settings in which practice occurs will continue to expand (e.g., urban planning, environmental design, education, industry), as will the stakeholders with whom occupational therapists interact (e.g., architects, engineers, management consultants, educators).

* *Evidence will demonstrate the link between environments, occupational performance, and health.* Occupational therapists need to continue generating evidence through research and practice that supports the links among environments, occupational performance, and health. Models of practice such as the PEO, EHP, MOHO, and PEOP need continued validation. Specific intervention strategies, including examples from this text, need to be evaluated systematically. For example, the effectiveness of health promotion initiatives and various strategies to create supportive work environments for people with mental illness need to be examined. Such research may not always take the form of randomized controlled trials. In health promotion, for example, an international work group on evaluation in health promotion suggested that such approaches do not reflect the nature of the initiatives (WHO European Working Group on Health Promotion Evaluation, 1998). Evidence needs to be generated to demonstrate the contributions of occupational therapy in a variety of practice arenas while respecting the underlying premises of the interventions.

* *Occupational therapy educational programs will include theory and practice that address environmental interventions,* such as universal design, health promotion, culture, community development, advocacy, and policy development. Although policy development has not received significant attention in this book, clearly for occupational therapists practicing at the level of groups, communities, and societies, it is vital to understand policy development processes. In one course offered already to student occupational

therapists, students research a piece of legislation and try to learn how advocacy groups are working for policy change (Renwick, 2000). This provides students with opportunities to understand the importance of institutional environments, and the potential influence modifying policies can have on environments and occupational performance.

STRATEGIES TO IMPLEMENT ENVIRONMENTAL INTERVENTIONS

There are numerous strategies that occupational therapists can adopt to translate the ideas presented in this book into their own practice. It is our hope that the book will be a source of ideas and inspiration for occupational therapists to use environments to enable occupational performance. To begin the process, we have generated some ideas to stimulate change.

Begin With Small Steps

Although making a shift in practice may eventually result in significant changes to the approach taken by occupational therapists, there are small steps that can be implemented to begin the process of exploring how practice can be approached differently. Occupational therapists could begin by changing the ways they initiate their interactions with new clients. For example, try out an assessment that is occupation-focused, such as the Canadian occupational performance measure (COPM) (Law et al., 1998), the occupational performance history interview (Kielhofner et al., 1998), the community adaptive planning assessment (Spencer & Davidson, 1998), or the test of playfulness (ToP) (Bundy, Nelson, Metzger, & Bingaman, 2001). Instead of adopting a new assessment, occupational therapists can explore new ways to conduct interviews that provide clients with opportunities to share their own perspective on the occupational performance issues of greatest importance to them.

Another strategy might be to identify and try out an environmental assessment that is appropriate to the specific practice area in which occupational therapists work, such as the SAFER Tool (Chui, Marshall, Oliver, & Letts, 2002), the readily achievable checklist (Adaptive Environments Centre, 1993), or the workplace environment impact scale (Moore-Corner, Olson, & Kielhofner, 1998).

Although these are first steps, assessment is a good place to begin creating change in practice. If information about occupational performance from the clients' perspectives and environments is gathered, it will almost certainly influence how the issues are analyzed and intervention is planned by occupational therapists.

Adopt a Model of Occupational Performance

It might also be appropriate to adopt a new model of practice that incorporates the environment, such as the PEO, EHP, MOHO, etc. This could begin simply by reading an overview of these theoretical approaches in an occupational therapy text such as *Willard and Spackman's Occupational Therapy* (Crepeau, Cohn, & Schell, 2003) or *Occupational Therapy: Enabling Functioning and Well-Being* (Christiansen & Baum, 1997). Select the model that either offers the most appeal or that others are using. Collaborate with other occupational therapists to select the best model for your area of practice. Begin by trying to use the terminology from the model in discussions with colleagues until everyone feels comfortable with the language.

Another way to use the models is to adopt them to guide an occupational analysis. An occupational analysis might be undertaken in place of a traditional activity analysis. By doing so, the client's performance is considered in its broadest context by focusing on occupational performance, rather than simply the task or activity being undertaken.

Modify Documentation Templates

Documentation is a powerful tool, both for the occupational therapist completing the documentation, the client(s), and readers of the notes. Messages about priorities and the nature of occupational therapy can be sent through documentation. Fearing (1993) shared one approach to occupational therapy documentation that is grounded in an occupational therapy process that begins with occupational performance. Although the occupational therapy process model described by Fearing has since been modified to incorporate environmental considerations as well as performance components of the person (Fearing et al., 1997), the initial ideas related to strategies for documenting the clinical reasoning of occupational therapy remain valuable.

If occupational therapy documentation always began with occupational performance issues, that would send a signal to all who read the notes. It would be clear that the starting point for occupational therapy is to identify the key issues from the client's perspective. This would be followed by an analysis that considers person, environment, and occupation factors that contribute to the issues. Environmental analyses could encompass all environmental aspects of relevance to practice, including physical, social, cultural, and institutional components. Intervention strategies that include those with an environmental focus would naturally follow such analyses.

Consider Measuring Intervention Outcomes

Once the model has been adopted, the need to focus on outcomes of intervention becomes apparent. Using the intervention approaches described in this text, the major outcome of interest will be occupational performance. Although intervention strategies may emphasize the environment or the person, changes to accessibility or policy or range of motion are relevant only when they contribute to improved occupational performance for clients. Thus, it may be necessary to measure outcomes related to the strategies, but also ensure that occupational performance issues have been adequately addressed.

Changing Practice to Reflect Changing Models

Rather than changing practice for all clients simultaneously, it may be more manageable to select one or two clients with whom to pilot the assessments or documentation changes. Not only is this more manageable for many clinicians, it also provides a way to demonstrate to yourself, clients, referral sources, and funders the difference that a change in practice model or outcome measure can make in demonstrating the value of enabling occupational performance through environments.

Changing practice may also mean approaching the work of occupational therapy differently at the operational level. In Chapter 3, Karen Whalley Hammell has argued that occupational therapists need to consider providing services outside of traditional business hours, if they are to meet client needs in more contextually relevant ways.

Another strategy to facilitate changing practice is to find a mentor or occupational therapy role model to assist in the process. Finding someone who is already practicing in ways that reflect changing models of practice may prove to be a source of inspiration and practical assistance. Alternatively, a group of occupational therapists interested in exploring models of practice and environmental interventions might form themselves into a team so that they can offer support and ideas to one another.

CONCLUSION

It is clear that occupational therapists have a role to play in using social, cultural, political, physical, and institutional environments to enable occupational performance. By beginning with a focus on occupational performance issues from the clients' perspectives, occupational therapists can plan a variety of interventions. Theories and models can be used to guide environmental interventions, which are powerful approaches to optimize performance. Occupational therapy practice can expand beyond traditional boundaries with a focus on occupational performance and environments. A number of strategies can be used to begin implementing the ideas shared in this text. It is our hope that these will provide insights and inspirations to many occupational therapists, so that environments can be used in intervention, and occupational performance optimized for the individuals, groups, and communities with whom occupational therapists work.

Study Questions

1. What are key themes about occupational therapy practice that thread throughout the book?

2. Name two strategies that you could implement to change your practice to reflect the themes described in this book.

3. You are an occupational therapist who has just been hired to provide services in a 15-bed general rehabilitation unit in a small city of 30,000. The unit has undergone significant staff changes over recent months, and has been without occupational therapy services for some time. On your first day, the ward clerk hands you the standard occupational therapy forms. Physical status is the first heading you see. You plan to propose a new form to the rehabilitation services director that you will use on the unit. Draft the major headings and subheadings that you feel should be included on the standard initial assessment form.

4. You are interested in health promotion, and have done some readings that support your idea that occupational therapists can be involved in community initiatives to promote health. What steps might you take to create a role for yourself in health promotion? How might you receive remuneration for such a role?

REFERENCES

Adaptive Environments Center. (1993). *Readily achievable checklist: A survey for accessibility.* Boston, MA: Author.

American Occupational Therapy Association (AOTA). (2002). Occupational therapy practice framework: Domain and process. *Am J Occup Ther, 56,* 609-639.

Bundy, A., Nelson, L., Metzger M., & Bingaman, K. (2001). Validity and reliability of a test of playfulness. *The Occupational Therapy Journal of Research, 21,* 277-293.

Canadian Association of Occupational Therapists (CAOT). (1997). *Enabling occupation: An occupational therapy perspective.* Ottawa, ON: CAOT Publications ACE.

Chui, T., Marshall, L., Oliver, R., & Letts, L. (2002). *The safety assessment of function and the environment for rehabilitation (SAFER) tool.* Toronto, ON: COTA Community Rehabilitation and Mental Health Services.

Christiansen, C., & Baum, C. (Eds.). (1997). *Occupational therapy: Enabling function and well-being* (2nd ed.). Thorofare, NJ: SLACK Incorporated.

Crepeau, E. B., Cohn, E. S., & Schell, B. A. B. (Eds.). (2003). *Willard and Spackman's occupational therapy* (10th ed.). Philadelphia, PA: Lippincott, Williams & Wilkins.

Dunn, W., Brown, C., & McGuigan, A. (1994). The ecology of human performance: A framework for considering the effect of context. *Am J Occup Ther, 48,* 595-607.

Fearing, V. G. (1993). Occupational therapists chart a course through the health record. *Can J Occup Ther, 60,* 232-240.

Fearing, V. G., Law, M., & Clark, J. (1997). An occupational performance process model: Fostering client and therapist alliances. *Can J Occup Ther, 64,* 7-15.

Hagedorn, R. (2000). *Tools for practice in occupational therapy: A structured approach to core skills and processes.* London, UK: Churchill Livingstone.

Kielhofner, G. (2002). *Model of human occupation: Theory and application* (3rd ed.). Philadelphia, PA: Lippincott Williams & Wilkins.

Kielhofner, G., Mallinson, T., Crawford, C., Nowak, M., Rigby, M., Henry, A., & Walens, D. (1998). *Occupational performance history interview II.* Chicago, IL: Model of Human Occupation Clearinghouse, University of Illinois at Chicago.

Law, M., Baptiste, S., Carswell, A., McColl, M. A., Polatajko, H., & Pollock, N. (1998). *Canadian occupational performance measure* (3rd ed.). Ottawa, ON: CAOT Publications ACE.

Law, M., Cooper, B., Strong, S., Stewart, D., Rigby, P. & Letts, L. (1996). The person-environment-occupation model: A transactive approach to occupational performance. *Can J Occup Ther, 63,* 186-192.

Moore-Corner, R., Olson, L., & Kielhofner, G. (1998). *Work environment impact scale.* Chicago, IL: Model of Human Occupation Clearinghouse, University of Illinois at Chicago.

Renwick, R. (2000). *OCT 122Y: Person-environment relationships.* Unpublished course outline, Department of Occupational Therapy, Faculty of Medicine, University of Toronto.

Spencer, J., & Davidson, H. (1998). The community adaptive planning assessment: A clinical tool for documenting future planning with clients. *Am J Occup Ther, 52,* 19-30.

WHO European Working Group on Health Promotion Evaluation (1998). *Health promotion evaluation: Recommendations to policymakers.* Geneva: World Health Organization.

World Health Organization (2001). *International Classification of Functioning, Disability and Health.* Geneva: Author.

Typology of Mental Health Vocational Programs

(Compiled from: Bond & Boyer, 1988; Bond, Drake, Mueser, & Becker, 1997; Cochrane, Goering, & Rogers, 1991)

Assertive Community Treatment Model (ACT): Developed by Stein and Test (1980), assertive community treatment involves a multidisciplinary team providing intensive, time-unlimited support and individualized assistance in the individual's home or living environments. This hybrid of treatment and rehabilitation assumes every client has a vocational goal to be addressed by the team.

Career Counseling Programs: A traditional approach involving exploration of values, interests, and goals after completion of a work adjustment program and before job seeking.

Career Placement Training Programs and Job Clubs: Programs aim to help clients develop effective job-seeking and marketing strategies (e.g., presentation, resume writing and interviewing techniques). Some offer a comprehensive program of career counseling, work adjustment training and marketing techniques. Often job clubs are offered as a follow-up for clients to share strategies and offer each other support.

"Choose-Get-Keep" Model: Operating within the Psychiatric Rehabilitation Practice model, this psycho-educational person-centred approach emphasizes client choice in selecting, obtaining, and maintaining jobs. Career planning occurs in vocational counseling type sessions with trained staff together with experiential community living skills training.

Cooperatives (includes Affirmative Businesses): Typically small business operations run directly by consumers in which profits are shared (Trainor & Tremblay, 1992).

Work Adjustment Training Programs: A traditional approach emphasizing employment preparation, training focuses on fundamental work skills and habits. It is offered in sheltered community enclaves, mobile work crews, or hospital-based locations. There is little reported evaluation.

Home Employment: Modeled after cottage industries, piecework is performed at home with materials delivered by a staff person who supervises production (Kates et al., 1989)

Individual Placement and Support (IPS): Clients are directly placed in competitive jobs with job-site training, flexible support, ongoing monitoring, and indefinite follow-up. Proponents point to the difficulties people with psychiatric disabilities have with transferring training, the need for work histories and the belief that anyone, irrespective of the severity of disability, can work given the proper support. See Chapter 9 regarding how recent research is supporting the use of this approach for diverse disadvantaged clients.

Job Coach Model: Wehman (1986) demonstrated the feasibility of an individual placement model in which clients were placed and then trained on the job, the reverse of the traditional train followed by place approach. At the worksite, job coaches provide intensive, time-unlimited support, and training of clients in their work roles. The on-site coaching is reduced over time according to each client's needs.

Part-Time Competitive Employment: Part-time competitive jobs found by staff for clients with direct and on-going support (e.g., wake-up calls, transportation, appropriate clothing)

Sheltered Work: Usually light industrial contract assembly work in a segregated, protected environment for token wages (Black, 1988).

Transitional Employment and Clubhouse Model: Clubhouses operate as a community with members having collective responsibility to operate a place for members to meet, socialize, live, learn, and work. To maintain the clubhouse, members participate in work units (Beard, Propst, & Malamud, 1982). Fountain House, one of the oldest clubhouses, extended work opportunities to pioneer what is referred to as *transitional employment*. Professional staff finds and negotiates temporary part-time jobs with community employers that match members' stamina and stress tolerance. Staff members act as job coaches and take full responsibility for client's orientation and training. If the client is absent, staff members guarantee a replacement, hence removing the barrier of employer's fearing excessive absenteeism. Throughout the process, staff provides support and the clubhouse operates as a home base. By providing a series of these time limited placements, participants gain work experience, increase self-confidence, and build a work history for resumes. This widely used approach by many treatment centres may or may not operate within a clubhouse model and has been shown to extend community tenure (Turkat & Buzzell, 1983). The clubhouse or treatment program rather than the participant controls the positions creating problems with full inclusion, belonging, and making significant relationships (Marrone, Balzell, & Gold, 1995).

REFERENCES

Beard, J. H., Propst, R. N., & Malamud, T. J. (1982). The Fountain House model of rehabilitation. *Psychosocial Rehabilitation Journal, 5*(1), 47-53.

Black, B. (1988). *Work and mental illness: Transitions to employment*. Baltimore: John Hopkins University.

Bond, G. R., & Boyer, S. (1988). Rehabilitation programs and outcomes. In J. A. Ciardiello & M. D. Bell (Eds.), *Vocational rehabilitation of persons with prolonged mental illness* (pp. 231-263). Baltimore: John Hopkins University.

Bond, G. R., Drake, R. E., Mueser, K. T., & Becker, D. R. (1997). An update on supported employment for people with severe mental illness. *Psychiatric Services, 48*(3), 335-346.

Cochrane, J. J., Goering, P., & Rogers, J. M. (1991). Vocational programs and services in Canada. *Canadian Journal of Community Mental Health, 10*(1), 51-63.

Kates, N., Woodside, H., Gavin, D., O'Callaghan, J., Young, S., Jones, B., & Case, P. (1989). Home employment: A work alternative for persons who are mentally ill. *Psychosocial Rehabilitation Journal, 12*(4), 66-69.

Marrone, J., Balzell, A., & Gold, M. (1995). Employment supports for people with mental illness. *Psychiatric Services, 46*(7), 707-711.

Stein, L. I., & Test, M. A. (1980). Alternative to mental health treatment: 1. Conceptual model, treatment program, and clinical evaluation. *Arch Gen Psychiatry, 37*, 392-397.

Trainor, J., & Tremblay, J. (1992). Consumer/Survivor businesses in Ontario: Challenging the rehabilitation model. *Canadian Journal of Community Mental Health, 11*(2), 65-77.

Turkat, D., & Buzzell, V. M. (1983). Recidivism and employment rates among psychosocial rehabilitation clients. *Hospital and Community Psychiatry, 4*(8), 741-742.

Wehman, P. (1986). Supported competitive employment for persons with severe disabilities. *Journal of Applied Rehabilitation Counselling, 17*, 24-29.

Explanation of Key Terms
Related to the Environment

Throughout this text, a number of words are used frequently that warrant a bit more explanation. This section provides more clarification of the meaning of these terms, and references to sources that provide more detailed insight about the concepts. These explanations represent an effort to describe how terms are typically used in this book.

Context: Many people, in this book and more generally, use the words environment and context interchangeably. Context implies the person within the environment (Dunn, Brown, & McGuigan, 1994) and is often described as including temporal components as well as physical, social, cultural, and institutional. The AOTA (2002) uses the term context in place of environment, and contexts are described as being cultural, physical, social, personal, spiritual, temporal, and virtual components.

Cultural Environment: The Canadian Association of Occupational Therapists (CAOT, 1997, p. 46) defines the cultural environment as "ethnic, racial, ceremonial, and routine practices, based on ethos and value system of particular groups." The American Occupational Therapy Association (AOTA, 1994, p. 1054) also includes customs, beliefs, behavior standards, and expectations accepted by the society (or a group) in which the individual is a member. In Chapter 6 of this book, Baptiste pays significant attention to the concept of culture: whether it is learned or inherited, how it is manifested, and notes that above all it is something that is shared.

Disability: Although there are many definitions and descriptions of disability, the WHO (2001) defines it as "an umbrella term for impairments, activity limitations or participation restrictions" (p. 3). Disability is an outcome of the relationship between a person and the environment, rather than being attributed exclusively to the individual. The ICF and its view of health and disability is further described in Chapter 1, and a number of contributing authors refer to this view. In Chapter 3, Hammell discusses models of disability (social and medical) and provides a critique of the ICF.

Environment: Although there are many ways to describe the environment, it is most generically defined as the "surroundings, conditions, and circumstances etc., in which a person lives" (*Oxford Modern English Dictionary*, 1996, p. 328). Law (1991) defines it as "those contexts (situations) that occur outside individuals and elicit response in them" (p. 175). The environment is not static, but continually changing. There are numerous ways to classify the environment, but it is generally considered to include physical, social, cultural, and and institutional components (CAOT, 1997).

Environmental Barriers: Any aspect of the environment that constrains satisfactory occupational performance. These can be physical (e.g., stairs impeding functional mobility of people using wheeled walkers), social (e.g., caregiver feeding a child with disabilities when the child can at least partially self-feed), cultural (e.g., expectations for women to take on homemaking roles and not continue educational endeavors), or institutional (e.g., policies in long-term care facilities that impede residents' ability to make decisions about the time of day they would like to complete self-care activities).

Environmental Supports: Any aspect of the environment that enables or encourages satisfactory occupational performance. Similar to environmental barriers, supports can be physical (e.g., universal design of public buildings to make access simple for all users), social (e.g., peer-support groups for people with mental illness in the workplace), cultural (e.g., expectations that aging family members will continue to be integrated into family life), and/or institutional (e.g., community organization meetings by policy always held in wheelchair accessible spaces with adequate sound systems).

Institutional Environment: The institutional environment includes organizational components (such as policies), legal, economic, and political aspects. As Hammell notes in Chapter 3 of this book, there is a distinction to be made between the institutions (buildings) within which many occupational therapists and their clients interact (e.g., hospitals, long-term care facilities), and the institutional environment within which the interactions between therapists and clients take place. Legal, political, and economic aspects of the institutional environment significantly influence the structures that guide how occupational therapists receive referrals and payment, and how they are supported to work with individuals, groups, and communities.

Occupational Performance: "Occupational performance is the outcome of the transaction of the person, environment, and occupation" (Law et al., 1996). It is experienced by a person when striving to complete an occupation within an environmental context. It is a complex and dynamic process. It has also been defined as "the ability to choose and satisfactorily perform meaningful occupations that are culturally defined, and appropriate for looking after one's self, enjoying life, and contributing to the social and economic fabric in a community" (CAOT, 1997, p. 45).

Person-Environment Fit: Originally, the term person-environment fit is rooted in environment-behavior studies, such as Baker & Intagliata (1982) and Kahana (1982). Other terms that are frequently considered to be synonymous with person-environment fit include congruence, adaptation, and coping. It is typically considered in terms of "goodness of fit" in relation to facilitating or hindering performance of an occupation.

Physical Environment: The physical environment consists of the built and natural aspects of the environment. It includes light, noise and other ambient characteristics (Christiansen & Baum, 1997, p. 601), the ground, buildings, furniture, assistive devices, technology, and natural settings within which people experience their lives (CAOT, 1997).

Press: The roots of the term press seem to be from the work of Lawton & Nahemow (1973), where it is used to describe the demand characteristics of the environment. Press consists of forces from the environment that demand or evoke a behavioral response from a person. Often press is described as a negative characteristic (i.e., as a demand), but it can also be conceived as a support (i.e., as a reminder or a cue).

Social Environment: The social environment includes social relationships in which people are involved; both intimate with family, friends, and co-workers, and those that are less personal (such as interactions in crowds of people) (Christiansen & Baum, 1997; Law et al., 1996). It includes the ways in which people relate to one another, understand social priorities, and organize themselves as members of a community (CAOT, 1997).

Transaction: Transactions can be differentiated from interactions, in their complex and dynamic natures. While interactions are typically understood as one-way relationships (e.g., environments influence people), transactions are considered to be two-way (environments influence people and people in turn influence their environments). A further extension of this idea describes that in fact, people and environments are elements of a whole rather than being separate entities (Altman & Rogoff, 1987). From an occupational therapy perspective, this would imply that the distinctions we make between person, occupation, and environment are useful to help us understand occupational performance, but are more artificial than real. Occupational performance can only be understood by understanding the PEO relationships as a whole (Law et al., 1996).

REFERENCES

Altman, I., & Rogoff, B. (1987). World views in psychology: Triat, interactional, organismic and transactional perspectives. In D. Sokols & I. Altman (Eds.), *Handbook of environmental psychology* (pp. 7-40). New York: Wiley.

American Occupational Therapy Association (AOTA). (1994). Uniform terminology for occupational therapy. *Am J Occup Ther, 48,* 1047-1054.

American Occupational Therapy Association (AOTA). (2002). Occupational therapy practice framework: Domain and process. *Am J Occup Ther, 56,* 609-639.

Baker, F., & Intagliata, J. (1982). Quality of life in the evaluation of community support systems. *Evaluation and Program Planning, 5,* 69-79.

Canadian Association of Occupational Therapists (CAOT). (1997). *Enabling occupation: An occupational therapy perspective.* Ottawa, ON: CAOT Publications ACE.

Christiansen, C., & Baum, C. (1997). Person-environment-occupational performance: A conceptual model for practice. In C. Christiansen & C. Baum (Eds.), *Occupational therapy: Enabling function and well-being* (2nd ed., pp. 46-70). Thorofare, NJ: SLACK Incorporated.

Dunn W., Brown, C., & McGuigan, A. (1994). The ecology of human performance: A framework for considering the effect of context. *Am J Occup Ther, 48,* 595-607.

Kahana, E. (1982). A congruence model of person-environment interaction. In M. P. Lawton, P. G. Windley, & T. D. Byers (Eds), *Aging and the environment: Theoretical approaches* (pp. 97-121). New York: Springer.

Law, M. (1991). The environment: A focus for occupational therapy. *Can J Occup Ther, 58,* 171-1179.

Law, M., Cooper, B., Strong, S., Stewart, D., Rigby, P., & Letts, L. (1996). The person-environment-occupation model: A transactive approach to occupational performance. *Can J Occup Ther, 63,* 9-23.

Lawton, M. P., & Nahemow, L. (1973). Ecology and the aging process. In C. Eisdorfer, & M. P. Lawton (Eds.), *The psychology of adult development and aging* (pp. 619-674). Washington, DC: American Psychological Association.

Oxford Modern English Dictionary. (2nd ed.). (1996). Oxford, UK: Oxford University.

World Health Organization (WHO). (2001). *ICF Introduction.* Retrieved November 13, 2002 from http://www.who.int/classification/icf/intros/ICF-Eng-Intro.pdf.

Index

Build Your Library

Along with this title, we publish numerous products on a variety of topics. We are sure that you will find the below titles to be an essential addition to your library. Order your copies today or contact us for a copy of our latest catalog for additional product information.

USING ENVIRONMENTS TO ENABLE OCCUPATIONAL PERFORMANCE

Lori Letts, PhD, OT Reg. (Ont.); Patty Rigby, MHSc, OT Reg. (Ont.); and Debra Stewart, MSc, OT Reg. (Ont.)

336 pp., Hard Cover, 2003, ISBN 1-55642-578-3, Order #35783, **$35.95**

Using Environments to Enable Occupational Performance is a unique text that specifically focuses on how environments (physical, social, cultural, institutional) can be used by occupational therapists to enable occupational performance with all types of clients. Each chapter contains "real world" scenarios from occupational therapists about how the environment can be used to optimize occupational performance. This feature is a beneficial element for both students and clinicians as they are learning these applications.

EVIDENCE-BASED REHABILITATION: A GUIDE TO PRACTICE

Mary Law, PhD, OT Reg. (Ont.)

384 pp., Soft Cover, 2002, ISBN 1-55642-453-1, Order #44531, **$39.95**

Evidence-Based Rehabilitation: A Guide to Practice is designed as an entry-level book on evidence-based practice in rehabilitation. Specifically written for rehabilitation practitioners, this exceptional text is not designed to teach students how to do research, but rather how to become critical consumers of research, therefore developing skills to ensure that their rehabilitation practice is based on the best evidence that is available.

PROBLEM-BASED LEARNING: A SELF-DIRECTED JOURNEY

Sue Baptiste, MHSc, OT Reg. (Ont.)

144 pp., Soft Cover, 2003, ISBN 1-55642-563-5, Order #35635, **$30.95**

Problem-Based Learning: A Self-Directed Journey is an interesting and unique text written for both occupational therapists and health professionals seeking to develop their understanding of this integral concept. Following the PBL process, the scenario for exploration is established by using "problem-based learning."

Contact Us

SLACK Incorporated, Professional Book Division
6900 Grove Road, Thorofare, NJ 08086
1-800-257-8290/1-856-848-1000, Fax: 1-856-853-5991
orders@slackinc.com or www.slackbooks.com

ORDER FORM

QUANTITY	TITLE	ORDER #	PRICE
	Using Environments to Enable Occupational Performance	35783	$35.95
	Evidence-Based Rehabilitation: A Guide to Practice	44531	$39.95
	Problem-Based Learning: A Self-Directed Journey	35635	$30.95

Subtotal	$
Applicable state and local tax will be added to your purchase	$
Handling	$4.50
Total	$

Name: _____

Address: _____

City: _____ State:_____ Zip: _____

Phone:_____ Fax: _____

Email: _____

• Check enclosed (Payable to SLACK Incorporated)_____

• Charge my: ___ MasterCard ___ VISA ___

Account #: _____

Exp. date: _____ Signature:_____

NOTE: *Prices are subject to change without notice.*
Shipping charges will apply.
Shipping and handling charges are non-refundable.

CODE: 328